Power Practice Problems

for the PE Exam, Third Edition

John Camara, PE

PPI2PASS.COM
A **KAPLAN** COMPANY

Register Your Book at ppi2pass.com

- Receive the latest exam news.
- Obtain exclusive exam tips and strategies.
- Receive special discounts.

Report Errors for This Book

PPI is grateful to every reader who notifies us of a possible error. Your feedback allows us to improve the quality and accuracy of our products. Report errata at **ppi2pass.com**.

POWER PRACTICE PROBLEMS FOR THE PE EXAM
Third Edition

Current release of this edition: 1

Release History

date	edition number	revision number	update
Feb 2016	2	1	New edition. Additional content. Code updates. Copyright update.
Oct 2017	2	2	Minor corrections. Minor cover updates.
Oct 2018	1	3	New edition. Code updates. Additional content. Copyright update.

© 2018 Kaplan, Inc. All rights reserved.

PPI
1250 Fifth Avenue, Belmont, CA 94002
(650) 593-9119
ppi2pass.com

ISBN: 978-1-59126-632-7

F E D C B A

Topics

Mathematics

Machinery and Devices

Basic Theory

Electronics

Field Theory

Special Applications

Circuit Theory

Measurement/ Instrumentation

Generation

Electrical Materials

Distribution

Codes and Standards

System Analysis

Professional

Protection and Safety

Table of Contents

Topic XIII: Electrical Materials

Topic XIV: Codes and Standards

Topic XV: Professional

Preface

While I wrote *Power Practice Problems for the PE Exam* with the National Council of Examiners for Engineering and Surveying (NCEES) Electrical and Computer: Power PE exam specifications in mind, the content of *Power Practice Problems* is not limited to them. This book also contains practice problems based on the fundamental concepts you need to build a complete understanding of power engineering. With an understanding of the basic problems that arise in electrical engineering, you will comprehend the majority of exam problems, as well as get a glimpse of real-world electrical engineering problems.

For this third edition, I revised problems so that this book is consistent with revisions made to the second edition of the *Power Reference Manual for the PE Exam* and the NCEES adopted codes and standards. (The actual codes and standards used to develop this book are listed in "Codes and References.")

I revised all problems related to and dependent on the 2017 edition of the *National Electrical Code* (NEC) and the 2017 edition of the *National Electrical and Safety Code* (NESC).

FOR WHOM THIS BOOK IS WRITTEN

Power Practice Problems is written for you. If you are an exam candidate, *Power Practice Problems* is an efficient resource for applying exam topics systematically and exhaustively. If you are a practicing power engineer, it functions as a companion to the *Power Reference Manual*. Finally, for the engineering student, it provides the opportunity to practice using your knowledge of the fundamentals of electrical engineering.

LOOKING FORWARD TO THE FUTURE

Future editions of this book will be very much shaped by what you and others want to see in it. I am braced for the influx of comments and suggestions from readers who (1) want more problems in some chapters, (2) want more detail in existing problems, and (3) think that continuing to include some problems is just plain lame. Computer hardware and programming change quickly, as does the grand unified (atomic) theory of everything. I expect several chapters to generate some kind of debate.

Should you find an error in this book, know that it is mine, and that I regret it. Beyond that, I hope two things happen. First, please let me know about the error by reporting it on the Errata section of the PPI website at **ppi2pass.com**. Second, I hope you learn something from the error—I know I will! I appreciate suggestions for improvement, additional questions, and recommendations for expansion so that new editions or similar texts will better meet your needs.

Read carefully, prepare well, and you will triumph over the exam!

John A. Camara, PE

Acknowledgments

It is with enduring gratitude that I thank Michael R. Lindeburg, PE, who, with the sixth editions of the *Electrical Engineering Reference Manual* and *Electrical Engineering Practice Problems*, allowed me to realize my dream of authoring significant engineering texts. In taking the reins from the titles' previous author, the late Raymond Yarbrough, I was entrusted with the responsibility of shaping those texts as the electrical and computer PE exam and the engineering profession evolved. It is my hope that *Power Practice Problems for the PE Exam*—which is based on *Electrical Engineering Practice Problems*—will match the comprehensive and cohesive quality of the best of PPI's family of engineering books, which I have admired for more than three decades.

The third edition of the *Power Practice Problems* would not have been possible without the professionalism and meticulous attention to detail shown by the responsive team at PPI, including Steve Shea, product manager; Sam Webster, product data manager; Ellen Nordman, publishing systems specialist; Scott Rutherford, copy editor; Tom Bergstrom, production specialist and technical drawings; Cathy Schrott, production services manager; and Grace Wong, director of Editorial Operations.

Michael Lindeburg provided Chapters 1–14 on mathematics, as well as the chapters on economic analysis, law, ethics, and engineering licensure. He also provided sections of the illumination chapter. These contributions saved me considerable time, and I am grateful to use material of this quality. John Goularte, Jr., taught me firsthand the applications of biomedical engineering, and the real-world practicality of this topic is attributable to him.

Additionally, the following individuals contributed in a variety of ways, from thoughtful reviews, criticisms, and corrections, to suggestions for additional material. I am grateful for their efforts, which resulted in enhanced coverage of the material and enriched my knowledge of each topic. Thank you to Chris Iannello, PhD; Thomas P. Barrera, PhD; Robert G. Henriksen, PhD; Glenn G. Butcher, DCS; Chen J. Chang, MSE; and Aubrey Clements, BSEE.

The knowledge expressed in this book represents years of training, instruction, and self-study in electrical, electronic, mechanical, nuclear, marine, space, and a variety of other types of engineering—all of which I find fascinating. I would not understand any of it were it not for the teachers, instructors, mentors, family, and friends who have spent countless hours with me and from whom I have learned so much. A few of them include Mary Avila, Capt. C. E. Ellis, Claude Estes, Capt. Karl Hasslinger, Jerome Herbeck, Jack Hunnicutt, Ralph Loya, Harry Lynch, Harold Mackey, Michael O'Neal, Mark Richwine, Charles Taylor, Abigail Thyarks, and Jim Triguerio. There are others who have touched my life in special ways: Jim, Marla, Lora, Todd, Tom, and Rick. There are many others, though here they will remain nameless, to whom I am indebted. Thanks.

A very special thanks to my son, Jac Camara, whose annotated periodic table became the basis for the Electrical Materials chapter. Additionally, our discussions about quantum mechanics and a variety of other topics helped stir the intellectual curiosity I thought was fading.

Thanks also to my daughter, Cassiopeia, who has been a constant source of inspiration, challenge, and pride.

Thanks to Shelly for providing insight into the intricacies of the NEC, and thanks to Wyatt Wade for teaching me the new math. Also, thanks to all those who have graced and improved my life in a myriad of ways: Georgia Ann, Becky Marlene, and Marcelle. And, last but not least, thanks to Miss DeDe.

Finally, thanks to my mother, Arlene, who made me a better person.

John A. Camara, PE

Codes and References

The information that was used to write and update this book was based on the exam specifications at the time of publication. However, as with engineering practice itself, the PE exam is not always based on the most current codes or cutting-edge technology. Similarly, codes, standards, and regulations adopted by state and local agencies often lag issuance by several years. It is likely that the codes that are most current, the codes that you use in practice, and the codes that are the basis of your exam will all be different. However, differences between code editions typically minimally affect the technical accuracy of this book, and the methodology presented remains valid. For more information about the variety of codes related to electrical engineering, refer to the following organizations and their websites.

American National Standards Institute (ansi.org)
Electronic Components Industry Association (ecianow.org)
Federal Communications Commission (fcc.gov)
Institute of Electrical and Electronics Engineers (ieee.org)
International Organization for Standardization (iso.org)
International Society of Automation (isa.org)
National Electrical Manufacturers Association (nema.org)
National Fire Protection Association (nfpa.org)

The PPI website (**ppi2pass.com**) provides the dates and editions of the codes, standards, and regulations on which NCEES has announced the PE exams are based. It is your responsibility to find out which codes are relevant to your exam.

The minimum recommended library of materials to bring to the Power PE exam consists of this book, any applicable code books, a standard handbook of electrical engineering, and one or two textbooks that cover fundamental circuit theory (both electrical and electronic).

CODES AND STANDARDS

47 CFR 73: *Code of Federal Regulations*, "Title 47—Telecommunication, Part 73—Radio Broadcast Services," 2018. Office of the Federal Register National Archives and Records Administration, Washington, DC.

IEEE/ASTM SI 10: *American National Standard for Metric Practice*, 2016. ASTM International, West Conshohocken, PA.

IEEE Std 141 (IEEE Red Book): *IEEE Recommended Practice for Electric Power Distribution for Industrial Plants*, 1993. The Institute of Electrical and Electronics Engineers, Inc., New York, NY.

IEEE Std 142 (IEEE Green Book): *IEEE Recommended Practice for Grounding of Industrial and Commercial Power Systems*, 2007.

IEEE Std 241 (IEEE Gray Book): *IEEE Recommended Practice for Electrical Power Systems in Commercial Buildings*, 1990.

IEEE Std 242 (IEEE Buff Book): *IEEE Recommended Practice for Protection and Coordination of Industrial and Commercial Power Systems*, 2001.

IEEE Std 399 (IEEE Brown Book): *IEEE Recommended Practice for Industrial and Commercial Power Systems Analysis*, 1997.

IEEE Std 446 (IEEE Orange Book): *IEEE Recommended Practice for Emergency and Standby Power Systems for Industrial and Commercial Applications*, 1995.

IEEE Std 493 (IEEE Gold Book): *IEEE Recommended Practice for the Design of Reliable Industrial and Commercial Power Systems*, 2007.

IEEE Std 551 (IEEE Violet Book): *IEEE Recommended Practice for Calculating Short-Circuit Currents in Industrial and Commercial Power Systems*, 2006.

IEEE Std 602 (IEEE White Book): *IEEE Recommended Practice for Electric Systems in Health Care Facilities*, 2007.

IEEE Std 739 (IEEE Bronze Book): *IEEE Recommended Practice for Energy Management in Industrial and Commercial Facilities*, 1995.

IEEE Std 902 (IEEE Yellow Book): *IEEE Guide for Maintenance, Operation, and Safety of Industrial and Commercial Power Systems*, 1998.

IEEE Std 1015 (IEEE Blue Book): *IEEE Recommended Practice for Applying Low-Voltage Circuit Breakers Used in Industrial and Commercial Power Systems*, 2006.

IEEE Std 1100 (IEEE Emerald Book): *IEEE Recommended Practice for Powering and Grounding Electronic Equipment*, 2005.

NEC (NFPA 70): *National Electrical Code*, 2017. National Fire Protection Association, Quincy, MA.

NESC (IEEE C2): *2017 National Electrical Safety Code*, 2017. The Institute of Electrical and Electronics Engineers, Inc., New York, NY.

NFPA 30: *Flammable and Combustible Liquids Code*, 2018. National Fire Protection Association, Quincy, MA.

NFPA 30B: *Code for the Manufacture and Storage of Aerosol Products*, 2019.

NFPA 70E: *Standard for Electrical Safety in the Workplace*, 2018.

NFPA 497: *Recommended Practice for the Classification of Flammable Liquids, Gases, or Vapors and of Hazardous (Classified) Locations for Electrical Installations in Chemical Process Areas*, 2017.

NFPA 499: *Recommended Practice for the Classification of Combustible Dusts and of Hazardous (Classified) Locations for Electrical Installations in Chemical Process Areas*, 2017.

REFERENCES

Anthony, Michael A. *NEC Answers*. New York, NY: McGraw-Hill. (*National Electrical Code* example applications textbook.)

Bronzino, Joseph D. *The Biomedical Engineering Handbook*. Boca Raton, FL: CRC Press. (Electrical and electronics handbook.)

Chemical Rubber Company. *CRC Standard Mathematical Tables and Formulae*. Boca Raton, FL: CRC Press. (General engineering reference.)

Croft, Terrell and Wilford I. Summers. *American Electricians' Handbook*. New York, NY: McGraw-Hill. (Power handbook.)

Earley, Mark W. et al. *National Electrical Code Handbook*, 2014 ed. Quincy, MA: National Fire Protection Association. (Power handbook.)

Fink, Donald G. and H. Wayne Beaty. *Standard Handbook for Electrical Engineers*. New York, NY: McGraw-Hill. (Power and electrical and electronics handbook.)

Grainger, John J. and William D. Stevenson, Jr. *Power System Analysis*. New York, NY: McGraw-Hill. (Power textbook.)

Horowitz, Stanley H. and Arun G. Phadke. *Power System Relaying*. Chichester, West Sussex: John Wiley & Sons, Ltd. (Power protection textbook.)

Huray, Paul G. *Maxwell's Equations*. Hoboken, NJ: John Wiley & Sons, Inc. (Power and electrical and electronics textbook.)

Jaeger, Richard C. and Travis Blalock. *Microelectronic Circuit Design*. New York, NY: McGraw-Hill Education. (Electronic fundamentals textbook.)

Lee, William C.Y. *Wireless and Cellular Telecommunications*. New York, NY: McGraw-Hill. (Electrical and electronics handbook.)

Marne, David J. *National Electrical Safety Code (NESC) 2012 Handbook*. New York, NY: McGraw-Hill Professional. (Power handbook.)

McMillan, Gregory K. and Douglas Considine. *Process/Industrial Instruments and Controls Handbook*. New York, NY: McGraw-Hill Professional. (Power and electrical and electronics handbook.)

Millman, Jacob and Arvin Grabel. *Microelectronics*. New York, NY: McGraw-Hill. (Electronic fundamentals textbook.)

Mitra, Sanjit K. *An Introduction to Digital and Analog Integrated Circuits and Applications*. New York, NY: Harper & Row. (Digital circuit fundamentals textbook.)

Parker, Sybil P., ed. *McGraw-Hill Dictionary of Scientific and Technical Terms*. New York, NY: McGraw-Hill. (General engineering reference.)

Plonus, Martin A. *Applied Electromagnetics*. New York, NY: McGraw-Hill. (Electromagnetic theory textbook.)

Rea, Mark S., ed. *The IESNA Lighting Handbook: Reference & Applications*. New York, NY: Illuminating Engineering Society of North America. (Power handbook.)

Shackelford, James F. and William Alexander, eds. *CRC Materials Science and Engineering Handbook*. Boca Raton, FL: CRC Press, Inc. (General engineering handbook.)

Van Valkenburg, M.E. and B.K. Kinariwala. *Linear Circuits*. Englewood Cliffs, NJ: Prentice-Hall. (AC/DC fundamentals textbook.)

Wildi, Theodore and Perry R. McNeill. *Electrical Power Technology*. New York, NY: John Wiley & Sons. (Power theory and application textbook.)

How to Use This Book

This book is a companion to the *Power Reference Manual*—a compendium of power engineering. Admittedly, despite its size, the *Reference Manual* cannot cover all the details of power engineering. Nevertheless, a thorough study of that reference manual and its equations, terms, diagrams, and tables, as well as the problem statements and solutions given in this Practice Problems book, will expose you to many electrical engineering topics. This broad overview of the field can benefit you in a number of ways, not the least of which is by thoroughly preparing you for the Power exam. Further, the range of subjects covered allows you to tailor your review toward the topics of your choice and to study at a level that suits your needs.

The philosophy guiding the preparation of this book is that all problems, regardless of how complex, are comprised of building blocks of simpler questions and concepts. The problems on the PE exam take an average of six minutes per problem. You should focus your study on the fundamental building blocks of knowledge for an electrical engineer so you can quickly evaluate and complete problems on the exam.

In this book, there are a variety of problems that range from "easy" to "difficult." Easy problems are intended to remind you of the fundamentals and bring you closer to the level of knowledge you had when college afforded you the opportunity to concentrate on fundamental subjects daily. Moderately difficult problems, meant to simulate the exam, can be answered in approximately six minutes. Longer, more complex essay problems are designed to help you explore more difficult concepts and show how complex problems may be broken into simpler pieces. Finally, difficult problems are intended to deepen your knowledge and increase your confidence.

Before you start studying, read about the format and content of the Power exam in the *Power Reference Manual* or on the PPI website at **ppi2pass.com**. Then, begin your studies with the fundamentals and progressively expand your review to include more specific topics. Keep in mind that the key to success on the exam is to practice solving problems.

Depending on your specific study methods, you may wish to study the "easier" material first to build confidence. Alternatively, you might delay reading the easy material until just prior to the exam. The choice is yours.

If you are using this text to practice a particular area of study, go directly to the topic of interest and begin. Any weakness noted while attempting to solve the problems should prompt you to review the applicable material in the *Reference Manual*.

Allow yourself three to six months to prepare. You will get the most benefit out of your exam preparation efforts if you make a plan and stick to it. Make sure, as you review chapters in the accompanying *Reference Manual*, that you also leave enough time to work the problems associated with those chapters. You should revisit practice problems during your exam preparation. Even if you successfully worked all the problems in a given chapter upon your initial review, you must maintain that knowledge as you work through unfamiliar or difficult material.

I am confident that with diligent study, you will find the electrical engineering knowledge you seek within these pages or within the pages of the *Reference Manual*. At the very least, this companion book will allow you to conduct an intelligent search for knowledge. I hope these texts serve you well. Enjoy the adventure of learning!

John A. Camara, PE

Topic I: Mathematics

Chapter

1 Systems of Units

PRACTICE PROBLEMS

1. Convert 250°F to degrees Celsius.

(A)   115°C

(B)   121°C

(C)   124°C

(D)   420°C

2. Convert the Stefan-Boltzmann constant (1.71×10^{-9} Btu/ft²-hr-°R⁴) from customary U.S. to SI units.

(A) 5.14×10^{-10} W/m²·K⁴

(B) 0.950×10^{-8} W/m²·K⁴

(C) 5.67×10^{-8} W/m²·K⁴

(D) 7.33×10^{-6} W/m²·K⁴

3. How many U.S. tons (2000 lbm per ton) of coal with a heating value of 13,000 Btu/lbm must be burned to provide as much energy as a complete nuclear conversion of 1 g of its mass? (Hint: Use Einstein's equation: $E = mc^2$.)

(A)   1.7 tons

(B)   14 tons

(C)   780 tons

(D)   3300 tons

4. What is the SI unit for force, N, in terms of more basic units?

(A)   kg·m/s²

(B)   kg·m²/s² - *Joule - N·m*

(C)   kg·m²/s³

(D)   kg/m·s²

5. What is the appropriate SI unit for magnetic flux density, **B**?

(A)   henry, H - *inductance*

(B)   siemens, S - *elec conductance*

(C)   tesla, T

(D)   weber, W - *magnetic flux*

SOLUTIONS

1. The conversion of temperature Fahrenheit (T_F) to Celsius (T_C) is

$$T_C = \left(\frac{5}{9}\right)(T_F - 32)$$
$$= \left(\frac{5}{9}\right)(250°F - 32°F)$$
$$= \boxed{121.1°C \quad (121°C)}$$

The answer is (B).

2. In customary U.S. units, the Stefan-Boltzmann constant, σ, is 1.71×10^{-9} Btu/ft²-hr-°R⁴.

Use the following conversion factors.

$$1 \text{ Btu/hr} = 0.2931 \text{ W}$$
$$1 \text{ ft} = 0.3048 \text{ m}$$
$$T_R = \frac{9}{5} T_K$$

Performing the conversion gives

$$\sigma = \left(1.71 \times 10^{-9} \frac{\text{Btu}}{\text{hr-ft}^2\text{-}°R^4}\right)\left(0.2931 \frac{\text{W}}{\frac{\text{Btu}}{\text{hr}}}\right)$$
$$\times \left(\frac{1 \text{ ft}}{0.3048 \text{ m}}\right)^2\left(\frac{9}{5}\frac{°R}{K}\right)^4$$
$$= \boxed{5.67 \times 10^{-8} \text{ W/m}^2\text{·K}^4}$$

The answer is (C).

3. The energy produced from the nuclear conversion of any quantity of mass is given as

$$E = mc^2$$

The speed of light, c, is 2.9979×10^8 m/s.

For a mass of 1 g (0.001 kg),

$$E = mc^2$$
$$= (0.001 \text{ kg})\left(2.9979 \times 10^8 \frac{\text{m}}{\text{s}}\right)^2$$
$$= 8.9874 \times 10^{13} \text{ J}$$

Convert to U.S. customary units with the conversion 1 Btu = 1055 J.

$$E = (8.9874 \times 10^{13} \text{ J})\left(\frac{1 \text{ Btu}}{1055 \text{ J}}\right)$$
$$= 8.5189 \times 10^{10} \text{ Btu}$$

The number, n, of tons of 13,000 Btu/lbm coal is

$$n \text{ tons} = \frac{8.5189 \times 10^{10} \text{ Btu}}{\left(13,000 \frac{\text{Btu}}{\text{lbm}}\right)\left(2000 \frac{\text{lbm}}{\text{ton}}\right)}$$
$$= \boxed{3276 \text{ tons} \quad (3300 \text{ tons})}$$

The answer is (D).

4. The newton, N, is the SI unit of force. Using primary dimensional analysis,

$$F = \frac{ML}{\theta^2}$$
$$= \boxed{\text{kg·m/s}^2}$$

The answer is (A).

5. The SI unit for magnetic flux density, **B**, is the tesla, T.

The answer is (C).

2 Energy, Work, and Power

PRACTICE PROBLEMS

1. A volume of 40×10^6 m^3 of water behind a dam is at an average height of 100 m above the input to electrical turbines. The turbines are 90% efficient. What is the approximate maximum potential energy converted to electrical energy?

(A) 3.9×10^{11} J

(B) 3.5×10^{13} J

(C) 3.9×10^{13} J

(D) 1.3×10^{14} J

2. A certain motor parameter is measured by the following expression.

$$\int \mathbf{T} \cdot d\theta$$

What is the motor parameter being measured?

(A) power

(B) thrust

(C) torque

(D) work

3. A linear spring, held by an electric coil, is designed to open a cooling valve in the event of an electrical failure. The conditions are as shown.

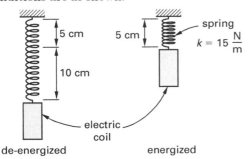

What work must the electric coil do to compress the spring?

(A) 0.018 N·m

(B) 0.075 N·m

(C) 750 N·m

(D) 75×10^3 N·m

4. Consider the illustration shown. An induction motor is rotating at 1800 rpm. The mass is concentrated on the circumference of the rotor. What is the approximate kinetic energy of the rotor?

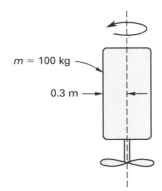

(A) 2.8×10^3 J

(B) 1.6×10^5 J

(C) 3.2×10^5 J

(D) 6.4×10^5 J

5. A rotating motor is calculated to have 20×10^4 J of rotational energy during operation. During fault analysis, this motor is expected to dissipate all this energy in one second through a ground. What power is provided to the ground site by the motor?

(A) 20 W

(B) 200 W

(C) 200 kW

(D) 400 kW

SOLUTIONS

1. Using the average height of 100 m, the potential energy, $E_{\text{available}}$, is given by the product of the mass of the water (its density, ρ_{H_2O}, times its volume), the acceleration of gravity, g, and the height of the dam above the turbines, z.

$$
\begin{aligned}
E_{\text{available}} &= mgz \\
&= (\rho_{H_2O} V_{\text{dam}}) gz \\
&= \left(1000\ \frac{\text{kg}}{\text{m}^3}\right)(40 \times 10^6\ \text{m}^3) \\
&\quad \times \left(9.81\ \frac{\text{m}}{\text{s}^2}\right)(100\ \text{m}) \\
&= 3.924 \times 10^{13}\ \text{J}
\end{aligned}
$$

At an efficiency, η, of 90%, the maximum energy output is

$$
\begin{aligned}
E_{\text{output}} &= \eta E_{\text{available}} \\
&= (0.90)(3.924 \times 10^{13}\ \text{J}) \\
&= \boxed{3.53 \times 10^{13}\ \text{J} \quad (3.5 \times 10^{13}\ \text{J})}
\end{aligned}
$$

The answer is (B).

2. The expression represents the work done in a rotational system by a variable torque.

$$
W = \int \mathbf{T} \cdot d\theta
$$

The answer is (D).

3. Because no value is given for the coil efficiency, assume 100%. Therefore, the work of the coil, W_{coil}, equals the work of a spring, W_{spring}, with spring constant, k, and positions, δ_{initial} and δ_{final}.

$$
W_{\text{coil}} = W_{\text{spring}} = \tfrac{1}{2}k(\delta_{\text{final}}^2 - \delta_{\text{initial}}^2)
$$

With no initial deflection, the equation becomes

$$
\begin{aligned}
W_{\text{spring}} &= \tfrac{1}{2}k\delta_{\text{final}}^2 \\
&= \left(\frac{1}{2}\right)\left(15\ \frac{\text{N}}{\text{m}}\right)(10\ \text{cm})^2\left(\frac{1\ \text{m}}{100\ \text{cm}}\right)^2 \\
&= \boxed{0.075\ \text{N·m}}
\end{aligned}
$$

The answer is (B).

4. The kinetic energy of the rotor is

$$
E = \tfrac{1}{2}mv^2 \qquad \text{[I]}
$$

The mass is m and the velocity is v.

The velocity is

$$
\begin{aligned}
v &= \omega r \\
&= \left| \frac{\left(1800\ \dfrac{\text{rev}}{\text{min}}\right)\left(\dfrac{2\pi}{\text{rev}}\right)}{\left(60\ \dfrac{\text{s}}{\text{min}}\right)} \right| (0.3\ \text{m}) \\
&= 56.55\ \text{m/s} \qquad \text{[II]}
\end{aligned}
$$

Substitute the result of Eq. II into Eq. I, along with the given information about the mass.

$$
\begin{aligned}
E &= \tfrac{1}{2}mv^2 \\
&= \left(\frac{1}{2}\right)(100\ \text{kg})\left(56.55\ \frac{\text{m}}{\text{s}}\right)^2 \\
&= \boxed{1.598 \times 10^5\ \text{J} \quad (1.6 \times 10^5\ \text{J})}
\end{aligned}
$$

This problem may also be solved using moment of inertia concepts.

The answer is (B).

5. During worst-case conditions, a motor's rotational energy will be converted to electrical energy and provided to the fault site. Power, in watts, is energy per unit time, or

$$
\begin{aligned}
P &= \frac{E}{t} \\
&= \frac{20 \times 10^4\ \text{J}}{1\ \text{s}} \\
&= \boxed{20 \times 10^4\ \text{W} \quad (200\ \text{kW})}
\end{aligned}
$$

The answer is (C).

3 Engineering Drawing Practice

PRACTICE PROBLEMS

1. Which of the following symbols represents an electrical connection, or tie-point, between one part of an electrical circuit and another?

(A) ⟶▷

(B) ⟶╫╷

(C) ⟶⪙

(D)

2. Which symbol represents a ground for safety purposes on a metallic enclosure?

(A) ⟶▷

(B) ⟶╫╷ Earth

(C) ⟶⪙ Chassis

(D) ⟶⟋⟍⟶ Thermal overloads

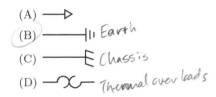

3. Which of the following symbols would be used in an electrical schematic to represent a MOSFET?

(A)

(B)

(C)

(D)

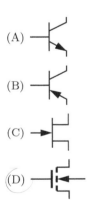

4. Which of the following symbols would be used in an electrical schematic to represent an NPN bipolar junction transistor?

(A) B C npn E

(B) pnp

(C)

(D)

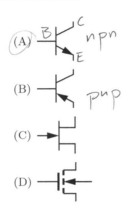

5. Which of the following symbols represents a controlled source?

(A) ⟶╫

(B) ⟶◯

(C) ⟶◇

(D)

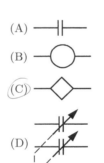

SOLUTIONS

1. A tie-point is represented in an electrical schematic by the symbol

The answer is (A).

2. An earth ground is used for safety. The symbol for an earth ground is

The answer is (B).

3. An *n*-channel enhancement (four-terminal) MOS-FET is symbolized by

The answer is (D).

4. An NPN bipolar junction transistor is represented by

A useful memory aid is that an NPN transistor's arrow is "<u>n</u>ot <u>p</u>ointing <u>in</u>" toward the base, giving NPN.

The answer is (A).

5. A controlled source is used in schematics of electronic models, and is represented by the diamond shape.

The answer is (C).

Algebra

PRACTICE PROBLEMS

1. Calculate the following sum.

$$\sum_{j=1}^{5}\left((j+1)^2 - 1\right)$$

(A) 15

(B) 24

(C) 35

(D) 85

2. If every 0.1 sec a quantity increases by 0.1% of its current value, calculate the doubling time.

(A) 14 sec

(B) 70 sec

(C) 690 sec

(D) 69,000 sec

3. What are the roots of the following equation?

$$x^2 + 5x + 4 = 0$$

(A) $-1, -4$

(B) $-1, 4$

(C) $1, -4$

(D) $-11, -14$

4. What is the base-10 logarithm of 115^2?

(A) 2.06

(B) 4.12

(C) 8.24

(D) 16.5

5. What is the exponential form of $\mathbf{Z} = 2 + j2$?

(A) $0.7e^{j2.83}$

(B) $e^{j0.78}$

(C) $0.78e^{j}$

(D) $2.83e^{j0.78}$

Mathematics

SOLUTIONS

1. Let $S_n = (j+1)^2 - 1$.

For $j = 1$, $S_1 = (1+1)^2 - 1 = 3$

For $j = 2$, $S_2 = (2+1)^2 - 1 = 8$

For $j = 3$, $S_3 = (3+1)^2 - 1 = 15$

For $j = 4$, $S_4 = (4+1)^2 - 1 = 24$

For $j = 5$, $S_5 = (5+1)^2 - 1 = 35$

Substituting the above expressions gives

$$\sum_{j=1}^{5} \left((j+1)^2 - 1 \right) = \sum_{j=1}^{5} S_j$$
$$= S_1 + S_2 + S_3 + S_4 + S_5$$
$$= 3 + 8 + 15 + 24 + 35$$
$$= \boxed{85}$$

The answer is (D).

2. Let n represent the number of elapsed periods of 0.1 sec, and let y_n represent the quantity present after n periods.

y_0 represents the initial quantity.

$$y_1 = 1.001 y_0$$
$$y_2 = 1.001 y_1 = (1.001)(1.001 y_0) = (1.001)^2 y_0$$

Therefore, by deduction,

$$y_n = (1.001)^n y_0$$

The expression for a doubling of the original quantity is

$$2 y_0 = y_n$$

Substitute for y_n.

$$2 y_0 = (1.001)^n y_0$$
$$2 = (1.001)^n$$

Take the logarithm of both sides.

$$\log 2 = \log 1.001^n$$
$$= n \log 1.001$$

Solve for n.

$$n = \frac{\log 2}{\log 1.001} = 693.5$$

Because each period, p, is 0.1 sec, the time is

$$t = np$$
$$= n(0.1 \text{ sec})$$
$$= (693.5)(0.1 \text{ sec})$$
$$= \boxed{69.35 \text{ sec} \quad (70 \text{ sec})}$$

The answer is (B).

3. The equation is in the quadratic form,

$$ax^2 + bx + c = 0$$

Quadratic equations can represent a second order response by electrical circuits. The roots are

$$x_1, x_2 = \frac{-b \pm \sqrt{b^2 - 4ac}}{2a}$$
$$= \frac{-5 \pm \sqrt{(5)^2 - (4)(1)(4)}}{(2)(1)}$$
$$= \boxed{-1, -4}$$

The answer is (A).

4. The property of logarithms used in this case is

$$\log x^a = a \log x$$

Substitute the given information.

$$\log 115^2 = 2 \log 115$$
$$= (2)(2.06)$$
$$= \boxed{4.12}$$

The answer is (B).

5. The exponential form of a complex number, **Z**, is

$$z = re^{j\theta}$$

The radius in the complex plane is r. The square root of -1 is j, and θ is the angle of the radius in the complex plane. Calculate the unknown values using

$$r = \sqrt{x_{real}^2 + y_{imaginary}^2}$$
$$= \sqrt{2^2 + 2^2}$$
$$= 2.83$$
$$\theta = \arctan\frac{y}{x}$$
$$= \arctan\frac{2}{2}$$
$$= 0.78 \text{ rad}$$

Therefore,

$$z = re^{j\theta}$$
$$= \boxed{2.83e^{j0.78}}$$

The answer is (D).

5 Linear Algebra

PRACTICE PROBLEMS

1. What is the determinant of matrix $\mathbf{A}$?

$$\mathbf{A} = \begin{bmatrix} 8 & 2 & 0 & 0 \\ 2 & 8 & 2 & 0 \\ 0 & 2 & 8 & 2 \\ 0 & 0 & 2 & 4 \end{bmatrix}$$

(A)   459

(B)   832

(C)   1552

(D)   1776

2. Use Cramer's rule to solve for the values of x, y, and z that simultaneously satisfy the following equations.

$$x + y = -4$$
$$x + z - 1 = 0$$
$$2z - y + 3x = 4$$

(A) $(x, y, z) = (3, 2, 1)$

(B) $(x, y, z) = (-3, -1, 2)$

(C) $(x, y, z) = (3, -1, -3)$

(D) $(x, y, z) = (-1, -3, 2)$

3. What value of x satisfies all three given equations?

$$4x + 6y - 8z = 2$$
$$6x - 2y - 4z = 8$$
$$8x - 14y - 12z = -14$$

(A)   0.10

(B)   0.20

(C)   0.33

(D)   3.0

4. What is the product of matrices $\mathbf{A}$ and $\mathbf{B}$?

$$\mathbf{A} = \begin{bmatrix} 1 & 2 & 5 \\ 3 & 4 & 7 \end{bmatrix}$$

$$\mathbf{B} = \begin{bmatrix} 2 & 9 \\ 6 & 2 \\ 3 & 4 \end{bmatrix}$$

(A) $\begin{bmatrix} 525 \\ 347 \end{bmatrix}$

(B) $\begin{bmatrix} 29 & 51 \\ 33 & 63 \end{bmatrix}$

(C) $\begin{bmatrix} 29 & 33 \\ 51 & 63 \end{bmatrix}$

(D)   cannot be multiplied

5. What is the determinant of matrix $\mathbf{A}$?

$$\mathbf{A} = \begin{bmatrix} 4 & 4 \\ 3 & 6 \end{bmatrix}$$

(A) -12

(B)   12

(C)   24

(D)   36

SOLUTIONS

1. Expand by cofactors of the first column, because there are two zeros in that column. (The first row could also have been used.)

$$D = 8 \begin{vmatrix} 8 & 2 & 0 \\ 2 & 8 & 2 \\ 0 & 2 & 4 \end{vmatrix} - 2 \begin{vmatrix} 2 & 0 & 0 \\ 2 & 8 & 2 \\ 0 & 2 & 4 \end{vmatrix} + 0 - 0$$

by first column:

$$\begin{vmatrix} 8 & 2 & 0 \\ 2 & 8 & 2 \\ 0 & 2 & 4 \end{vmatrix} = (8)\big((8)(4) - (2)(2)\big) \\ -(2)\big((2)(4) - (2)(0)\big) \\ = 208$$

by first column:

$$\begin{vmatrix} 2 & 0 & 0 \\ 2 & 8 & 2 \\ 0 & 2 & 4 \end{vmatrix} = (2)\big((8)(4) - (2)(2)\big) = 56$$

$$D = (8)(208) - (2)(56) = \boxed{1552}$$

The answer is (C).

2. Rearrange the equations.

$$\begin{aligned} x &+ y & &= -4 \\ x & &+ z &= 1 \\ 3x &- y &+ 2z &= 4 \end{aligned}$$

Write the set of equations in matrix form: $\mathbf{AX} = \mathbf{B}$.

$$\begin{bmatrix} 1 & 1 & 0 \\ 1 & 0 & 1 \\ 3 & -1 & 2 \end{bmatrix} \begin{bmatrix} x \\ y \\ z \end{bmatrix} = \begin{bmatrix} -4 \\ 1 \\ 4 \end{bmatrix}$$

Find the determinant of the matrix $\mathbf{A}$.

$$\begin{aligned} |\mathbf{A}| &= \begin{vmatrix} 1 & 1 & 0 \\ 1 & 0 & 1 \\ 3 & -1 & 2 \end{vmatrix} \\ &= 1 \begin{vmatrix} 0 & 1 \\ -1 & 2 \end{vmatrix} - 1 \begin{vmatrix} 1 & 0 \\ -1 & 2 \end{vmatrix} + 3 \begin{vmatrix} 1 & 0 \\ 0 & 1 \end{vmatrix} \\ &= (1)\big((0)(2) - (1)(-1)\big) \\ &\quad -(1)\big((1)(2) - (-1)(0)\big) \\ &\quad +(3)\big((1)(1) - (0)(0)\big) \\ &= (1)(1) - (1)(2) + (3)(1) \\ &= 1 - 2 + 3 \\ &= 2 \end{aligned}$$

Find the determinant of the substitutional matrix $\mathbf{A}_1$.

$$\begin{aligned} |\mathbf{A}_1| &= \begin{vmatrix} -4 & 1 & 0 \\ 1 & 0 & 1 \\ 4 & -1 & 2 \end{vmatrix} \\ &= -4 \begin{vmatrix} 0 & 1 \\ -1 & 2 \end{vmatrix} - 1 \begin{vmatrix} 1 & 0 \\ -1 & 2 \end{vmatrix} + 4 \begin{vmatrix} 1 & 0 \\ 0 & 1 \end{vmatrix} \\ &= (-4)\big((0)(2) - (1)(-1)\big) \\ &\quad -(1)\big((1)(2) - (-1)(0)\big) \\ &\quad +(4)\big((1)(1) - (0)(0)\big) \\ &= (-4)(1) - (1)(2) + (4)(1) \\ &= -4 - 2 + 4 \\ &= -2 \end{aligned}$$

Find the determinant of the substitutional matrix $\mathbf{A}_2$.

$$\begin{aligned} |\mathbf{A}_2| &= \begin{vmatrix} 1 & -4 & 0 \\ 1 & 1 & 1 \\ 3 & 4 & 2 \end{vmatrix} \\ &= 1 \begin{vmatrix} 1 & 1 \\ 4 & 2 \end{vmatrix} - 1 \begin{vmatrix} -4 & 0 \\ 4 & 2 \end{vmatrix} + 3 \begin{vmatrix} -4 & 0 \\ 1 & 1 \end{vmatrix} \\ &= (1)\big((1)(2) - (4)(1)\big) \\ &\quad -(1)\big((-4)(2) - (4)(0)\big) \\ &\quad +(3)\big((-4)(1) - (1)(0)\big) \\ &= (1)(-2) - (1)(-8) + (3)(-4) \\ &= -2 + 8 - 12 \\ &= -6 \end{aligned}$$

Find the determinant of the substitutional matrix $\mathbf{A}_3$.

$$|\mathbf{A}_3| = \begin{vmatrix} 1 & 1 & -4 \\ 1 & 0 & 1 \\ 3 & -1 & 4 \end{vmatrix}$$

$$= 1 \begin{vmatrix} 0 & 1 \\ -1 & 4 \end{vmatrix} - 1 \begin{vmatrix} 1 & -4 \\ -1 & 4 \end{vmatrix} + 3 \begin{vmatrix} 1 & -4 \\ 0 & 1 \end{vmatrix}$$

$$= (1)\big((0)(4) - (-1)(1)\big)$$

$$\quad - (1)\big((1)(4) - (-1)(-4)\big)$$

$$\quad + (3)\big((1)(1) - (0)(-4)\big)$$

$$= (1)(1) - (1)(0) + (3)(1)$$

$$= 1 - 0 + 3$$

$$= 4$$

Use Cramer's rule.

$$x = \frac{|\mathbf{A}_1|}{|\mathbf{A}|} = \frac{-2}{2} = \boxed{-1}$$

$$y = \frac{|\mathbf{A}_2|}{|\mathbf{A}|} = \frac{-6}{2} = \boxed{-3}$$

$$z = \frac{|\mathbf{A}_3|}{|\mathbf{A}|} = \frac{4}{2} = \boxed{2}$$

The answer is (D).

3. Use Cramer's rule to solve for x. The determinant of the coefficient matrix is

$$|\mathbf{A}| = \begin{vmatrix} 4 & 6 & -8 \\ 6 & -2 & -4 \\ 8 & -14 & -12 \end{vmatrix}$$

$$= 4 \begin{vmatrix} -2 & -4 \\ -14 & -12 \end{vmatrix} - 6 \begin{vmatrix} 6 & -8 \\ -14 & -12 \end{vmatrix} + 8 \begin{vmatrix} 6 & -8 \\ -2 & -4 \end{vmatrix}$$

$$= (4)\big((-2)(-12) - (-4)(-14)\big)$$

$$\quad - (6)\big((6)(-12) - (-8)(-14)\big)$$

$$\quad + (8)\big((6)(-4) - (-2)(-8)\big)$$

$$= (4)(24 - 56) - (6)(-72 - 112)$$

$$\quad + (8)(-24 - 16)$$

$$= 656$$

The determinant of the substitutional matrix is

$$|\mathbf{A}_1| = \begin{vmatrix} 2 & 6 & -8 \\ 8 & -2 & -4 \\ -14 & -14 & -12 \end{vmatrix}$$

$$= 1968$$

Therefore,

$$x = \frac{|\mathbf{A}_1|}{|\mathbf{A}|} = \frac{1968}{656} = 3$$

$$= \boxed{3}$$

The answer is (D).

4. Let the product matrix be $\mathbf{C}$.

$$\mathbf{C} = \begin{bmatrix} c_{11} & c_{12} \\ c_{21} & c_{22} \end{bmatrix} = \mathbf{AB}$$

$$= \begin{bmatrix} 1 & 2 & 5 \\ 3 & 4 & 7 \end{bmatrix} \begin{bmatrix} 2 & 9 \\ 6 & 2 \\ 3 & 4 \end{bmatrix}$$

$$c_{11} = (1)(2) + (2)(6) + (5)(3)$$

$$\quad = 2 + 12 + 15$$

$$\quad = 29$$

$$c_{12} = (1)(9) + (2)(2) + (5)(4)$$

$$\quad = 9 + 4 + 20$$

$$\quad = 33$$

$$c_{21} = (3)(2) + (4)(6) + (7)(3)$$

$$\quad = 6 + 24 + 21$$

$$\quad = 51$$

$$c_{22} = (3)(9) + (4)(2) + (7)(4)$$

$$\quad = 27 + 8 + 28$$

$$\quad = 63$$

Therefore,

$$\mathbf{C} = \boxed{\begin{bmatrix} 29 & 33 \\ 51 & 63 \end{bmatrix}}$$

The answer is (C).

5. The determinant of any 2×2 matrix is

$$\begin{vmatrix} \mathbf{A} \end{vmatrix} = \begin{vmatrix} a & c \\ b & d \end{vmatrix} = ad - bc$$

In this case,

$$|\mathbf{A}| = \begin{vmatrix} 4 & 4 \\ 3 & 6 \end{vmatrix} = (4)(6) - (4)(3)$$

$$= 24 - 12$$

$$= \boxed{12}$$

The answer is (B).

6 Vectors

PRACTICE PROBLEMS

1. What is the unit vector for $\mathbf{V} = 2\mathbf{i} + 4\mathbf{j} + 4\mathbf{k}$?

(A) $1\mathbf{i} + 2\mathbf{j} + 2\mathbf{k}$

(B) $(1/18)\mathbf{i} + (1/6)\mathbf{j} + (1/6)\mathbf{k}$

(C) $(2/3)\mathbf{i} + (1/3)\mathbf{j} + (1/3)\mathbf{k}$

(D) $(1/3)\mathbf{i} + (2/3)\mathbf{j} + (2/3)\mathbf{k}$

2. What is the dot product of the two vectors given?

$$\mathbf{V}_1 = 2\,\mathbf{i} + 2\,\mathbf{j} + 4\mathbf{k}$$
$$\mathbf{V}_2 = 2\,\mathbf{i} + 2\,\mathbf{j} + 1\mathbf{k}$$

(A) $4\angle 35°$

(B) 4

(C) $12\angle 35°$

(D) 12

3. Which of the following vectors is orthogonal to $\mathbf{V}_1$ and $\mathbf{V}_2$?

$$\mathbf{V}_1 = \mathbf{i} + \mathbf{j} + 2\mathbf{k}$$
$$\mathbf{V}_2 = \mathbf{i} + 2\mathbf{j}$$

(A) $\mathbf{i} - 2\mathbf{j} + 2\mathbf{k}$

(B) $-4\mathbf{i} + 2\mathbf{j} + \mathbf{k}$

(C) $-4\mathbf{i} - 2\mathbf{j} - \mathbf{k}$

(D) $4\mathbf{i} - 2\mathbf{j} + \mathbf{k}$

4. The magnitude of the vector cross product corresponds to which of the following?

(A) area

(B) line

(C) projection

(D) volume

5. Which of the following vectors is orthogonal to $\mathbf{V}_1 = 2\mathbf{i} + 4\mathbf{j}$?

(A) $2\mathbf{i} - 4\mathbf{j}$

(B) $4\mathbf{i} - 2\mathbf{j}$

(C) $-2\mathbf{i} - 4\mathbf{j}$

(D) $2\mathbf{i} + 4\mathbf{j}$

SOLUTIONS

1. The unit vector **a** for vector **V** is

$$\mathbf{a} = \frac{\mathbf{V}}{|\mathbf{V}|}$$
$$= \frac{2\mathbf{i} + 4\mathbf{j} + 4\mathbf{k}}{\sqrt{2^2 + 4^2 + 4^2}}$$
$$= \frac{2\mathbf{i} + 4\mathbf{j} + 4\mathbf{k}}{6}$$
$$= \boxed{(1/3)\mathbf{i} + (2/3)\mathbf{j} + (2/3)\mathbf{k}}$$

The answer is (D).

2. The dot product is

$$\mathbf{V}_1 \cdot \mathbf{V}_2 = |\mathbf{V}_1||\mathbf{V}_2|\cos\phi = V_{1x}V_{2x} + V_{1y}V_{2y} + V_{1z}V_{2z}$$
$$= (2)(2) + (2)(2) + (4)(1)$$
$$= \boxed{12}$$

The answer is (D).

3. An orthogonal vector is found from the cross product. The order of the cross product will change the perpendicular found. Using $\mathbf{V}_1 \times \mathbf{V}_2$ gives

$$\mathbf{V}_1 \times \mathbf{V}_2 = \begin{vmatrix} \mathbf{i} & 1 & 1 \\ \mathbf{j} & 1 & 2 \\ \mathbf{k} & 2 & 0 \end{vmatrix}$$
$$= \mathbf{i}\begin{vmatrix} 1 & 2 \\ 2 & 0 \end{vmatrix} - \mathbf{j}\begin{vmatrix} 1 & 1 \\ 2 & 0 \end{vmatrix} + \mathbf{k}\begin{vmatrix} 1 & 1 \\ 1 & 2 \end{vmatrix}$$
$$= \mathbf{i}(0 - 4) - \mathbf{j}(0 - 2) + \mathbf{k}(2 - 1)$$
$$= \boxed{-4\mathbf{i} + 2\mathbf{j} + \mathbf{k}}$$

The answer is (B).

4. The magnitude of the vector cross product for two vectors, say $\mathbf{V}_1 \times \mathbf{V}_2$, corresponds to the area of a parallelogram with $\mathbf{V}_1$ and $\mathbf{V}_2$ as its sides.

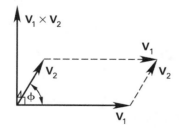

The answer is (A).

5. A vector is orthogonal to another if the dot product of the two vectors equals zero.

Checking answer (A) gives

$$\mathbf{V}_1 \cdot \mathbf{A} = V_{1x}\mathbf{A}_{1x} + V_{2y}\mathbf{A}_{2y}$$
$$= (2)(2) + (4)(-4)$$
$$= -12$$

Because the dot product result is not zero, **A** is not orthogonal.

Checking answer (B) gives

$$\mathbf{V}_1 \cdot \mathbf{B} = V_{1x}\mathbf{B}_{1x} + V_{2y}\mathbf{B}_{2y}$$
$$= (2)(4) + (4)(-2)$$
$$= 0$$

$$\boxed{4\mathbf{i} - 2\mathbf{j} \text{ is orthogonal to } \mathbf{V}_1 = 2\mathbf{i} + 4\mathbf{j}.}$$

The answer is (B).

7 Trigonometry

PRACTICE PROBLEMS

1. A 5 lbm (5 kg) block sits on a 20° incline without slipping.

(a) Draw the free-body diagram with respect to axes parallel and perpendicular to the surface of the incline.

(b) Determine the magnitude of the frictional force on the block.

   (A)   1.71 lbf (16.8 N)

   (B)   3.35 lbf (32.9 N)

   (C)   4.70 lbf (46.1 N)

   (D)   5.00 lbf (49.1 N)

2. Consider the unit circle shown.

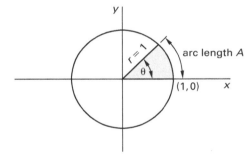

What is the value of the arc length, A, in radians, if the shaded area is $\theta/2$?

   (A) θ

   (B) $\pi/4$

   (C) $\pi/2$

   (D) $r\theta$

3. If the cosine of the angle θ is given as C, what is the tangent of the angle θ?

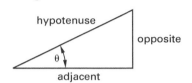

   (A) $\dfrac{C \times \text{hypotenuse}}{\text{opposite}}$

   (B) $\dfrac{\text{adjacent}}{\text{opposite}}$

   (C) $\dfrac{\text{opposite} \times \text{adjacent}}{C \times \text{hypotenuse}}$

   (D) $\dfrac{\text{opposite}}{C \times \text{hypotenuse}}$

4. What is the value of the sum of the squares of the sine and cosine of an angle θ?

   (A)   0

   (B) $\pi/4$

   (C)   1

   (D) 2π

5. What is the approximate value of tan 10° in radians?

   (A)   1.00

   (B)   10.0°

   (C)   0.0500

   (D)   0.175

SOLUTIONS

1. (a)

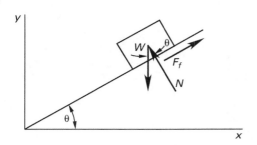

Customary U.S. Solution

(b) The mass of the block, m, is 5 lbm.

The angle of inclination, θ, is 20°. The weight is

$$W = \frac{mg}{g_c}$$

$$= \frac{(5 \text{ lbm})\left(32.2 \ \dfrac{\text{ft}}{\text{sec}^2}\right)}{32.2 \ \dfrac{\text{lbm-ft}}{\text{lbf-sec}^2}}$$

$$= 5 \text{ lbf}$$

The frictional force is

$$F_f = W \sin\theta$$

$$= (5 \text{ lbf})\sin 20°$$

$$= \boxed{1.71 \text{ lbf}}$$

The answer is (A).

SI Solution

(b) The mass of the block, m, is 5 kg.

The angle of inclination, θ, is 20°. The gravitational force is

$$W = mg$$

$$= (5 \text{ kg})\left(9.81 \ \frac{\text{m}}{\text{s}^2}\right)$$

$$= 49.1 \text{ N}$$

The frictional force is

$$F_f = W \sin\theta$$

$$= (49.1 \text{ N})\sin 20°$$

$$= \boxed{16.8 \text{ N}}$$

The answer is (A).

2. The area of a sector with a central angle of θ radians is $\theta/2$ units for a unit circle. The arc length is therefore θ radians.

The answer is (A).

3. The cosine is

$$C = \frac{\text{adjacent}}{\text{hypotenuse}} \qquad [\text{I}]$$

The tangent is

$$T = \frac{\text{opposite}}{\text{adjacent}} \qquad [\text{II}]$$

Substituting the value of adjacent from Eq. I into Eq. II gives

$$T = \frac{\text{opposite}}{\text{adjacent}} = \boxed{\frac{\text{opposite}}{C \times \text{hypotenuse}}}$$

The answer is (D).

4. For any angle θ, the following is true.

$$\cos^2\theta + \sin^2\theta = \boxed{1}$$

The answer is (C).

5. For angles less than approximately 10°, the following is valid.

$$\sin\theta \approx \tan\theta \approx \theta\big|_{\theta \leq 10°}$$

The value of θ must be expressed in radians. Converting the 10° to radians gives

$$\theta_{\text{radians}} = 10°\left(\frac{2\pi \text{ rad}}{360°}\right)$$

$$= 0.05\pi \text{ rad}$$

$$= \boxed{0.175 \text{ rad}}$$

The answer is (D).

8 Analytic Geometry

PRACTICE PROBLEMS

1. The diameter of a sphere and of the base of a cone are equal. What percentage of that diameter must the cone's height be, so that both volumes are equal?

(A) 133%

(B) 150%

(C) 166%

(D) 200%

2. Consider the graph shown. What is the slope of the line?

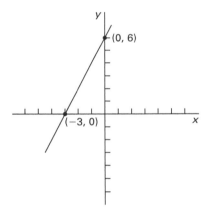

(A) $-1/2$

(B) $1/2$

(C) 2

(D) 4

3. What type of symmetry is shown?

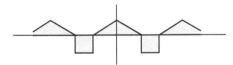

(A) even

(B) odd

(C) rotational

(D) quadratic

4. What expression in the spherical coordinate system represents the z coordinate in the Cartesian coordinate system?

(A) z

(B) $r \sin \theta$

(C) $r \cos \phi$

(D) $r \sin \phi \sin \theta$

5. What geometric figure does the following equation represent?

$$(x-2)^2 + (y-2)^2 + (z-2)^2 = 100$$

(A) circle

(B) ellipse

(C) helix

(D) sphere

SOLUTIONS

1. Let d be the diameter of the sphere and the base of the cone.

The volume of the sphere is

$$V_{\text{sphere}} = \frac{4}{3}\pi r^3 = \frac{4}{3}\pi\left(\frac{d}{2}\right)^3$$
$$= \frac{\pi}{6}d^3$$

The volume of the circular cone is

$$V_{\text{cone}} = \frac{1}{3}\pi r^2 h = \frac{1}{3}\pi\left(\frac{d}{2}\right)^2 h$$
$$= \frac{\pi}{12}d^2 h$$

The volumes of the sphere and cone are equal.

$$V_{\text{cone}} = V_{\text{sphere}}$$
$$\frac{\pi}{12}d^2 h = \frac{\pi}{6}d^3$$
$$h = 2d$$

The cone's height must be 200% of its diameter.

The answer is (D).

2. The two-point form of a line is

$$\frac{y - y_1}{x - x_1} = \frac{y_2 - y_1}{x_2 - x_1}$$

Substitute for the two points given.

$$\frac{y - 6}{x - 0} = \frac{0 - 6}{-3 - 0}$$
$$\frac{y - 6}{x} = 2$$
$$y - 6 = 2x$$
$$y = 2x + 6$$

The equation is in the slope intercept form of a line,

$$y = mx + b$$

The slope value is m. Therefore, in this case, the slope is 2.

The answer is (C).

3. The symmetry is such that the value of y is identical for any $\pm\,x$ value. This is even symmetry.

The answer is (A).

4. In the spherical system

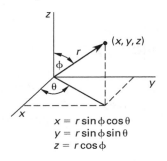

$$x = r\sin\phi\cos\theta$$
$$y = r\sin\phi\sin\theta$$
$$z = r\cos\phi$$

The value of z is $r\cos\phi$.

The answer is (C).

5. The general equation for a sphere that is centered at (h, k, l) is

$$(x - h)^2 + (y - k)^2 + (z - l)^2 = r^2$$

The equation given is in this form. The equation represents a sphere centered at $(2,2,2)$.

The answer is (D).

 **Differential Calculus**

PRACTICE PROBLEMS

1. What are the values of a, b, and c in the following expression, such that $n(\infty) = 100$, $n(0) = 10$, and $dn(0)/dt = 0.5$?

$$n(t) = \frac{a}{1 + be^{ct}}$$

(A) $a = 10$, $b = 9$, $c = 1$

(B) $a = 100$, $b = 10$, $c = 1.5$

(C) $a = 100$, $b = 9$, $c = -0.056$

(D) $a = 1000$, $b = 10$, $c = -0.056$

2. Find all minima, maxima, and inflection points for

$$y = x^3 - 9x^2 - 3$$

(A)

    maximum at $x = 0$
    inflection at $x = 3$
    minimum at $x = 6$

(B)

    maximum at $x = 0$
    inflection at $x = -3$
    minimum at $x = -6$

(C)

    maximum at $x = 0$
    inflection at $x = 3$
    minimum at $x = 6$

(D)

    maximum at $x = 3$
    inflection at $x = 0$
    minimum at $x = -3$

3. What is the derivative of the following equation?

$$f(t) = 170 \cos 377t$$

(A) $120 \sin 377t$

(B) $170 \sin 377t$

(C) $-170 \sin 377t$

(D) $-64{,}000 \sin 377t$

4. What is the gradient of the following function?

$$f(x, y) = x^2 - 2y^2 + 2x + y$$

(A) $(2x - 2)\mathbf{i} - (4y + 1)\mathbf{j}$

(B) $(2x + 2)\mathbf{i} - (-4y + 1)\mathbf{j}$

(C) $(-2x - 2)\mathbf{i} + (-4y + 1)\mathbf{j}$

(D) $(2x + 2)\mathbf{i} + (-4y + 1)\mathbf{j}$

5. What is the divergence of the following vector function?

$$\mathbf{F}(x, y, z) = 2xz\mathbf{i} + e^x y^2 \mathbf{j} + x^2 y \mathbf{k}$$

(A) $(2z + e^x + 2x)\mathbf{i} + (2ye^x + x^2)\mathbf{j} + 2x\mathbf{k}$

(B) $2z + 2ye^x$

(C) $2z + 2ye^x + 1$

(D) $2z\mathbf{i} + 2ye^x\mathbf{j} + \mathbf{k}$

Mathematics

SOLUTIONS

1. If c is positive, then $n(\infty)$ is zero, which is contrary to the data given. Therefore, c is less than or equal to zero. If c is zero, then $n(\infty) = a/(1 + b) = 100$, which is possible depending on the values of a and b. However, $n(0) = a/(1 + b)$ would also equal 100, which is contrary to the given data. Therefore, c is not zero. c must be less than zero.

Because $c < 0$, then $n(\infty) = a$, so $\boxed{a = 100.}$

Applying the condition $t = 0$ gives

$$n(0) = \frac{a}{1 + b} = 10$$

Because $a = 100$,

$$n(0) = \frac{100}{1 + b}$$
$$100 = (10)(1 + b)$$
$$10 = 1 + b$$
$$\boxed{b = 9}$$

Substitute the results for a and b into the expression.

$$n(t) = \frac{100}{1 + 9e^{ct}}$$

Take the first derivative.

$$\frac{d}{dt} n(t) = \left(\frac{100}{\left(1 + 9e^{ct}\right)^2} \right)(-9ce^{ct})$$

Apply the initial condition.

$$\frac{d}{dt} n(0) = \left(\frac{100}{\left(1 + 9e^{c(0)}\right)^2} \right)(-9ce^{c(0)}) = 0.5$$

$$\left(\frac{100}{(1 + 9)^2} \right)(-9c) = 0.5$$

$$(1)(-9c) = 0.5$$

$$c = \frac{-0.5}{9}$$

$$= \boxed{-0.056}$$

Substitute the terms a, b, and c into the expression.

$$n(t) = \frac{100}{1 + 9e^{-0.056t}}$$

The answer is (C).

2. Determine the critical points by taking the first derivative of the function and setting it equal to zero.

$$\frac{dy}{dx} = 3x^2 - 18x = 3x(x - 6)$$
$$3x(x - 6) = 0$$
$$x(x - 6) = 0$$

The critical points are located at $x = 0$ and $x = 6$.

Determine the inflection points by setting the second derivative equal to zero. Take the second derivative.

$$\frac{d^2y}{dx^2} = \left(\frac{d}{dx} \right) \left(\frac{dy}{dx} \right) = \frac{d}{dx}(3x^2 - 18x)$$
$$= 6x - 18$$

Set the second derivative equal to zero.

$$\frac{d^2y}{dx^2} = 0 = 6x - 18 = 6(x - 3)$$
$$6(x - 3) = 0$$
$$x - 3 = 0$$
$$x = 3$$

$$\boxed{\text{This inflection point is at } x = 3.}$$

Determine the local maximum and minimum by substituting the critical points into the expression for the second derivative.

At the critical point $x = 0$,

$$\left. \frac{d^2y}{dx^2} \right|_{x=0} = 6(x - 3) = (6)(0 - 3)$$
$$= -18$$

Because $-18 < 0$, $\boxed{x = 0 \text{ is a local maximum.}}$

At the critical point $x = 6$,

$$\left. \frac{d^2y}{dx^2} \right|_{x=6} = 6(x - 3) = (6)(6 - 3)$$
$$= 18$$

Because $18 > 0$, $\boxed{x = 6 \text{ is a local minimum.}}$

The answer is (A).

3. The derivative of $f(t)$ is

$$f'(t) = 170\frac{d\cos 377t}{dt}$$

$$= (170)(-\sin 377t)\left(\frac{d(377t)}{dt}\right)$$

$$= (-170)\sin 377t\left[377\frac{dt}{dt}\right]$$

$$= \boxed{-64{,}090\sin 377t \quad (-64{,}000\sin 377t)}$$

The answer is (D).

4. The gradient of $f(x, y)$ is

$$\boldsymbol{\nabla}f(x,\ y) = \frac{\partial f(x,\ y)}{\partial x}\,\mathbf{i} + \frac{\partial f(x,\ y)}{\partial y}\mathbf{j}$$

$$= \frac{\partial(x^2 - 2y^2 + 2x + y)}{\partial x}\,\mathbf{i}$$

$$+\frac{\partial(x^2 - 2y^2 + 2x + y)}{\partial y}\mathbf{j}$$

$$= \boxed{(2x+2)\,\mathbf{i} + (-4y+1)\mathbf{j}}$$

The answer is (D).

5. The divergence is a scalar, given by

$$\text{div } \mathbf{F} = \boldsymbol{\nabla}\cdot\mathbf{F} = \frac{\partial F_x}{\partial x} + \frac{\partial F_y}{\partial y} + \frac{\partial F_z}{\partial z}$$

$$= \frac{\partial(2xz)}{\partial x} + \frac{\partial(e^x y^2)}{\partial y} + \frac{\partial(x^2 y)}{\partial z}$$

$$= 2z + 2e^x y + 0$$

$$= \boxed{2z + 2ye^x}$$

The answer is (B).

10 Integral Calculus

PRACTICE PROBLEMS

1. What is the result of the indefinite integration shown?

$$\int \sin \theta \, d\theta$$

(A) $-\cos \theta + C$

(B) $\cos \theta + C$

(C) $-\sin \theta + C$

(D) $\dfrac{-\sin^2 \theta}{2}$

2. What is the value of the following definite integral?

$$\int_0^{\pi/2} \sin x \, dx$$

(A) -1

(B) 0

(C) 1

(D) $\pi/2$

3. Consider the following square wave function.

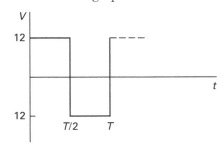

The period of the function is 0.0167 s. What is the value of a_0? That is, what is the value of the first term in a Fourier series representing the function?

(A) -6

(B) 0

(C) $+6$

(D) $+12$

4. What is the area between the curve shown, $y_1 = e^x$, and the straight line, $y_2 = x + 1$, shown on the interval $x = [0, 2]$?

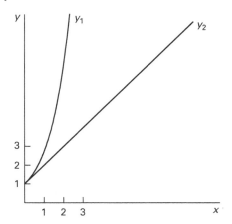

(A) 2.4

(B) 3.4

(C) 5.4

(D) 7.4

5. What is the approximate length of the curve $y = x^2$ from $x = 0$ to $x = 5$?

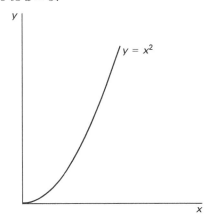

(A) 10

(B) 17

(C) 25

(D) 170

SOLUTIONS

1. The indefinite integral of the sine function is

$$\int \sin\theta \, d\theta = \boxed{-\cos\theta + C}$$

C represents a constant.

The answer is (A).

2. The definite integral is

$$\int_0^{\pi/2} \sin x \, dx = -\cos x \Big|_0^{\pi/2}$$
$$= -\cos\frac{\pi}{2} - (-\cos 0)$$
$$= 0 + 1$$
$$= \boxed{1}$$

The answer is (C).

3. The Fourier series is

$$f(t) = \tfrac{1}{2}a_0 + a_1\cos\omega t + a_2\cos 2\omega t + \cdots$$
$$+ b_1\sin\omega t + b_2\sin 2\omega t + \cdots$$

Normalizing by dividing all frequencies by ω gives

$$f(t) = \tfrac{1}{2}a_0 + a_1\cos t + a_2\cos 2t + \cdots$$
$$+ b_1\sin t + b_2\sin 2t + \cdots$$

The first term, $(1/2)a_0$, is the average value of the function. Calculate the value of a_0.

$$a_0 = \frac{2}{T}\int_0^T f(t)\,dt$$
$$= \frac{2}{T}\left(\int_0^{T/2} 12\,dt + \int_{T/2}^T (-12)\,dt\right)$$
$$= \frac{2}{T}\left(12t\Big|_0^{T/2} - 12t\Big|_{T/2}^T\right)$$
$$= \frac{2}{T}\left(\left(12\frac{T}{2} - (12)(0)\right) + \left((-12T) - (-12)\frac{T}{2}\right)\right)$$
$$= \frac{2}{T}\left(12\frac{T}{2} - 12T + 12\frac{T}{2}\right)$$
$$= \frac{2}{T}(12T - 12T)$$
$$= \frac{2}{T}(0)$$
$$= \boxed{0}$$

The average value, $(1/2)a_0$, equals zero, which can also be determined by inspection of the waveform.

The answer is (B).

4. The area bounded by the two functions is given by the following.

$$A = \int_a^b \big(y_1(x) - y_2(x)\big)\,dx$$
$$= \int_0^2 \big(e^x - (x+1)\big)\,dx = \int_0^2 (e^x - x - 1)\,dx$$
$$= \int_0^2 e^x\,dx - \int_0^2 x\,dx - \int_0^2 dx$$
$$= e^x\Big|_0^2 - \frac{x^2}{2}\Big|_0^2 - x\Big|_0^2$$
$$= (e^2 - e^0) - \left(\frac{2^2}{2} - \frac{0^2}{2}\right) - (2 - 0)$$
$$= e^2 - 1 - 2 - 2 = e^2 - 5 = 7.389 - 5$$
$$= \boxed{2.389 \quad (2.4)}$$

The answer is (A).

5. The length of the curve may be found from

$$L_{\text{curve}} = \int_a^b \sqrt{1 + \big(f'(x)\big)^2}\,dx \qquad \text{[I]}$$

Because $f(x) = x^2$, the derivative $f'(x) = 2x$. Substitute the derivative function and the given limits into Eq. I.

$$L_{\text{curve}} = \int_0^5 \sqrt{1 + (2x)^2}\,dx$$
$$= \int_0^5 \sqrt{1 + 4x^2}\,dx \qquad \text{[II]}$$

Use a table of integrals to determine the following.

$$\int \sqrt{a^2 + x^2}\,dx = \tfrac{1}{2}\Big((x\sqrt{a^2 + x^2})$$
$$+ a^2\ln(x + \sqrt{a^2 + x^2})\Big)$$

Put Eq. II into the form of the solution from the table of integrals and solve.

$$L_{\text{curve}} = \int_0^5 \sqrt{1 + 4x^2}\, dx = 2\int_0^5 \sqrt{\left(\frac{1}{2}\right)^2 + x^2}\, dx$$

$$= (2)\left(\left(\frac{1}{2}\right)\left(\frac{x\sqrt{\left(\frac{1}{2}\right)^2 + x^2} + \left(\frac{1}{2}\right)^2}{\times \ln\left(x + \sqrt{\left(\frac{1}{2}\right)^2 + x^2}\right)}\right)\Bigg|_0^5\right)$$

$$= (2)\left(\begin{array}{l}\left(\frac{1}{2}\right)\left(\begin{array}{l}5\sqrt{\left(\frac{1}{2}\right)^2 + 5^2} + \left(\frac{1}{2}\right)^2 \\ \times \ln\left(5 + \sqrt{\left(\frac{1}{2}\right)^2 + 5^2}\right)\end{array}\right) \\ -\left(\frac{1}{2}\right)\left(\begin{array}{l}0\sqrt{\left(\frac{1}{2}\right)^2 + 0^2} + \left(\frac{1}{2}\right)^2 \\ \times \ln\left(0 + \sqrt{\left(\frac{1}{2}\right)^2 + 0^2}\right)\end{array}\right)\end{array}\right)$$

$$= \boxed{25.8 \quad (25)}$$

The answer is (C).

11 Differential Equations

PRACTICE PROBLEMS

1. Solve the following differential equation for y.

$$y'' - 4y' - 12y = 0$$

(A) $A_1 e^{6x} + A_2 e^{-2x}$

(B) $A_1 e^{-6x} + A_2 e^{2x}$

(C) $A_1 e^{6x} + A_2 e^{2x}$

(D) $A_1 e^{-6x} + A_2 e^{-2x}$

2. Solve the following differential equation for y.

$$y' - y = 2xe^{2x} \qquad y(0) = 1$$

(A) $y = 2e^{-2x}(x-1) + 3e^{-x}$

(B) $y = 2e^{2x}(x-1) + 3e^{x}$

(C) $y = -2e^{-2x}(x-1) + 3e^{-x}$

(D) $y = 2e^{2x}(x-1) + 3e^{-x}$

3. The oscillation exhibited by the top story of a certain building in free motion is given by the following differential equation.

$$x'' + 2x' + 2x = 0 \qquad x(0) = 0 \qquad x'(0) = 1$$

(a) What is x as a function of time?

(A) $e^{-2t}\sin t$

(B) $e^{t}\sin t$

(C) $e^{-t}\sin t$

(D) $e^{-t}\sin t + e^{-t}\cos t$

(b) What is the building's fundamental natural frequency of vibration?

(A) $\frac{1}{2}$

(B) 1

(C) $\sqrt{2}$

(D) 2

(c) What is the amplitude of oscillation?

(A) 0.32

(B) 0.54

(C) 1.7

(D) 6.6

(d) What is x as a function of time if a lateral wind load is applied with a form of $\sin t$?

(A) $\frac{2}{5}e^{-t}\cos t + \frac{6}{5}e^{-t}\sin t$

(B) $\frac{2}{5}e^{t}\cos t + \frac{6}{5}e^{t}\sin t$

(C) $\frac{6}{5}e^{-t}\cos t + \frac{2}{5}e^{-t}\sin t - \frac{2}{5}\cos t + \frac{2}{5}\sin t$

(D) $\frac{2}{5}e^{-t}\cos t + \frac{6}{5}e^{-t}\sin t - \frac{2}{5}\cos t + \frac{1}{5}\sin t$

4. A 90 lbm (40 kg) bag of a chemical is accidentally dropped into an aerating lagoon. The chemical is water soluble and nonreacting. The lagoon is 120 ft (35 m) in diameter and filled to a depth of 10 ft (3 m). The aerators circulate and distribute the chemical evenly throughout the lagoon.

Water enters the lagoon at a rate of 30 gal/min (115 L/min). Fully mixed water is pumped into a reservoir at a rate of 30 gal/min (115 L/min).

The established safe concentration of this chemical is 1 ppb (part per billion). Approximately how many days will it take for the concentration of the discharge water to reach this level?

(A)   25 days

(B)   50 days

(C)   100 days

(D)   200 days

5. A tank contains 100 gal (100 L) of brine made by dissolving 60 lbm (60 kg) of salt in pure water. Salt water with a concentration of 1 lbm/gal (1 kg/L) enters the tank at a rate of 2 gal/min (2 L/min). A well-stirred mixture is drawn from the tank at a rate of 3 gal/min (3 L/min). Find the mass of salt in the tank after 1 hr.

(A)   13 lbm (13 kg)

(B)   37 lbm (37 kg)

(C)   43 lbm (43 kg)

(D)   51 lbm (51 kg)

SOLUTIONS

1. Obtain the characteristic equation by replacing each derivative with a polynomial term of equal degree.

$$r^2 - 4r - 12 = 0$$

Factor the characteristic equation.

$$(r - 6)(r + 2) = 0$$

The roots are $r_1 = 6$ and $r_2 = -2$.

Since the roots are real and distinct, the solution is

$$y = A_1 e^{r_1 x} + A_2 e^{r_2 x}$$
$$= \boxed{A_1 e^{6x} + A_2 e^{-2x}}$$

The answer is (A).

2. The equation is a first-order linear differential equation of the form

$$y' + p(x)y = g(x)$$
$$p(x) = -1$$
$$g(x) = 2xe^{2x}$$

The integration factor $u(x)$ is given by

$$u(x) = e^{\left(\int p(x)\,dx\right)}$$
$$= e^{\left(\int (-1)\,dx\right)}$$
$$= e^{-x}$$

The closed form of the solution is given by

$$y = \left(\frac{1}{u(x)}\right)\left(\int u(x)g(x)\,dx + C\right)$$
$$= \left(\frac{1}{e^{-x}}\right)\left(\int (e^{-x})(2xe^{2x})\,dx + C\right)$$
$$= e^x\left(2(xe^x - e^x) + C\right)$$
$$= e^x\left(2e^x(x - 1) + C\right)$$
$$= 2e^{2x}(x - 1) + Ce^x$$

Apply the initial condition $y(0) = 1$ to obtain the integration constant C.

$$y(0) = 2e^{(2)(0)}(0 - 1) + Ce^0 = 1$$
$$(2)(1)(-1) + C(1) = 1$$
$$-2 + C = 1$$
$$C = 3$$

Substituting in the value for the integration constant, C, the solution is

$$y = \boxed{2e^{2x}(x-1) + 3e^x}$$

The answer is (B).

3. (a) The differential equation is a homogeneous second-order linear differential equation with constant coefficients. Write the characteristic equation.

$$r^2 + 2r + 2 = 0$$

This is a quadratic equation of the form $ar^2 + br + c = 0$, where a equals 1, b equals 2, and c equals 2.

Solve for r.

$$
\begin{aligned}
r &= \frac{-b \pm \sqrt{b^2 - 4ac}}{2a} \\
&= \frac{-2 \pm \sqrt{(2)^2 - (4)(1)(2)}}{(2)(1)} \\
&= \frac{-2 \pm \sqrt{4 - 8}}{2} \\
&= \frac{-2 \pm \sqrt{-4}}{2} \\
&= \frac{-2 \pm 2\sqrt{-1}}{2} \\
&= -1 \pm \sqrt{-1} \\
&= -1 \pm i \\
r_1 &= -1 + i \text{ and } r_2 = -1 - i
\end{aligned}
$$

Because the roots are imaginary and of the form $\alpha + i\omega$ and $\alpha - i\omega$, where α equals -1 and ω equals 1, the general form of the solution is given by

$$
\begin{aligned}
x(t) &= A_1 e^{\alpha t} \cos \omega t + A_2 e^{\alpha t} \sin \omega t \\
&= A_1 e^{-1t} \cos(1t) + A_2 e^{-1t} \sin(1t) \\
&= A_1 e^{-t} \cos t + A_2 e^{-t} \sin t
\end{aligned}
$$

Apply the initial conditions $x(0) = 0$ and $x'(0) = 1$ to solve for A_1 and A_2.

First, apply the initial condition $x(0) = 0$.

$$
\begin{aligned}
x(t) &= A_1 e^0 \cos 0 + A_2 e^0 \sin 0 = 0 \\
&\quad A_1(1)(1) + A_2(1)(0) = 0 \\
&\quad\quad\quad\quad\quad\quad A_1 = 0
\end{aligned}
$$

Substituting, the solution of the differential equation becomes

$$x(t) = A_2 e^{-t} \sin t$$

To apply the second initial condition, take the first derivative.

$$
\begin{aligned}
x'(t) &= \frac{d}{dt}(A_2 e^{-t} \sin t) \\
&= A_2 \frac{d}{dt}(e^{-t} \sin t) \\
&= A_2 \left(\sin t \frac{d}{dt}(e^{-t}) + e^{-t} \frac{d}{dt} \sin t \right) \\
&= A_2 (\sin t(-e^{-t}) + e^{-t} \cos t) \\
&= A_2 (e^{-t})(-\sin t + \cos t)
\end{aligned}
$$

Apply the initial condition, $x'(0) = 1$.

$$
\begin{aligned}
x(0) &= A_2(e^0)(-\sin 0 + \cos 0) = 1 \\
&\quad A_2(1)(0 + 1) = 1 \\
&\quad\quad\quad\quad A_2 = 1
\end{aligned}
$$

The solution is

$$
\begin{aligned}
x(t) &= A_2 e^{-t} \sin t \\
&= (1)e^{-t} \sin t \\
&= \boxed{e^{-t} \sin t}
\end{aligned}
$$

The answer is (C).

(b) To determine the natural frequency, set the damping term to zero. The equation has the form

$$x'' + 2x = 0$$

This equation has a general solution of the form

$$x(t) = x_0 \cos \omega t + \left(\frac{v_0}{\omega} \right) \sin \omega t$$

ω is the natural frequency.

Given the equation $x'' + 2x = 0$, the characteristic equation is

$$
\begin{aligned}
r^2 + 2 &= 0 \\
r &= \sqrt{-2} \\
&= \pm\sqrt{2}\, i
\end{aligned}
$$

Because the roots are imaginary and of the form $\alpha + i\omega$ and $\alpha - i\omega$, where α equals 0 and ω equals $\sqrt{2}$, the general form of the solution is given by

$$\begin{aligned}
x(t) &= A_1 e^{\alpha t} \cos \omega t + A_2 e^{\alpha t} \sin \omega t \\
&= A_1 e^{0t} \cos \sqrt{2}\, t + A_2 e^{0t} \sin \sqrt{2}\, t \\
&= A_1(1)\cos \sqrt{2}\, t + A_2(1)\sin \sqrt{2}\, t \\
&= A_1 \cos \sqrt{2}\, t + A_2 \sin \sqrt{2}\, t
\end{aligned}$$

Apply the initial conditions $x(0) = 0$ and $x'(0) = 1$ to solve for A_1 and A_2. Applying the initial condition $x(0) = 0$ gives

$$\begin{aligned}
x(0) = A_1 \cos\big((\sqrt{2})(0)\big) + A_2 \sin\big((\sqrt{2})(0)\big) &= 0 \\
A_1 \cos 0 + A_2 \sin 0 &= 0 \\
A_1(1) + A_2(0) &= 0 \\
A_1 &= 0
\end{aligned}$$

Substituting, the solution of the different equation becomes

$$x(t) = A_2 \sin \sqrt{2}\, t$$

To apply the second initial condition, take the first derivative.

$$\begin{aligned}
x'(t) &= \frac{d}{dt}(A_2 \sin \sqrt{2}\, t) \\
&= A_2 \sqrt{2} \, \cos \sqrt{2}\, t
\end{aligned}$$

Apply the second initial condition, $x'(0) = 1$.

$$\begin{aligned}
x'(0) = A_2 \sqrt{2} \, \cos\big((\sqrt{2})(0)\big) &= 1 \\
A_2 \sqrt{2} \, \cos 0 &= 1 \\
A_2(\sqrt{2})(1) &= 1 \\
A_2 \sqrt{2} &= 1 \\
A_2 = \frac{1}{\sqrt{2}} &= \frac{\sqrt{2}}{2}
\end{aligned}$$

Substituting, the undamped solution becomes

$$x(t) = \frac{\sqrt{2}}{2} \sin \sqrt{2}\, t$$

The undamped natural frequency is

$$\boxed{\omega = \sqrt{2}}$$

The answer is (C).

(c) The amplitude of the oscillation is the maximum displacement.

Take the derivative of the solution, $x(t) = e^{-t} \sin t$.

$$\begin{aligned}
x'(t) &= \frac{d}{dt} e^{-t} \sin t \\
&= (\sin t)\frac{d}{dt}(e^{-t}) + (e^{-t})\frac{d}{dt}\sin t \\
&= \sin t(-e^{-t}) + (e^{-t})\cos t \\
&= e^{-t}(\cos t - \sin t)
\end{aligned}$$

The maximum displacement occurs at $x'(t) = 0$.

Because $e^{-t} \neq 0$ except as t approaches infinity,

$$\begin{aligned}
\cos t - \sin t &= 0 \\
\tan t &= 1 \\
t &= \arctan 1 \\
&= 0.785 \text{ rad}
\end{aligned}$$

At time $t = 0.785$ rad, the displacement is at a maximum. Substitute into the original solution to obtain a value for the maximum displacement.

$$\begin{aligned}
x(0.785) &= e^{-0.785} \sin 0.785 \\
&= 0.32
\end{aligned}$$

The amplitude is $\boxed{0.32.}$

The answer is (A).

(d) (An alternative solution using Laplace transforms follows this solution.) The application of a lateral wind load with the form $\sin t$ revises the differential equation to the form

$$x'' + 2x' + 2x = \sin t$$

Express the solution as the sum of the complementary x_c and particular x_p solutions.

$$x(t) = x_c(t) + x_p(t)$$

From part (a),

$$x_c(t) = A_1 e^{-t} \cos t + A_2 e^{-t} \sin t$$

The general form of the particular solution is given by

$$x_p(t) = x^s(A_3 \cos t + A_4 \sin t)$$

Determine the value of s; check to see if the terms of the particular solution solve the homogeneous equation.

Examine the term $A_3 \cos t$.

Take the first derivative.

$$\frac{d}{dt}(A_3 \cos t) = -A_3 \sin t$$

Take the second derivative.

$$\frac{d}{dt}\left(\frac{d}{dt}(A_3 \cos t)\right) = \frac{d}{dt}(-A_3 \sin t)$$
$$= -A_3 \cos t$$

Substitute the terms into the homogeneous equation.

$$x'' + 2x' + 2x = -A_3 \cos t + 2(-A_3 \sin t)$$
$$+ 2(-A_3 \cos t)$$
$$= A_3 \cos t - 2A_3 \sin t$$
$$\neq 0$$

Except for the trivial solution $A_3 = 0$, the term $A_3 \cos t$ does not solve the homogeneous equation.

Examine the second term $A_4 \sin t$.

Take the first derivative.

$$\frac{d}{dt}(A_4 \sin t) = A_4 \cos t$$

Take the second derivative.

$$\frac{d}{dt}\left(\frac{d}{dt}(A_4 \sin t)\right) = \frac{d}{dt}(A_4 \cos t)$$
$$= -A_4 \sin t$$

Substitute the terms into the homogeneous equation.

$$x'' + 2x' + 2x = -A_4 \sin t + 2(A_4 \cos t)$$
$$+ 2(A_4 \sin t)$$
$$= A_4 \sin t + 2A_4 \cos t$$
$$\neq 0$$

Except for the trivial solution $A_4 = 0$, the term $A_4 \sin t$ does not solve the homogeneous equation.

Neither of the terms satisfies the homogeneous equation $s = 0$; therefore, the particular solution is of the form

$$x_p(t) = A_3 \cos t + A_4 \sin t$$

Use the method of undetermined coefficients to solve for A_3 and A_4. Take the first derivative.

$$x'_p(t) = \frac{d}{dt}(A_3 \cos t + A_4 \sin t)$$
$$= -A_3 \sin t + A_4 \cos t$$

Take the second derivative.

$$x''_p(t) = \frac{d}{dt}\left(\frac{d}{dt}(A_3 \cos t + A_4 \sin t)\right)$$
$$= \frac{d}{dt}(-A_3 \sin t + A_4 \cos t)$$
$$= -A_3 \cos t - A_4 \sin t$$

Substitute the expressions for the derivatives into the differential equation.

$$x'' + 2x' + 2x = (-A_3 \cos t - A_4 \sin t)$$
$$+ 2(-A_3 \sin t + A_4 \cos t)$$
$$+ 2(A_3 \cos t + A_4 \sin t)$$
$$= \sin t$$

Rearranging terms gives

$$(-A_3 + 2A_4 + 2A_3)\cos t$$
$$+ (-A_4 - 2A_3 + 2A_4)\sin t = \sin t$$
$$(A_3 + 2A_4)\cos t + (-2A_3 + A_4)\sin t = \sin t$$

Equating coefficients gives

$$A_3 + 2A_4 = 0$$
$$-2A_3 + A_4 = 1$$

Multiplying the first equation by 2 and adding equations gives

$$2A_3 + 4A_4 = 0$$
$$+ (-2A_3 + A_4) = 1$$
$$\overline{ 5A_4 = 1 \text{ or } A_4 = \frac{1}{5}}$$

From the first equation for $A_4 = \frac{1}{5}$, $A_3 + (2)(\frac{1}{5}) = 0$ and $A_3 = -\frac{2}{5}$.

Substituting for the coefficients, the particular solution becomes

$$x_p(t) = -\frac{2}{5}\cos t + \frac{1}{5}\sin t$$

Combining the complementary and particular solutions gives

$$x(t) = x_c(t) + x_p(t)$$
$$= A_1 e^{-t} \cos t + A_2 e^{-t} \sin t - \frac{2}{5}\cos t + \frac{1}{5}\sin t$$

Apply the initial conditions to solve for the coefficients A_1 and A_2, then apply the first initial condition, $x(0) = 0$.

$$x(t) = A_1 e^0 \cos 0 + A_2 e^0 \sin 0$$
$$- \frac{2}{5} \cos 0 + \frac{1}{5} \sin 0 = 0$$

$$A_1(1)(1) + A_2(1)(0) + \left(-\frac{2}{5}\right)(1) + \left(\frac{1}{5}\right)(0) = 0$$

$$A_1 - \frac{2}{5} = 0$$

$$A_1 = \frac{2}{5}$$

Substituting for A_1, the solution becomes

$$x(t) = \frac{2}{5} e^{-t} \cos t + A_2 e^{-t} \sin t - \frac{2}{5} \cos t + \frac{1}{5} \sin t$$

Take the first derivative.

$$x'(t) = \frac{d}{dt}\left(\frac{2}{5} e^{-t} \cos t + A_2 e^{-t} \sin t\right)$$
$$+ \left(-\frac{2}{5}\cos t + \frac{1}{5} \sin t\right)$$
$$= \left(\frac{2}{5}\right)(-e^{-t} \cos t - e^{-t} \sin t)$$
$$+ A_2(-e^{-t} \sin t + e^{-t} \cos t)$$
$$+ \left(-\frac{2}{5}\right)(-\sin t) + \frac{1}{5} \cos t$$

Apply the second initial condition, $x'(0) = 1$.

$$x'(0) = \frac{2}{5}(-e^0 \cos 0 - e^0 \sin 0)$$
$$+ A_2(-e^0 \sin 0 + e^0 \cos 0)$$
$$+ \left(-\frac{2}{5}\right)(-\sin 0) + \frac{1}{5} \cos 0$$
$$= 1$$

$$\left(\frac{2}{5}\right)(-(1)(1) - (1)(0))$$
$$+ A_2\big(-(1)(0) + (1)(1)\big) + \left(-\frac{2}{5}\right)(0) + \left(\frac{1}{5}\right)(1) = 1$$
$$\left(\frac{2}{5}\right)(-1) + A_2(1) + \left(\frac{1}{5}\right) = 1$$

$$A_2 = \frac{6}{5}$$

Substituting for A_2, the solution becomes

$$x(t) = \boxed{\begin{array}{c} \frac{2}{5} e^{-t} \cos t + \frac{6}{5} e^{-t} \sin t \\ -\frac{2}{5}\cos t + \frac{1}{5} \sin t \end{array}}$$

The answer is (D).

(d) *Alternate Solution:*

Use the Laplace transform method.

$$x' + 2x' + 2x = \sin t$$
$$\mathcal{L}(x') + 2\mathcal{L}(x') + 2\mathcal{L}(x) = \mathcal{L}\sin t$$
$$s^2\mathcal{L}(x) - 1 + 2s\mathcal{L}(x) + 2\mathcal{L}(x) = \frac{1}{s^2 + 1}$$
$$\mathcal{L}(x)(s^2 + 2s + 2) - 1 = \frac{1}{s^2 + 1}$$
$$\mathcal{L}(x) = \frac{1}{s^2 + 2s + 2} + \frac{1}{(s^2 + 1)(s^2 + 2s + 2)}$$
$$= \frac{1}{(s+1)^2 + 1} + \frac{1}{(s^2 + 1)(s^2 + 2s + 2)}$$

Use partial fractions to expand the second term.

$$\frac{1}{(s^2 + 1)(s^2 + 2s + 2)} = \frac{A_1 + B_1 s}{s^2 + 1} + \frac{A_2 + B_2 s}{s^2 + 2s + 2}$$

Cross multiply.

$$\frac{A_1 + B_1 s}{s^2 + 1} + \frac{A_2 + B_2 s}{s^2 + 2s + 2}$$
$$= \frac{\begin{array}{c}A_1 s^2 + 2A_1 s + 2A_1 + B_1 s^3 + 2B_1 s^2 \\ + 2B_1 s + A_2 s^2 + A_2 + B_2 s^3 + B_2 s\end{array}}{(s^2 + 1)(s^2 + 2s + 2)}$$
$$= \frac{\begin{array}{c}s^3(B_1 + B_2) + s^2(A_1 + A_2 + 2B_1) \\ + s(2A_1 + 2B_1 + B_2) + 2A_1 + A_2\end{array}}{(s^2 + 1)(s^2 + 2s + 2)}$$

Compare numerators to obtain the following four simultaneous equations.

$$B_1 + B_2 = 0$$
$$A_1 + A_2 + 2B_1 \qquad = 0$$
$$2A_1 \qquad + 2B_1 + B_2 = 0$$
$$2A_1 + A_2 \qquad = 1$$

Use Cramer's rule to find A_1.

$$A_1 = \frac{\begin{vmatrix} 0 & 0 & 1 & 1 \\ 0 & 1 & 2 & 0 \\ 0 & 0 & 2 & 1 \\ 1 & 1 & 0 & 0 \end{vmatrix}}{\begin{vmatrix} 0 & 0 & 1 & 1 \\ 1 & 1 & 2 & 0 \\ 2 & 0 & 2 & 1 \\ 2 & 1 & 0 & 0 \end{vmatrix}} = \frac{-1}{-5} = \frac{1}{5}$$

The rest of the coefficients are found similarly.

$$A_1 = \frac{1}{5}$$

$$A_2 = \frac{3}{5}$$

$$B_1 = -\frac{2}{5}$$

$$B_2 = \frac{2}{5}$$

Then,

$$\mathcal{L}(x) = \frac{1}{(s+1)^2+1} + \frac{\frac{1}{5}}{s^2+1} + \frac{-\frac{2}{5}s}{s^2+1}$$
$$+ \frac{\frac{3}{5}}{s^2+2s+2} + \frac{\frac{2}{5}s}{s^2+2s+2}$$

Take the inverse transform.

$$x(t) = \mathcal{L}^{-1}\{\mathcal{L}(x)\}$$
$$= e^{-t}\sin t + \frac{1}{5}\sin t - \frac{2}{5}\cos t + \frac{3}{5}e^{-t}\sin t$$
$$+ \frac{2}{5}(e^{-t}\cos t - e^{-t}\sin t)$$
$$= \boxed{\frac{2}{5}e^{-t}\cos t + \frac{6}{5}e^{-t}\sin t - \frac{2}{5}\cos t + \frac{1}{5}\sin t}$$

The answer is (D).

4. *Customary U.S. Solution*

The differential equation is given as

$$m'(t) = a(t) = \frac{m(t)o(t)}{V(t)}$$

$a(t) =$ rate of addition of chemical

$m(t) =$ mass of chemical at time t

$o(t) =$ volumetric flow out of the lagoon (30 gal/min)

$V(t) =$ volume in the lagoon at time t

$d =$ diameter of the lagoon

$z =$ depth of the lagoon

Water flows into the lagoon at a rate of 30 gal/min, and a water-chemical mix flows out of the lagoon at a rate of 30 gal/min. Therefore, the volume of the lagoon at time t is equal to the initial volume.

$$V(t) = \frac{\pi}{4}d^2 z$$
$$= \left(\frac{\pi}{4}\right)(120 \text{ ft})^2(10 \text{ ft})$$
$$= 113{,}097 \text{ ft}^3$$

Use a conversion factor of 7.48 gal/ft³.

$$o(t) = \frac{30 \frac{\text{gal}}{\text{min}}}{7.48 \frac{\text{gal}}{\text{ft}^3}}$$
$$= 4.01 \text{ ft}^3/\text{min}$$

Substituting into the general form of the differential equation gives

$$m'(t) = a(t) - \frac{m(t)o(t)}{V(t)}$$
$$= 0 \text{ lbm} - m(t)\left(\frac{4.01 \frac{\text{ft}^3}{\text{min}}}{113{,}097 \text{ ft}^3}\right)$$
$$= -\left(\frac{3.985 \times 10^{-5}}{\text{min}}\right)m(t)$$

$$m'(t) + \left(\frac{3.985 \times 10^{-5}}{\text{min}}\right)m(t) = 0$$

The differential equation of the problem has a characteristic equation.

$$r + \frac{3.55 \times 10^{-5}}{\text{min}} = 0$$
$$r = -3.55 \times 10^{-5}/\text{min}$$

The general form of the solution is

$$m(t) = A e^{rt}$$

Substituting for the root, r, gives

$$m(t) = A e^{(-3.55 \times 10^{-5}/\text{min})t}$$

Apply the initial condition $m(0) = 90$ lbm at time $t = 0$ min.

$$m(0) = A e^{(-3.55 \times 10^{-5}/\text{min})(0)} = 90 \text{ lbm}$$
$$A e^0 = 90 \text{ lbm}$$
$$A = 90 \text{ lbm}$$

Therefore,

$$m(t) = (90 \text{ lbm}) e^{(-3.55 \times 10^{-5}/\text{min})t}$$

Solve for t.

$$\frac{m(t)}{90 \text{ lbm}} = e^{(-3.55 \times 10^{-5}/\text{min})t}$$

$$\ln \frac{m(t)}{90 \text{ lbm}} = \ln e^{(-3.55 \times 10^{-5}/\text{min})t}$$

$$= \frac{-3.55 \times 10^{-5}}{\text{min}} t$$

$$t = \frac{\ln \dfrac{m(t)}{90 \text{ lbm}}}{\dfrac{-3.55 \times 10^{-5}}{\text{min}}}$$

The initial mass of the water in the lagoon is

$$m_i = V\rho$$
$$= (113{,}097 \text{ ft}^3)\left(62.4 \ \frac{\text{lbm}}{\text{ft}^3}\right)$$
$$= 7.06 \times 10^6 \text{ lbm}$$

The final mass of chemicals is achieved at a concentration of 1 ppb or

$$m_f = \frac{m_i}{1 \times 10^9} = \frac{7.06 \times 10^6 \text{ lbm}}{1 \times 10^9}$$
$$= 7.06 \times 10^{-3} \text{ lbm}$$

Find the time required to achieve a mass of 7.06×10^{-3} lbm.

$$t = \left(\frac{\ln \dfrac{m(t)}{90 \text{ lbm}}}{\dfrac{-3.55 \times 10^{-5}}{\text{min}}} \right)\left(\frac{1 \text{ hr}}{60 \text{ min}}\right)\left(\frac{1 \text{ day}}{24 \text{ hr}}\right)$$

$$= \left(\frac{\ln \dfrac{7.06 \times 10^{-3} \text{ lbm}}{90 \text{ lbm}}}{\dfrac{-3.55 \times 10^{-5}}{\text{min}}} \right)\left(\frac{1 \text{ hr}}{60 \text{ min}}\right)\left(\frac{1 \text{ day}}{24 \text{ hr}}\right)$$

$$= \boxed{185 \text{ days} \quad (200 \text{ days})}$$

The answer is (D).

SI Solution

The differential equation is given as

$$m'(t) = a(t) - \frac{m(t)o(t)}{V(t)}$$

$a(t) =$ rate of addition of chemical
$m(t) =$ mass of chemical at time t
$o(t) =$ volumetric flow out of the lagoon
$\quad$ (115 L/min)
$V(t) =$ volume in the lagoon at time t
$d =$ diameter of the lagoon
$z =$ depth of the lagoon

Water flows into the lagoon at a rate of 115 L/min, and a water-chemical mix flows out of the lagoon at a rate of 115 L/min. Therefore, the volume of the lagoon at time t is equal to the initial volume.

$$V(t) = \frac{\pi}{4} d^2 z$$
$$= \left(\frac{\pi}{4}\right)(35 \text{ m})^2(3 \text{ m})$$
$$= 2886 \text{ m}^3$$

Using a conversion factor of 1 m³/1000 L gives

$$o(t) = \left(115 \ \frac{\text{L}}{\text{min}}\right)\left(\frac{1 \text{ m}^3}{1000 \text{ L}}\right)$$
$$= 0.115 \text{ m}^3/\text{min}$$

Substitute into the general form of the differential equation.

$$m'(t) = a(t) - \frac{m(t)\,o(t)}{V(t)}$$

$$= 0 \text{ lbm} - m(t)\left(\frac{0.115 \ \frac{\text{m}^3}{\text{min}}}{2886 \ \text{m}^3}\right)$$

$$= -\left(\frac{3.985 \times 10^{-5}}{\text{min}}\right)m(t)$$

$$m'(t) + \left(\frac{3.985 \times 10^{-5}}{\text{min}}\right)m(t)$$

$$= 0$$

The differential equation of the problem has the following characteristic equation.

$$r + \frac{3.985 \times 10^{-5}}{\text{min}} = 0$$

$$r = -3.985 \times 10^{-5}/\text{min}$$

The general form of the solution is

$$m(t) = A\,e^{rt}$$

Substituting in for the root, r, gives

$$m(t) = A\,e^{(-3.985 \times 10^{-5}/\text{min})t}$$

Apply the initial condition $m(0) = 40$ kg at time $t = 0$ min.

$$m(0) = A\,e^{(-3.985 \times 10^{-5}/\text{min})(0)} = 40 \text{ kg}$$

$$A\,e^0 = 40 \text{ kg}$$

$$A = 40 \text{ kg}$$

Therefore,

$$m(t) = (40 \text{ kg})\,e^{(-3.985 \times 10^{-5}/\text{min})t}$$

Solve for t.

$$\frac{m(t)}{40 \text{ kg}} = e^{(-3.985 \times 10^{-5}/\text{min})t}$$

$$\ln\left(\frac{m(t)}{40 \text{ kg}}\right) = \ln e^{(-3.985 \times 10^{-5}/\text{min})t}$$

$$= \left(\frac{-3.985 \times 10^{-5}}{\text{min}}\right)t$$

$$t = \ln\frac{\dfrac{m(t)}{40 \text{ kg}}}{\dfrac{-3.985 \times 10^{-5}}{\text{min}}}$$

The initial mass of water in the lagoon is

$$m_i = V\rho$$

$$= (2886 \text{ m}^3)\left(1000 \ \frac{\text{kg}}{\text{m}^3}\right)$$

$$= 2.866 \times 10^6 \text{ kg}$$

The final mass of chemicals is achieved at a concentration of 1 ppb or

$$m_f = \frac{2.886 \times 10^6 \text{ kg}}{1 \times 10^9}$$

$$= 2.886 \times 10^{-3} \text{ kg}$$

Find the time required to achieve a mass of 2.886×10^{-3} kg.

$$t = \left(\frac{\ln\dfrac{m(t)}{40 \text{ kg}}}{\dfrac{-3.985 \times 10^{-5}}{\text{min}}}\right)\left(\frac{1 \text{ h}}{60 \text{ min}}\right)\left(\frac{1 \text{ d}}{24 \text{ h}}\right)$$

$$= \left(\frac{\ln\left(\dfrac{2.886 \times 10^{-3} \text{ kg}}{40 \text{ kg}}\right)}{\dfrac{-3.985 \times 10^{-5}}{\text{min}}}\right)\left(\frac{1 \text{ h}}{60 \text{ min}}\right)\left(\frac{1 \text{ d}}{24 \text{ h}}\right)$$

$$= \boxed{166 \text{ d} \quad (200 \text{ d})}$$

The answer is (D).

5. Let

$m(t) =$ mass of salt in tank at time t

$m_0 = 60$ mass units

$m'(t) =$ rate at which salt content is changing

Two mass units of salt enter each minute, and three volumes leave each minute. The amount of salt leaving each minute is

$$\left(3 \ \frac{\text{vol}}{\text{min}}\right)\left(\text{concentration in } \frac{\text{mass}}{\text{vol}}\right)$$

$$= \left(3 \ \frac{\text{vol}}{\text{min}}\right)\left(\frac{\text{salt content}}{\text{volume}}\right)$$

$$= \left(3 \ \frac{\text{vol}}{\text{min}}\right)\left(\frac{m(t)}{100 - t}\right)$$

$$m'(t) = 2 - (3)\left(\frac{m(t)}{100 - t}\right)$$

Alternately,

$$m'(t) + \frac{3m(t)}{100 - t} = 2 \text{ mass/min}$$

This is a first-order linear differential equation. The integrating factor is

$$m = e^3 \int \frac{dt}{100 - t}$$

$$= e^{(3)\left(-\ln(100-t)\right)}$$

$$= (100 - t)^{-3}$$

$$m(t) = (100 - t)^3\left[2 \int \frac{dt}{(100 - t)^3} + k\right]$$

$$= 100 - t + k(100 - t)^3$$

But $m = 60$ mass units at time $t = 0$ min, so k is -0.00004.

$$m(t) = 100 - t - (0.00004)(100 - t)^3$$

At time $t = 60$ min,

$$m = 100 - 60 \text{ min} - (0.00004)(100 - 60 \text{ min})^3$$

$$= \boxed{37.44 \text{ mass units} \quad (37 \text{ mass units})}$$

The answer is (B).

12 Probability and Statistical Analysis of Data

PRACTICE PROBLEMS

1. Four military recruits whose respective shoe sizes are 7, 8, 9, and 10 report to the supply clerk to be issued boots. The supply clerk selects one pair of boots in each of the four required sizes and hands them at random to the recruits.

(a) What is the probability that all recruits will receive boots of an incorrect size?

   (A) 0.25

   (B) 0.38

   (C) 0.45

   (D) 0.61

(b) What is the probability that exactly three recruits will receive boots of the correct size?

   (A) 0.0

   (B) 0.063

   (C) 0.17

   (D) 0.25

2. The time taken by a toll taker to collect the toll from vehicles crossing a bridge is an exponential distribution with a mean of 23 sec. What is the probability that a random vehicle will be processed in 25 sec or more (i.e., will take longer than 25 sec)?

   (A) 0.17

   (B) 0.25

   (C) 0.34

   (D) 0.52

3. The number of cars entering a toll plaza on a bridge during the hour after midnight follows a Poisson distribution with a mean of 20.

(a) What is the probability that 17 cars will pass through the toll plaza during that hour on any given night?

   (A) 0.076

   (B) 0.12

   (C) 0.16

   (D) 0.23

(b) What is the probability that three or fewer cars will pass through the toll plaza at that hour on any given night?

   (A) 0.0000032

   (B) 0.0019

   (C) 0.079

   (D) 0.11

4. A mechanical component exhibits a negative exponential failure distribution with a mean time to failure, MTTF, of 1000 hr. What is the maximum operating time such that the reliability remains above 99%?

   (A) 3.3 hr

   (B) 5.6 hr

   (C) 8.1 hr

   (D) 10 hr

5. A survey field crew measures one leg of a traverse four times. The following results are obtained.

repetition	measurement	direction
1	1249.529	forward
2	1249.494	backward
3	1249.384	forward
4	1249.348	backward

The crew chief is under orders to obtain readings with confidence limits of 90%.

(a) Which readings are acceptable?

   (A)   No readings are acceptable.

   (B)   Two readings are acceptable.

   (C)   Three readings are acceptable.

   (D)   All four readings are acceptable.

(b) Which readings are NOT acceptable?

   (A)   No readings are unacceptable.

   (B)   One reading is unacceptable.

   (C)   Two readings are unacceptable.

   (D)   All four readings are unacceptable.

(c) Explain how to determine which readings are NOT acceptable.

   (A)   Readings inside the 90% confidence limits are unacceptable.

   (B)   Readings outside the 90% confidence limits are unacceptable.

   (C)   Readings outside the upper 90% confidence limit are unacceptable.

   (D)   Readings outside the lower 90% confidence limit are unacceptable.

(d) What is the most probable value of the distance?

   (A)   1249.399

   (B)   1249.410

   (C)   1249.439

   (D)   1249.452

(e) What is the error in the most probable value (at 90% confidence)?

   (A)   0.080

   (B)   0.11

   (C)   0.14

   (D)   0.19

(f) If the distance is one side of a square traverse whose sides are all equal, what is the most probable closure error?

   (A)   0.14

   (B)   0.20

   (C)   0.28

   (D)   0.35

(g) What is the probable error of part (f) expressed as a fraction?

   (A)   1:17,600

   (B)   1:14,200

   (C)   1:12,500

   (D)   1:10,900

(h) What is the order of accuracy of the closure?

   (A)   first order

   (B)   second order

   (C)   third order

   (D)   fourth order

(i) Define accuracy, and distinguish it from precision.

   (A)   If an experiment can be repeated with identical results, the results are considered accurate.

   (B)   If an experiment has a small bias, the results are considered precise.

   (C)   If an experiment is precise, it cannot also be accurate.

   (D)   If an experiment is unaffected by experimental error, the results are accurate.

(j) Give an example of systematic error.

   (A)   measuring a river's depth as a motorized ski boat passes by

   (B)   using a steel tape that is too short to measure consecutive distances

   (C)   locating magnetic north near a large iron ore deposit along an overland route

   (D)   determining local wastewater BOD after a toxic spill

6. Prior to the use of radar speed control on a road, California law requires a statistical analysis of the average speed driven there by motorists. The following speeds (all in mph) were observed in a random sample of 40 cars.

44, 48, 26, 25, 20, 43, 40, 42, 29, 39, 23, 26, 24, 47, 45, 28, 29, 41, 38, 36, 27, 44, 42, 43, 29, 37, 34, 31, 33, 30, 42, 43, 28, 41, 29, 36, 35, 30, 32, 31

(a) Tabulate the frequency distribution of the data.

(b) Draw the frequency histogram.

(c) Draw the frequency polygon.

(d) Tabulate the cumulative frequency distribution.

(e) Draw the cumulative frequency graph.

(f) What is the upper quartile speed?

(A) 30 mph

(B) 35 mph

(C) 40 mph

(D) 45 mph

(g) What is the median speed?

(A) 31 mph

(B) 33 mph

(C) 35 mph

(D) 37 mph

(h) What is the standard deviation of the sample data?

(A) 2.1 mph

(B) 6.1 mph

(C) 6.8 mph

(D) 7.4 mph

(i) What is the sample standard deviation?

(A) 7.5 mph

(B) 18 mph

(C) 35 mph

(D) 56 mph

(j) What is most nearly the sample variance?

(A) $56 \text{ mi}^2/\text{hr}^2$

(B) $320 \text{ mi}^2/\text{hr}^2$

(C) $1230 \text{ mi}^2/\text{hr}^2$

(D) $3140 \text{ mi}^2/\text{hr}^2$

7. A spot speed study is conducted for a stretch of roadway. During a normal day, the speeds were found to be normally distributed with a mean of 46 mph and a standard deviation of 3 mph.

(a) What is the 50th percentile speed?

(A) 39 mph

(B) 43 mph

(C) 46 mph

(D) 49 mph

(b) What is the 85th percentile speed?

(A) 47.1 mph

(B) 48.3 mph

(C) 49.1 mph

(D) 52.7 mph

(c) What speed is two standard deviations above the mean?

(A) 47.2 mph

(B) 49.3 mph

(C) 51.1 mph

(D) 52.0 mph

(d) The daily average speeds for the same stretch of roadway on consecutive normal days were determined by sampling 25 vehicles each day. What speed is two standard deviations above the mean?

(A) 46.6 mph

(B) 47.2 mph

(C) 52.0 mph

(D) 54.7 mph

8. The diameters of bolt holes drilled in structural steel members are normally distributed with a mean of 0.502 in and a standard deviation of 0.005 in. Holes are out of specification if their diameters are less than 0.497 in or more than 0.507 in.

(a) What is the probability that a hole chosen at random will be out of specification?

(A) 0.16

(B) 0.22

(C) 0.32

(D) 0.68

(b) What is the probability that 2 holes out of a sample of 15 will be out of specification?

(A) 0.074

(B) 0.12

(C) 0.15

(D) 0.32

9. 100 bearings were tested to failure. The average life was 1520 hr, and the standard deviation was 120 hr. The manufacturer claims a 1600 hr life. Evaluate using confidence limits of 95% and 99%.

(A) The claim is accurate at both 95% and 99% confidence.

(B) The claim is inaccurate only at 95%.

(C) The claim is inaccurate only at 99%.

(D) The claim is inaccurate at both 95% and 99% confidence.

10. (a) Find the best equation for a line passing through the points given. (b) Find the correlation coefficient.

x	y
400	370
800	780
1250	1210
1600	1560
2000	1980
2500	2450
4000	3950

11. Find the best equation for a line passing through the points given.

s	t
20	43
18	141
16	385
14	1099

12. The number of vehicles lining up behind a flashing railroad crossing has been observed for five trains of different lengths, as given. What is the mathematical formula that relates the two variables?

no. of cars in train	no. of vehicles
2	14.8
5	18.0
8	20.4
12	23.0
27	29.9

13. The following yield data are obtained from five identical treatment plants. (a) Develop a mathematical equation to correlate the yield and average temperature. (b) What is the correlation coefficient?

treatment plant	average temperature (T)	average yield (Y)
1	207.1	92.30
2	210.3	92.58
3	200.4	91.56
4	201.1	91.63
5	203.4	91.83

14. The following data are obtained from a soil compaction test. What is the mathematical formula that relates the two variables?

x	y
−1	0
0	1
1	1.4
2	1.7
3	2
4	2.2
5	2.4
6	2.6
7	2.8
8	3

15. Two resistances, the meter resistance and a shunt resistor, are connected in parallel in an ammeter. Most of the current passing through the meter goes through the shunt resistor. In order to determine the accuracy of the resistance of shunt resistors being manufactured for a line of ammeters, a manufacturer tests a sample of 100 shunt resistors. The numbers of shunt resistors with the resistance indicated (to the nearest hundredth of an ohm) are as follows.

0.200 Ω, 1; 0.210 Ω, 3; 0.220 Ω, 5; 0.230 Ω, 10; 0.240 Ω, 17; 0.250 Ω, 40; 0.260 Ω, 13; 0.270 Ω, 6; 0.280 Ω, 3; 0.290 Ω, 2

(a) What is the mean resistance?

(A) 0.235 Ω

(B) 0.247 Ω

(C) 0.251 Ω

(D) 0.259 Ω

(b) What is the sample standard deviation?

(A) 0.00030 Ω

(B) 0.010 Ω

(C) 0.016 Ω

(D) 0.24 Ω

(c) What is the median resistance?

(A) 0.221 Ω

(B) 0.244 Ω

(C) 0.252 Ω

(D) 0.259 Ω

(d) What is most nearly the sample variance?

(A) 0.00027 Ω²

(B) 0.0083 Ω²

(C) 0.011 Ω²

(D) 0.016 Ω²

SOLUTIONS

1. (a) There are $4! = 24$ different possible outcomes. By enumeration, the number of completely wrong combinations is 9, so that

$$p\{\text{all wrong}\} = \frac{9}{24} = \boxed{0.375 \quad (0.38)}$$

The answer is (B).

(b) If three recruits get the correct size, the fourth recruit will also, because there will be only one pair remaining.

$$p\{\text{exactly } 3\} = \boxed{0}$$

The answer is (A).

2. For an exponential distribution function, the mean is given as

$$\mu = \frac{1}{\lambda}$$

For a mean of 23,

$$\mu = 23 = \frac{1}{\lambda}$$
$$\lambda = 0.0435$$

For an exponential distribution function,

$$p\{X < x\} = F(x) = 1 - e^{-\lambda x}$$
$$p\{x > X\} = 1 - p\{X < x\}$$
$$= 1 - F(x)$$
$$= 1 - (1 - e^{-\lambda x})$$
$$= e^{-\lambda x}$$

The probability of a random vehicle being processed in 25 sec or more is

$$p\{x > 25\} = e^{-(0.0435)(25)}$$
$$= e^{-1.0875}$$
$$= \boxed{0.337 \quad (0.34)}$$

The answer is (C).

3. (a) The distribution is a Poisson distribution with an average of $\lambda = 20$. The probability for a Poisson distribution is

$$p\{x\} = f(x) = \frac{e^{-\lambda}\lambda^x}{x!}$$

The probability of 17 cars is

$$p\{x=17\} = f(17) = \frac{e^{-20}(20)^{17}}{17!}$$
$$= \boxed{0.076 \quad (7.6\%)}$$

The answer is (A).

(b) The probability of three or fewer cars is

$$p\{x \leq 3\} = p\{x=0\} + p\{x=1\} + p\{x=2\}$$
$$+ p\{x=3\}$$
$$= f(0) + f(1) + f(2) + f(3)$$
$$= \frac{e^{-20}(20)^0}{0!} + \frac{e^{-20}(20)^1}{1!}$$
$$+ \frac{e^{-20}(20)^2}{2!} + \frac{e^{-20}(20)^3}{3!}$$
$$= 2 \times 10^{-9} + 4.1 \times 10^{-8}$$
$$+ 4.12 \times 10^{-7} + 2.75 \times 10^{-6}$$
$$= \boxed{3.2 \times 10^{-6} \quad (0.0000032)}$$

The answer is (A).

4. To simplify the notation,

$$\lambda = \frac{1}{\text{MTTF}}$$
$$\frac{1}{1000 \text{ hr}} = 0.001 \text{ hr}^{-1}$$

The reliability function is

$$R\{t\} = e^{-\lambda t/\text{hr}} = e^{-0.001t/\text{hr}}$$

Because the reliability is greater than 99%,

$$e^{-0.001t/\text{hr}} > 0.99$$
$$\ln(e^{-0.001t/\text{hr}}) > \ln 0.99$$
$$\frac{-0.001t}{\text{hr}} > \ln 0.99$$
$$t < (-1000 \text{ hr}) \ln 0.99$$
$$\boxed{t < 10.05 \text{ hr} \quad (10 \text{ hr})}$$

The maximum operating time such that the reliability remains above 99% is 10 hr.

The answer is (D).

5. Find the average.

$$\bar{x} = \frac{\sum x_i}{n}$$
$$= \frac{1249.529 + 1249.494 + 1249.384 + 1249.348}{4}$$
$$= 1249.439$$

Because the sample population is small, use the sample standard deviation.

$$s = \sqrt{\frac{\sum(x_i - \bar{x})^2}{n-1}}$$
$$= \sqrt{\frac{\begin{array}{l}(1249.529 - 1249.439)^2 + (1249.494 - 1249.439)^2 \\ + (1249.384 - 1249.439)^2 + (1249.348 - 1249.439)^2\end{array}}{4-1}}$$
$$= 0.08647$$

From the standard deviation table, a 90% confidence limit falls within $1.645s$ of $\bar{x}$.

$$1249.439 \pm (1.645)(0.08647) = 1249.439 \pm 0.142$$

Therefore, (1249.297, 1249.581) is the 90% confidence range.

(a) $\boxed{\text{By observation, all the readings fall within the 90\% confidence range, so all are acceptable.}}$

The answer is (D).

(b) $\boxed{\text{No readings are unacceptable.}}$

The answer is (A).

(c) $\boxed{\text{Readings outside the 90\% confidence limits are unacceptable.}}$

The answer is (B).

(d) The unbiased estimate of the most probable distance is $\boxed{1249.439.}$

The answer is (C).

(e) The error for the 90% confidence range is $\boxed{0.14.}$

The answer is (C).

(f) If the surveying crew places a marker, measures a distance x, places a second marker, and then measures the same distance x back to the original marker, the ending point should coincide with the original marker. If, due to measurement errors, the ending and starting points do not coincide, the difference is the closure error.

In this example, the survey crew moves around the four sides of a square, so there are two measurements in the x-direction and two measurements in the y-direction. If the errors E_1 and E_2 are known for two measurements, x_1 and x_2, the error associated with the sum or difference $x_1 \pm x_2$ is

$$E\{x_1 \pm x_2\} = \sqrt{E_1^2 + E_2^2}$$

In this case, the error in the x-direction is

$$E_x = \sqrt{(0.1422)^2 + (0.1422)^2}$$
$$= 0.2011$$

The error in the y-direction is calculated the same way and is also 0.2011. E_x and E_y are combined by the Pythagorean theorem to yield

$$E_{\text{closure}} = \sqrt{(0.2011)^2 + (0.2011)^2}$$
$$= \boxed{0.2844 \quad (0.28)}$$

The answer is (C).

(g) In surveying, error may be expressed as a fraction of one or more legs of the traverse. Assume that the total of all four legs is to be used as the basis.

$$E = \frac{E_{\text{closure}}}{\sum_i x_i} = \frac{0.2844}{(4)(1249)} = \boxed{\frac{1}{17,567} \quad (1{:}17,600)}$$

The answer is (A).

(h) In surveying, a class 1 third-order error is smaller than 1/10,000. The error of 1/17,567 is smaller than the third-order error; therefore, the error is within the third-order accuracy.

The answer is (C).

(i) An experiment is accurate if it is unchanged by experimental error. Precision is concerned with the repeatability of the experimental results. If an experiment is repeated with identical results, the experiment is said to be precise. However, it is possible to have a highly precise experiment with a large bias.

The answer is (D).

(j) A systematic error is one that is always present and is unchanged from sample to sample. For example, a steel tape that is 0.02 ft short introduces a systematic error.

The answer is (B).

6. For (a) and (d), tabulate the frequency distribution data.

(The lowest speed is 20 mph and the highest speed is 48 mph; therefore, the range is 28 mph. Choose 10 cells with a width of 3 mph.)

midpoint	interval (mph)	frequency	cumulative frequency	cumulative percent
21	20–22	1	1	3
24	23–25	3	4	10
27	26–28	5	9	23
30	29–31	8	17	43
33	32–34	3	20	50
36	35–37	4	24	60
39	38–40	3	27	68
42	41–43	8	35	88
45	44–46	3	38	95
48	47–49	2	40	100

(b)

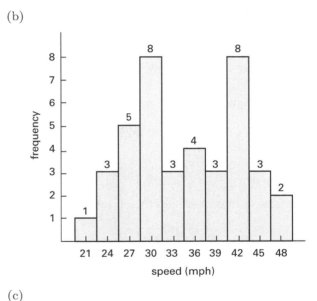

(c)

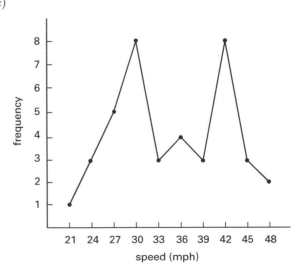

(e)

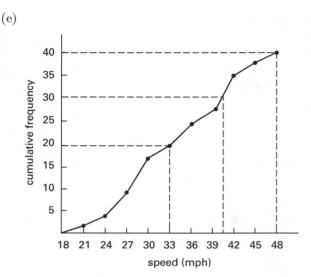

(f) From the cumulative frequency graph in part (e), the upper quartile speed occurs at 30 cars or 75%, which corresponds to approximately 40 mph.

The answer is (C).

(g) The mode occurs at two speeds (the frequency at each of the two speeds is 8): 30 mph and 42 mph.

The median occurs at 50% (or 20 cars) and, from the cumulative frequency chart, corresponds to 33 mph.

$$\sum x_i = 1390 \text{ mi/hr}$$
$$n = 40$$

The answer is (B).

(h) The standard deviation of the sample data is

$$\sigma = \sqrt{\frac{\sum x^2}{n} - \mu^2}$$
$$\sum x^2 = 50{,}496 \text{ mi}^2/\text{hr}^2$$

The mean is computed as

$$\overline{x} = \frac{\sum x_i}{n}$$
$$= \frac{1390 \dfrac{\text{mi}}{\text{hr}}}{40}$$
$$= \boxed{34.75 \text{ mph} \quad (35 \text{ mph})}$$

Use the sample mean as an unbiased estimator of the population mean, μ.

$$\sigma = \sqrt{\frac{\sum x^2}{n} - \mu^2}$$
$$= \sqrt{\frac{50{,}496 \dfrac{\text{mi}^2}{\text{hr}^2}}{40} - \left(34.75 \dfrac{\text{mi}}{\text{hr}}\right)^2}$$
$$= \boxed{7.405 \text{ mph} \quad (7.4 \text{ mph})}$$

The answer is (D).

(i) The sample standard deviation is

$$s = \sqrt{\frac{\sum x^2 - \dfrac{\left(\sum x\right)^2}{n}}{n - 1}}$$
$$= \sqrt{\frac{50{,}496 \dfrac{\text{mi}^2}{\text{hr}^2} - \dfrac{\left(1390 \dfrac{\text{mi}}{\text{hr}}\right)^2}{40}}{40 - 1}}$$
$$= \boxed{7.500 \text{ mph} \quad (7.5 \text{ mph})}$$

The answer is (A).

(j) The sample variance is given by the square of the sample standard deviation.

$$s^2 = \left(7.500 \dfrac{\text{mi}}{\text{hr}}\right)^2$$
$$= \boxed{56.25 \text{ mi}^2/\text{hr}^2 \quad (56 \text{ mi}^2/\text{hr}^2)}$$

The answer is (A).

7. (a) The 50th percentile speed is the mean speed,

$$\boxed{46 \text{ mph.}}$$

The answer is (C).

(b) The 85th percentile speed is the speed that is exceeded by only 15% of the measurements. Because this is a normal distribution, App. 12.A can be used.

15% in the upper tail corresponds to 35% between the mean and the 85th percentile. This occurs at approximately 1.04σ. The 85th percentile speed is

$$x_{85\%} = \mu + 1.04\sigma$$
$$= 46 \ \frac{\text{mi}}{\text{hr}} + (1.04)\left(3 \ \frac{\text{mi}}{\text{hr}}\right)$$
$$= \boxed{49.12 \text{ mph} \quad (49.1 \text{ mph})}$$

The answer is (C).

(c) The upper two standard deviation speed, 2σ, is

$$x_{2\sigma} = \mu + 2\sigma = 46 \ \frac{\text{mi}}{\text{hr}} + (2)\left(3 \ \frac{\text{mi}}{\text{hr}}\right)$$
$$= \boxed{52 \text{ mph}}$$

The answer is (D).

(d) According to the central limit theorem, the mean of the average speeds is the same as the distribution mean, and, for a sample size of K measurements, the standard deviation of sample means is

$$\sigma_{\overline{x}} = \frac{\sigma_x}{\sqrt{K}} = \frac{3 \ \frac{\text{mi}}{\text{hr}}}{\sqrt{25}}$$
$$= 0.6 \text{ mph}$$
$$\overline{x}_{2\sigma} = \mu + 2\sigma_{\overline{x}} = 46 \ \frac{\text{mi}}{\text{hr}} + (2)\left(0.6 \ \frac{\text{mi}}{\text{hr}}\right)$$
$$= \boxed{47.2 \text{ mph}}$$

The answer is (B).

8. (a) From Eq. 12.45,

$$z = \frac{x_o - \mu}{\sigma}$$
$$z_{\text{upper}} = \frac{0.507 \text{ in} - 0.502 \text{ in}}{0.005 \text{ in}} = +1$$

From App. 12.A, the area outside $z = +1$ is

$$0.5 - 0.3413 = 0.1587$$

Because these are symmetrical limits, $z_{\text{lower}} = -1$.

$$\text{total fraction defective} = (2)(0.1587)$$
$$= \boxed{0.3174 \quad (0.32)}$$

The answer is (C).

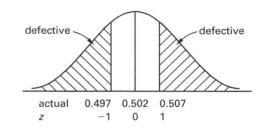

(b) This is a binomial problem.

$$p = p\{\text{defective}\} = 0.3174$$
$$q = 1 - p = 0.6826$$

From Eq. 12.28,

$$p\{x\} = f(x) = \left(\frac{n}{x}\right)\hat{p}^{\,x}\,\hat{q}^{\,n-x}$$
$$f(2) = \left(\frac{15}{2}\right)(0.3174)^2(0.6826)^{13}$$
$$= \left(\frac{15!}{13!\,2!}\right)(0.3174)^2(0.6826)^{13}$$
$$= \boxed{0.0739 \quad (0.074)}$$

The answer is (A).

9. This is a typical hypothesis test of two sample population means. The two populations are the original population the manufacturer used to determine the 1600 hr average life value and the new population the sample was taken from. The mean ($\overline{x} = 1520$ hr) of the sample and its standard deviation ($s = 120$ hr) are known, but the mean and standard deviation of a population of average lifetimes are unknown.

Assume that the average lifetime population mean and the sample mean are identical.

$$\overline{x} = \mu = 1520 \text{ hr}$$

The standard deviation of the average lifetime population is

$$\sigma_{\overline{x}} = \frac{s}{\sqrt{n}} = \frac{120 \text{ hr}}{\sqrt{100}} = 12 \text{ hr}$$

The manufacturer can be reasonably sure that the claim of a 1600 hr average life is justified if the average test life is near 1600 hr. "Reasonably sure" must be evaluated based on acceptable probability of being incorrect. If the manufacturer is willing to be wrong with a 5% probability, then a 95% confidence level is required.

Since the direction of bias is known, a one-tailed test is required. To determine if the mean has shifted downward, test the hypothesis that 1600 hr is within the 95% limit of a distribution with a mean of 1520 hr and a standard deviation of 12 hr. From a standard normal table, 5% of a standard normal distribution is outside of z equals 1.645.

$$\text{confidence limit} = \bar{x} + z\sigma_x$$

Therefore,

$$95\% \text{ confidence limit} = 1520 \text{ hr} + (1.645)(12 \text{ hr})$$
$$= 1540 \text{ hr}$$

The manufacturer can be 95% certain that the average lifetime of the bearings is less than 1600 hr.

If the manufacturer is willing to be wrong with a probability of only 1%, then a 99% confidence limit is required. From the normal table, $z = 2.33$, and the 99% confidence limit is

$$99\% \text{ confidence limit} = 1520 \text{ hr} + (2.33)(12 \text{ hr})$$
$$= 1548 \text{ hr}$$

> The manufacturer can be 99 % certain that the average bearing life is less than 1600 hr.

The answer is (D).

10. (a) Plot the data points to determine if the relationship is linear.

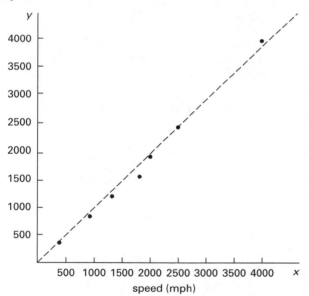

The data appear to be essentially linear. The slope, m, and the y-intercept, b, can be determined using linear regression.

The individual terms are

$$n = 7$$
$$\sum x_i = 400 + 800 + 1250 + 1600 + 2000 + 2500$$
$$ + 4000$$
$$= 12{,}550$$
$$\left(\sum x_i\right)^2 = (12{,}550)^2$$
$$= 1.575 \times 10^8$$
$$\bar{x} = \frac{\sum x_i}{n}$$
$$= \frac{12{,}550}{7}$$
$$= 1792.9$$
$$\sum x_i^2 = (400)^2 + (800)^2 + (1250)^2 + (1600)^2$$
$$ + (2000)^2 + (2500)^2 + (4000)^2$$
$$= 3.117 \times 10^7$$

Similarly,

$$\sum y_i = 370 + 780 + 1210 + 1560 + 1980$$
$$ + 2450 + 3950$$
$$= 12{,}300$$

$$\left(\sum y_i\right)^2 = (12{,}300)^2 = 1.513 \times 10^8$$
$$\bar{y} = \frac{\sum y_i}{n} = \frac{12{,}300}{7} = 1757.1$$
$$\sum y_i^2 = (370)^2 + (780)^2 + (1210)^2 + (1560)^2$$
$$ + (1980)^2 + (2450)^2 + (3950)^2$$
$$= 3.017 \times 10^7$$

Also,

$$\sum x_i y_i = (400)(370) + (800)(780) + (1250)(1210)$$
$$ + (1600)(1560) + (2000)(1980)$$
$$ + (2500)(2450) + (4000)(3950)$$
$$= 3.067 \times 10^7$$

The slope is

$$m = \frac{n\sum x_i y_i - \sum x_i \sum y_i}{n\sum x_i^2 - \left(\sum x_i\right)^2}$$
$$= \frac{(7)(3.067 \times 10^7) - (12{,}550)(12{,}300)}{(7)(3.117 \times 10^7) - (12{,}550)^2}$$
$$= 0.994$$

The y-intercept is

$$b = \bar{y} - m\bar{x}$$
$$= 1757.1 - (0.994)(1792.9)$$
$$= -25.0$$

The least squares equation of the line is

$$y = mx + b$$
$$= \boxed{0.994x - 25.0}$$

(b) The correlation coefficient is

$$r = \frac{n\sum x_i y_i - \sum x_i \sum y_i}{\sqrt{\left(n\sum x_i^2 - (\sum x_i)^2\right)\left(n\sum y_i^2 - (\sum y_i)^2\right)}}$$

$$= \frac{(7)(3.067 \times 10^7) - (12{,}500)(12{,}300)}{\sqrt{\begin{array}{l}\left((7)(3.117 \times 10^7) - (12{,}500)^2\right)\\ \times\left((7)(3.017 \times 10^7) - (12{,}300)^2\right)\end{array}}}$$

$$\asymp \boxed{1.00}$$

11. Plotting the data shows that the relationship is nonlinear.

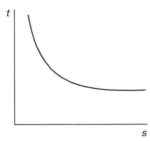

This appears to be an exponential with the form

$$t = ae^{bs}$$

Take the natural log of both sides.

$$\ln t = \ln(ae^{bs})$$
$$= \ln a + \ln(e^{bs})$$
$$= \ln a + bs$$

But, $\ln a$ is just a constant, c.

$$\ln t = c + bs$$

Make the transformation $R = \ln t$.

$$R = c + bs$$

s	R
20	3.76
18	4.95
16	5.95
14	7.00

This is linear.

$$n = 4$$
$$\sum s_i = 20 + 18 + 16 + 14 = 68$$
$$\bar{s} = \frac{\sum s}{n} = \frac{68}{4} = 17$$
$$\sum s_i^2 = (20)^2 + (18)^2 + (16)^2 + (14)^2 = 1176$$
$$\left(\sum s_i\right)^2 = (68)^2 = 4624$$
$$\sum R_i = 3.76 + 4.95 + 5.95 + 7.00 = 21.66$$
$$\bar{R} = \frac{\sum R_i}{n} = \frac{21.66}{4} = 5.415$$
$$\sum R_i^2 = (3.76)^2 + (4.95)^2 + (5.95)^2 + (7.00)^2$$
$$= 123.04$$
$$\left(\sum R_i\right)^2 = (21.66)^2 = 469.16$$
$$\sum s_i R_i = (20)(3.76) + (18)(4.95) + (16)(5.95)$$
$$\qquad + (14)(7.00)$$
$$= 357.5$$

The slope, b, of the transformed line is

$$b = \frac{n\sum s_i R_i - \sum s_i \sum R_i}{n\sum s_i^2 - \left(\sum s_i\right)^2}$$
$$= \frac{(4)(357.5) - (68)(21.66)}{(4)(1176) - (68)^2}$$
$$= -0.536$$

The intercept is

$$c = \bar{R} - b\bar{s}$$
$$= 5.415 - (-0.536)(17)$$
$$= 14.527$$

The transformed equation is

$$R = c + bs$$
$$= 14.527 - 0.536s$$
$$\boxed{\ln t = 14.527 - 0.536s}$$

12. The first step is to graph the data.

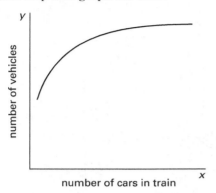

It is assumed that the relationship between the variables has the form $y = a + b \log x$. Therefore, the variable change $z = \log x$ is made, resulting in the following set of data.

z	y
0.301	14.8
0.699	18.0
0.903	20.4
1.079	23.0
1.431	29.9

$$\sum z_i = 4.413$$
$$\sum y_i = 106.1$$
$$\sum z_i^2 = 4.6082$$
$$\sum y_i^2 = 2382.2$$
$$\left(\sum z_i\right)^2 = 19.475$$
$$\left(\sum y_i\right)^2 = 11{,}257.2$$
$$\bar{z} = 0.8826$$
$$\bar{y} = 21.22$$
$$\sum z_i y_i = 103.06$$
$$n = 5$$

The slope is

$$m = \frac{n\sum z_i y_i - \sum z_i \sum y_i}{n\sum z_i^2 - \left(\sum z_i\right)^2}$$
$$= \frac{(5)(103.06) - (4.413)(106.1)}{(5)(4.6082) - 19.475}$$
$$= 13.20$$

The y-intercept is

$$b = \bar{y} - m\bar{z}$$
$$= 21.22 - (13.20)(0.8826)$$
$$= 9.570$$

The resulting equation is

$$y = 9.570 + 13.20z$$

The relationship between x and y is approximately

$$\boxed{y = 9.570 + 13.20\log x}$$

This is not an optimal correlation, as better correlation coefficients can be obtained if other assumptions about the form of the equation are made. For example, $y = 9.1 + 4\sqrt{x}$ has a better correlation coefficient.

13. (a) Plot the data to verify that they are linear.

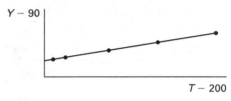

x	y
$T - 200$	$Y - 90$
7.1	2.30
10.3	2.58
0.4	1.56
1.1	1.63
3.4	1.83

step 1:

$$\sum x_i = 22.3 \qquad \sum y_i = 9.9$$
$$\sum x_i^2 = 169.43 \qquad \sum y_i^2 = 20.39$$
$$\left(\sum x_i\right)^2 = 497.29 \qquad \left(\sum y_i\right)^2 = 98.01$$
$$\bar{x} = \frac{22.3}{5} = 4.46 \qquad \bar{y} = 1.98$$
$$\sum x_i y_i = 51.54$$

step 2: From Eq. 12.73, the slope is

$$m = \frac{n\sum(x_i y_i) - \left(\sum x_i\right)\left(\sum y_i\right)}{n\sum x_i^2 - \left(\sum x_i\right)^2}$$
$$= \frac{(5)(51.54) - (22.3)(9.9)}{(5)(169.43) - 497.29}$$
$$= 0.1055$$

step 3: From Eq. 12.74, the y-intercept is

$$b = \bar{y} - m\bar{x} = 1.98 - (0.1055)(4.46)$$
$$= 1.509$$

The equation of the line is

$$y = 0.1055x + 1.509$$
$$Y - 90 = 0.1055(T - 200) + 1.509$$
$$Y = \boxed{0.1055\,T + 70.409}$$

(b) *step 4:* Use Eq. 12.75 to get the correlation coefficient.

$$r = \frac{n\sum(x_i y_i) - \left(\sum x_i\right)\left(\sum y_i\right)}{\sqrt{\left(n\sum x_i^2 - \left(\sum x_i\right)^2\right)\left(n\sum y_i^2 - \left(\sum y_i\right)^2\right)}}$$

$$= \frac{(5)(51.54) - (22.3)(9.9)}{\sqrt{\left((5)(169.43) - 497.29\right)\left((5)(20.39) - 98.01\right)}}$$

$$= \boxed{0.995}$$

14. Plot the data to see if they are linear.

This looks like it could be of the form

$$y = a + b\sqrt{x}$$

However, when x is negative (as in the first point), the function is imaginary. Try shifting the curve to the right, replacing x with $x + 1$.

$$y = a + bz$$
$$z = \sqrt{x + 1}$$

z	y
0	0
1	1
1.414	1.4
1.732	1.7
2	2
2.236	2.2
2.45	2.4
2.65	2.6
2.83	2.8
3	3

Because $y \approx z$, the relationship is

$$\boxed{y = \sqrt{x + 1}}$$

In this problem, the answer was found with some good guesses. Usually, regression would be necessary.

15. (a)

$R(\Omega)$	f	$fR(\Omega)$	$fR^2(\Omega^2)$
0.200	1	0.200	0.0400
0.210	3	0.630	0.1323
0.220	5	1.100	0.2420
0.230	10	2.300	0.5290
0.240	17	4.080	0.9792
0.250	40	10.000	2.5000
0.260	13	3.380	0.8788
0.270	6	1.620	0.4374
0.280	3	0.840	0.2352
0.290	2	0.580	0.1682
	100	24.730	6.1421

$$\overline{R} = \frac{\sum fR}{\sum f} = \frac{24.730\ \Omega}{100} = \boxed{0.2473\ \Omega \quad (0.247\ \Omega)}$$

The answer is (B).

(b) The sample standard deviation is given by Eq. 12.63.

$$s = \sqrt{\frac{\sum fR^2 - \dfrac{\left(\sum fR\right)^2}{n}}{n - 1}}$$

$$= \sqrt{\frac{6.1421\ \Omega^2 - \dfrac{(24.73\ \Omega)^2}{100}}{99}}$$

$$= \boxed{0.0163\ \Omega \quad (0.016\ \Omega)}$$

The answer is (C).

(c) 50% of the observations are below the median resistance, R_{median}. 36 observations are below 0.240, so 14 more are needed.

$$R_{median} = R_{36} + f_{interval}\,\Delta R_{interval}$$

$$= 0.240\ \Omega + \left(\frac{14}{40}\right)(0.250\ \Omega - 0.240\ \Omega)$$

$$= \boxed{0.2435\ \Omega \quad (0.244\ \Omega)}$$

The answer is (B).

(d) The sample variable is

$$s^2 = (0.0163\ \Omega)^2$$

$$= \boxed{0.0002656\ \Omega^2 \quad (0.00027\ \Omega^2)}$$

The answer is (A).

13 Computer Mathematics

PRACTICE PROBLEMS

1. What is the base-10 value of the base-2 number 1010?

(A) 3_{10}

(B) 7_{10}

(C) 10_{10}

(D) 20_{10}

2. What is the value of 144_{10} in base-2?

(A) $1\,001\,000_2$

(B) $1\,010\,000_2$

(C) $10\,010\,000_2$

(D) $10\,010\,001_2$

3. What is the value of 144_{10} in base-8, or octal, numbering?

(A) 22_8

(B) 202_8

(C) 220_8

(D) 221_8

4. What is the base-10 value of the hexadecimal number $(3C1)_{16}$?

(A) 192_{10}

(B) 768_{10}

(C) 961_{10}

(D) $15\,360_{10}$

5. What is the one's complement value of $101\,011_2$?

(A) $10\,010_2$

(B) $10\,100_2$

(C) $10\,111_2$

(D) $101\,011_2$

SOLUTIONS

1. For a positional numbering system,

$$(a_n a_{n-1} \cdots a_2 a_1 a_0) = a_n b^n + a_{n-1} b_{n-1} + \cdots$$
$$+ a_2 b^2 + a_1 b + a_0$$

Therefore, for $b = 2$,

$$(1010)_2 = (1)(2^3) + (0)(2^2) + (1)(2) + 0$$
$$= 8 + 0 + 2 + 0$$
$$= \boxed{10_{10}}$$

The answer is (C).

2. Using the remainder method gives

$$144/2 = 72 \quad \text{remainder } 0$$
$$72/2 = 36 \quad \text{remainder } 0$$
$$36/2 = 18 \quad \text{remainder } 0$$
$$18/2 = 9 \quad \text{remainder } 0$$
$$9/2 = 4 \quad \text{remainder } 1$$
$$4/2 = 2 \quad \text{remainder } 0$$
$$2/2 = 1 \quad \text{remainder } 0$$
$$1/2 = 0 \quad \text{remainder } 1$$

Reading up the remainder values from the last division remainder of 1 to the first remainder gives a base-2 value of $10\,010\,000_2$. The binary representation of 144_{10} is therefore $\boxed{10\,010\,000_2}$.

The answer is (C).

3. Using the remainder method gives

$$144/8 = 18 \quad \text{remainder } 0$$
$$18/8 = 2 \quad \text{remainder } 2$$
$$2/8 = 0 \quad \text{remainder } 2$$

Reading up the remainder values from the last division remainder of 1 to the first remainder gives a base-8 value of 220_8. The octal representation of 144_{10} is $\boxed{220_8.}$

The answer is (C).

4. For a positional numbering system,

$$(a_n a_{n-1} \cdots a_2 a_1 a_0) = a_n b^n + a_{n-1} b_{n-1} + \cdots$$
$$+ a_2 b^2 + a_1 b + a_0$$

Therefore, in base-10 representation, for $b = 16$,

$$(3C1)_{16} = (3)(16)^2 + (12)(16) + 1$$
$$= 768 + 192 + 1$$
$$= \boxed{961_{10}}$$

The answer is (C).

5. The one's complement is

$$N_1^x = N_2^x - 1$$

The one's complement is found by subtracting 1 from the two's complement value. Or, the one's complement can be found by reversing all the digits in the base-2 number. That is, all 0's become 1's and vice versa. Therefore, for $101\,011_2$, the one's complement is $010\,100_2$ or, dropping the leading 0, $\boxed{10\,100_2.}$

The answer is (B).

14 Numerical Analysis

PRACTICE PROBLEMS

1. A function is given as $y = 3x^{0.93} + 4.2$. What is the percent error if the value of y at $x = 2.7$ is found by using straight-line interpolation between $x = 2$ and $x = 3$?

(A) 0.060%

(B) 0.18%

(C) 2.5%

(D) 5.4%

2. Given the following data points, find y by straight-line interpolation for $x = 2.75$.

x	y
1	4
2	6
3	2
4	-14

(A) 2.1

(B) 2.4

(C) 2.7

(D) 3.0

3. Using the bisection method, find all the roots of $f(x) = 0$, to the nearest 0.000005.

$$f(x) = x^3 + 2x^2 + 8x - 2$$

4. What is most nearly the root of the following equation?

$$f(x) = x^3 - 4x - 5$$

Use Newton's method with $x_0 = 2$.

(A) 2.47

(B) 2.60

(C) 2.63

(D) 16.7

5. What is the primary DISADVANTAGE of performing a straight-line interpolation—as opposed to performing a polynomial interpolation—using either the Lagrange method or the Newton method?

(A) Calculations are complicated.

(B) Curvature is ignored.

(C) Iterations are required.

(D) The process is slow.

SOLUTIONS

1. The actual value at $x = 2.7$ is

$$y(x) = 3x^{0.93} + 4.2$$
$$y(2.7) = (3)(2.7)^{0.93} + 4.2$$
$$= 11.756$$

At $x = 3$,

$$y(3) = (3)(3)^{0.93} + 4.2$$
$$= 12.534$$

At $x = 2$,

$$y(2) = (3)(2)^{0.93} + 4.2$$
$$= 9.916$$

Use straight-line interpolation.

$$\frac{x_2 - x}{x_2 - x_1} = \frac{y_2 - y}{y_2 - y_1}$$
$$\frac{3 - 2.7}{3 - 2} = \frac{12.534 - y}{12.534 - 9.916}$$
$$y = 11.749$$

The relative error is given by

$$\frac{\begin{array}{l}\text{actual value}\\ -\text{predicted value}\end{array}}{\text{actual value}} = \frac{11.756 - 11.749}{11.756}$$
$$= \boxed{0.000595 \quad (0.060\%)}$$

The answer is (A).

2. Let $x_1 = 2$; from the table of data points, $y_1 = 6$. Let $x_2 = 3$; from the table of data points, $y_2 = 2$.

Let $x = 2.75$. By straight-line interpolation,

$$\frac{x_2 - x}{x_2 - x_1} = \frac{y_2 - y}{y_2 - y_1}$$
$$\frac{3 - 2.75}{3 - 2} = \frac{2 - y}{2 - 6}$$
$$\boxed{y = 3}$$

The answer is (D).

3. $f(x) = x^3 + 2x^2 + 8x - 2$

Try to find an interval in which there is a root.

x	$f(x)$
0	-2
1	9

A root exists in the interval $[0, 1]$.

Try $x = \left(\dfrac{1}{2}\right)(0 + 1) = 0.5$.

$$f(0.5) = (0.5)^3 + (2)(0.5)^2 + (8)(0.5) - 2 = 2.625$$

A root exists in $[0, 0.5]$.

Try $x = 0.25$.

$$f(0.25) = (0.25)^3 + (2)(0.25)^2 + (8)(0.25) - 2 = 0.1406$$

A root exists in $[0, 0.25]$.

Try $x = 0.125$.

$$f(0.125) = (0.125)^3 + (2)(0.125)^2 + (8)(0.125) - 2$$
$$= -0.967$$

A root exists in $[0.125, 0.25]$.

Try $x = \left(\dfrac{1}{2}\right)(0.125 + 0.25) = 0.1875$.

Continuing,

$$f(0.1875) = -0.42 \quad [0.1875, 0.25]$$
$$f(0.21875) = -0.144 \quad [0.21875, 0.25]$$
$$f(0.234375) = -0.002 \quad [\text{This is close enough.}]$$

One root is $x_1 \approx \boxed{0.234375.}$

Try to find the other two roots. Use long division to factor the polynomial.

$$
\begin{array}{r}
x^2 + \ 2.234375x \ +8.52368 \\
x - 0.234375 \overline{) \ x^3 + \quad\quad 2x^2 + \quad\quad 8x - 2} \\
-(x^3 - \ 0.234375x^2 \ ) \\
\hline
2.234375x^2 + \quad\quad 8x \\
-(2.234375x^2 - \ 0.52368x \ ) \\
\hline
8.52368x - 2 \\
- \ (8.52368x - 1.9977) \\
\hline
\approx 0
\end{array}
$$

Use the quadratic equation to find the roots of $x^2 + 2.234375x + 8.52368$.

$$x_2, x_3 = \frac{-2.234375 \pm \sqrt{(2.234375)^2 - (4)(1)(8.52368)}}{(2)(1)}$$

$$= \boxed{-1.117189 \pm j2.697327} \quad \text{[both imaginary]}$$

4. Newton's method for finding a root is

$$x_{n+1} = g(x_n) = x_n - \frac{f(x_n)}{f'(x_n)}$$

The function and its first derivative are

$$f(x) = x^3 - 4x - 5$$
$$f'(x) = 3x^2 - 4$$

The first iteration, with $n = 0$, gives

$$x_0 = 2$$
$$f(x_0) = f(2) = 2^3 - (4)(2) - 5 = -5$$
$$f'(x_0) = f'(2) = (3)(2)^2 - 4 = 8$$
$$x_1 = x_0 - \frac{f(x_0)}{f'(x_0)}$$
$$= 2 - \left(\frac{-5}{8}\right)$$
$$= 2.625$$

The second iteration, with $n = 1$, gives

$$x_1 = 2.625$$
$$f(x_1) = f(2.625) = 2.625^3 - (4)(2.625) - 5 = 2.588$$
$$f'(x_1) = f'(2.625) = (3)(2.625)^2 - 4 = 16.672$$
$$x_2 = x_1 - \frac{f(x_1)}{f'(x_1)}$$
$$= 2.625 - \left(\frac{2.588}{16.672}\right)$$
$$= \boxed{2.470 \quad (2.47)}$$

The iterations may be continued to the desired accuracy.

The answer is (A).

5. The straight-line interpolation technique ignores the effects of curvature. Lagrangian and Newtonian polynomial methods of interpolation do not.

The answer is (B).

15 Advanced Engineering Mathematics

PRACTICE PROBLEMS

1. What is the correct description of the following differential equation?

$$2y''(t) + 3y'(t) + 1 = f(t)$$

(A) linear, first order, homogeneous

(B) nonlinear, first order, homogeneous

(C) linear, second order, nonhomogeneous

(D) nonlinear, second order, nonhomogeneous

2. What is the characteristic equation for the following differential equation?

$$3y''(t) + 5y'(t) + 6y(t) = 24t$$

(A) $3\lambda^2 + 5\lambda + 6 = 0$

(B) $3\lambda^2 + 5\lambda + 6 = 24$

(C) $3\lambda^2 + 5\lambda = 18$

(D) $3\lambda^2 + 5\lambda = 24$

3. Which of the following is primarily used to analyze periodic signals when $s = j\omega$?

(A) Fourier series

(B) Fourier transforms

(C) Laplace transforms

(D) z-transforms

4. What is the Laplace transform of the following?

$$x(t - t_0)$$

(A) $X(s)$

(B) $e^{-st}X(s)$

(C) e^{-st_0}

(D) $e^{-st_0}X(s)$

5. What type of function is shown?

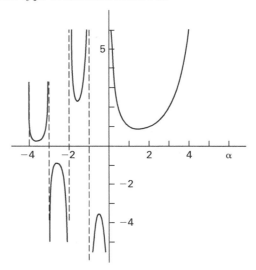

(A) delta function

(B) error function

(C) gamma function

(D) unit finite impulse function

SOLUTIONS

1. All the coefficients of the terms are constants, therefore, the equation is linear. Because there is a second derivative, the equation is of second order. Because the forcing function, $f(t)$, exists, the implication is that the value of $f(t)$ is nonzero. Therefore, the equation is nonhomogeneous.

The answer is (C).

2. The characteristic equation is

$$\boxed{3\lambda^2 + 5\lambda + 6 = 0}$$

The characteristic equation is used to solve the homogeneous differential equation.

The answer is (A).

3. Fourier series and Laplace transforms are used to analyze periodic signals. The Laplace transform is more general in nature.

$$\mathcal{L}\big(f(t)\big)\bigg|_{s=j\omega} = F(s)\bigg|_{s=j\omega} = \mathcal{F}\big(f(t)\big)$$

$$\boxed{\text{Therefore, the Fourier series is used.}}$$

The Fourier series is equivalent to the Laplace transform when $s = j\omega$, that is, when a transient condition is not present. When $s = \sigma + j\omega$, a transient condition is present, represented by the σ term, and the Fourier series is not equivalent to the Laplace transform.

The answer is (A).

4. The Laplace transform of a time-shifted function is

$$\boxed{e^{-st_0}X(s)}$$

The answer is (D).

5. The illustration shows the factorial function, more commonly called the gamma function.

$$\Gamma(\alpha) = \int_0^\infty e^{-t}t^{(\alpha-1)}dt \quad [\alpha > 0]$$

The answer is (C).

Topic II: Basic Theory

Chapter

16 Electromagnetic Theory

PRACTICE PROBLEMS

1. What is the magnitude of the electric force between a proton and electron 0.046 nm apart in free space?

(A) 9.651×10^{-19} N

(B) -5.018×10^{-18} N

(C) 5.018×10^{-18} N

(D) 1.090×10^{-7} N

2. What is the magnitude of the electric field strength for a proton in free space at 0.046 nm?

(A) 3.047×10^{-30} N/C

(B) 1.440×10^{-9} N/C

(C) 31.30 N/C

(D) 6.805×10^{11} N/C

3. What is the magnitude of the electric field intensity for a parallel plate capacitor with a potential of 12 V applied and a distance of 0.3 cm between plates?

(A) 0.12 V/m

(B) 0.40 V/m

(C) 40 V/m

(D) 4000 V/m

4. During design discussion, quartz is picked as the dielectric for certain capacitors. What is the expected permittivity?

(A) 1.8×10^{-12} C^2/N·m^2

(B) 8.8×10^{-12} C^2/N·m^2

(C) 4.4×10^{-11} C^2/N·m^2

(D) 8.8×10^{-8} C^2/N·m^2

5. What is the magnitude of the electric flux density 0.5 m from a wire with a uniform line charge of 30×10^{-5} C/m?

(A) 9.5×10^{-5} C/m^2

(B) 1.9×10^{-4} C/m^2

(C) 6.0×10^{-4} C/m^2

(D) 1.7×10^{-3} C/m^2

SOLUTIONS

1. The magnitude of the electric force is given by Coulomb's law. From Eq. 16.3,

$$F = \frac{|Q||Q|}{4\pi\epsilon_0 r^2}$$

$$= \frac{(1.6022 \times 10^{-19}\text{ C})(1.6022 \times 10^{-19}\text{ C})}{4\pi\left(8.854 \times 10^{-12}\ \dfrac{\text{C}^2}{\text{N}\cdot\text{m}^2}\right)(0.046 \times 10^{-9}\text{ m})^2}$$

$$= \boxed{1.090 \times 10^{-7}\text{ N}}$$

0.046 nm is the approximate distance between the proton and electron in the hydrogen atom.

The answer is (D).

2. From Eq. 16.6, the magnitude of the electric field strength, E, is

$$E = \frac{Q}{4\pi\epsilon_0 r^2}$$

$$= \frac{1.6022 \times 10^{-19}\text{ C}}{4\pi\left(8.854 \times 10^{-12}\ \dfrac{\text{C}^2}{\text{N}\cdot\text{m}^2}\right)(0.046 \times 10^{-9}\text{ m})^2}$$

$$= \boxed{6.805 \times 10^{11}\text{ N/C}}$$

The answer is (D).

3. From Table 16.6, the electric field intensity for a parallel plate capacitor (ignoring fringing effects) is

$$E = \frac{V}{r}$$

$$= \frac{12\text{ V}}{(0.3\text{ cm})\left(\dfrac{1\text{ m}}{100\text{ cm}}\right)}$$

$$= \boxed{4000\text{ V/m}}$$

The answer is (D).

4. From Table 16.7,

$$\epsilon_r = 5$$

From Eq. 16.11, the permittivity is

$$\epsilon = \epsilon_0 \epsilon_r$$

$$= \left(8.854 \times 10^{-12}\ \dfrac{\text{C}^2}{\text{N}\cdot\text{m}^2}\right)(5)$$

$$= \boxed{4.4 \times 10^{-11}\text{ C}^2/\text{N}\cdot\text{m}^2}$$

The answer is (C).

5. From Eq. 16.21, the electric flux density for a uniform line charge is

$$D = \frac{\rho_l}{2\pi r}$$

$$= \frac{30 \times 10^{-5}\ \dfrac{\text{C}}{\text{m}}}{2\pi(0.5\text{ m})}$$

$$= \boxed{9.5 \times 10^{-5}\text{ C/m}^2}$$

30×10^{-5} C/m correlates approximately with 30 A using an electron speed of 10^5 m/s, which is the thermal limit for electrons in silicon at room temperature.

The answer is (A).

17 Electronic Theory

PRACTICE PROBLEMS

1. A silicon surface has a work function of 4.8 V. If no kinetic energy will be imparted to the electron, what frequency of incoming radiation is required to free an electron from the surface?

   (A) 5.6×10^4 Hz

   (B) 1.2×10^{15} Hz

   (C) 3.5×10^{23} Hz

   (D) 7.2×10^{33} Hz

2. For germanium at room temperature, what is the conductivity of the holes?

   (A) 7.6×10^{-7} S/m

   (B) 7.6×10^{-1} S/m

   (C) 4.8×10^{14} S/m

   (D) 9.5×10^{14} S/m

3. How many valence electrons does a single atom of either silicon or germanium have?

   (A)   2

   (B)   4

   (C)   6

   (D)   10

4. In an n-type silicon crystal, the number of electrons is

   (A)   equal to the number of holes

   (B)   greater than the number of holes

   (C)   less than the number of holes

   (D)   equal to the square of the number of holes

$n = electrons$
$p = holes$

$n_i^2 = np$

5. The n_i of gallium arsenide is 1.1×10^7 cm^{-3}. If the hole concentration is 1.5×10^2 cm^{-3}, what is the electron concentration?

   (A) 1.5×10^2 cm^{-3}

   (B) 7.3×10^4 cm^{-3}

   (C) 1.1×10^7 cm^{-3}

   (D) 8.1×10^{11} cm^{-3}

SOLUTIONS

1. From Eq. 17.6,

$$h\nu = \varphi + \tfrac{1}{2}m\mathrm{v}^2$$

From App. 17.A, the work function for silicon is 4.8 V. Substituting into the equation gives

$$\frac{\varphi}{h} = \frac{(4.8 \text{ V})\left(1.6022 \times 10^{-19}\dfrac{\text{C}}{\text{electron}}\right)}{6.6256 \times 10^{-34} \text{ J·s}}$$

$$= \boxed{1.2 \times 10^{15} \text{ Hz}}$$

The answer is (B).

2. From Eq. 17.8,

$$\sigma_h = \rho_h \mu_h$$

From App. 17.A, the intrinsic electron carrier density (concentration) is

$$n_i = 2.5 \times 10^{13}\,\text{cm}^{-3} = n_{\text{holes}}$$

The electric unit charge of the holes is

$$q_h = 1.6022 \times 10^{-19} \text{ C}$$

Because there are equal numbers of electrons and holes in an intrinsic semiconductor,

$$\rho_h = n_i q_h = (2.5 \times 10^{13}\,\text{cm}^{-3})(1.6022 \times 10^{-19} \text{ C})$$

$$= 4.0 \times 10^{-6} \text{ C/cm}^3$$

The mobility of the holes is given in App. 17.A as

$$\mu_h = 1900 \text{ cm}^2/\text{V·s}$$

The conductivity of the holes is calculated by multiplying the charge density by the mobility.

$$\sigma = \rho_h \mu_h = \left(4.0 \times 10^{-6}\frac{\text{C}}{\text{cm}^3}\right)\left(1900\,\frac{\text{cm}^2}{\text{V·s}}\right)\left(100\,\frac{\text{cm}}{\text{m}}\right)$$

$$= \boxed{7.6 \times 10^{-1} \text{ S/m}}$$

The answer is (B).

3. Silicon has an electronic structure of

$$1s^2 2s^2 2p^6 3s^2 3p^2$$

Germanium has an electronic structure of

$$1s^2 2s^2 2p^6 3s^2 3p^6 3d^{10} 4s^2 4p^2$$

$$\boxed{\text{Both have four valence electrons in the outer shell.}}$$

The answer is (B).

4. An *n*-type crystal is doped with donor impurities. This increases the charge carrier density of electrons and thereby reduces the number of holes. By Eq. 17.13,

$$np = n_i^2 = \text{constant}$$

The answer is (B).

5. Using the law of mass action given by Eq. 17.13,

$$np = n_i^2$$

$$n = \frac{n_i^2}{p} = \frac{(1.1 \times 10^7 \text{ cm}^{-3})^2}{1.5 \times 10^2 \text{ cm}^{-3}}$$

$$= \boxed{8.1 \times 10^{11} \text{ cm}^{-3}}$$

The answer is (D).

18 Communication Theory

PRACTICE PROBLEMS

1. A 60 Hz signal is to be transformed into a digital signal. Determine the ideal sampling rate.

2. Determine the Nyquist interval for a 400 Hz signal.

3. A signal has an equivalent rectangular bandwidth of 6 MHz. Determine the response time.

4. A 32-bit computer system has equal probabilities for each of its information elements, x_i, or bits. Determine this probability.

5. Describe entropy as used in information theory.

6. If signal power is doubled at a transmitter, what is the decibel increase?

    (A) 0.3 dB

    (B) 2 dB

    (C) 3 dB

    (D) 20 dB

7. A communication signal uses a bandwidth, BW, of 2 kHz at 101 MHz. What is the classification of this signal?

    (A) narrowband, VLF

    (B) voice-grade, HF

    (C) voice-grade, VHF

    (D) wideband, UHF

8. A communication system uses the left-most bit as an even parity bit. Which of the following messages is in error?

    (A) 11101

    (B) 00101

    (C) 10001

    (D) 01011

9. A 5 kΩ resistor operating in a 6 MHz bandwidth is designed for a temperature service range of $-10°C$ to $60°C$. What is the maximum thermal noise voltage generated?

    (A) 9.71×10^{-6} V

    (B) 1.14×10^{-5} V

    (C) 2.35×10^{-5} V

    (D) 2.46×10^{-5} V

10. What is the shot noise current for a microphone with a 16 kHz bandwidth drawing 500 mA?

    (A) 2.60×10^{-12} A

    (B) 2.53×10^{-8} A

    (C) 2.56×10^{-8} A

    (D) 5.06×10^{-8} A

SOLUTIONS

1. The ideal sampling rate for a signal of frequency f_I is given by the Nyquist rate, f_S.

From Eq. 18.1,

$$f_S = 2f_I$$
$$= (2)(60 \text{ Hz})$$
$$= \boxed{120 \text{ Hz}}$$

In practice, the rate would have to be greater than this to avoid a zero output, which would occur if the initial sampling occurred at a value of zero for the analog signal.

2. From Eq. 18.2, the Nyquist interval, T_S, is

$$T_S = \frac{1}{2f_I}$$

The information signal, or message signal, frequency is given as 400 Hz. Therefore,

$$T_S = \frac{1}{(2)(400 \text{ Hz})}$$
$$= \boxed{0.00125 \text{ s} \quad (1.25 \text{ ms})}$$

3. From Eq. 18.4, the bandwidth, $\Delta\Omega$, and duration or response time, Δt, are related by the following uncertainty relationship.

$$\Delta t \Delta\Omega = 2\pi$$

Rearranging and substituting,

$$\Delta t = \frac{2\pi}{\Delta\Omega} = \frac{2\pi}{6 \times 10^6 \text{ Hz}}$$
$$= \boxed{1.047 \times 10^{-6} \text{ s}}$$

4. The information content of a message is given as 32 bits. The information content is related to the probability as follows.

From Eq. 18.6,

$$I(x_i) = \log_2 \frac{1}{p(x_i)}$$

Rearranging and substituting,

$$p(x_i) = \frac{1}{\text{antilog}_2 I(x_i)}$$
$$= \frac{1}{\text{antilog}_2(32)}$$
$$= \boxed{\frac{1}{4.295 \times 10^9}}$$

5. Entropy is the mean or average value of a message. It can be considered a measure of the disorder of a system, or of uncertainty in a message. For a given message of alphabet size A, the entropy is determined by Eq. 18.7.

$$H(X) = \sum_{i=1}^{A} p(x_i) I(x_i)$$

Here $p(x_i)$ is the probability of the piece of information x_i. Entropy is normally given in base 2, in units of bits. In base e, the unit is the nat.

6. The decibel is defined by the following ratio of power, P_{dB}.

$$P_{dB} = 10 \log_{10} \frac{P_2}{P_1}$$

Here, P_1 is a reference power and P_2 is the signal power. Because the signal power is doubled, P_2/P_1 equals 2.

$$P_{dB} = 10 \log_{10} 2$$
$$= \boxed{3 \text{ dB}}$$

The answer is (C).

7. Classification can mean a number of things. Given the possible choices, this signal is a voice-grade channel (BW is 300 Hz to 4 KHz) that is in the VHF range (30 MHz to 300 MHz).

The answer is (C).

Basic Theory

8. The message is contained in the bits to the right of the first bit in the signal, the parity bit. The parity represents an even or odd total of ones. The following table is helpful.

option	message	parity	signal
(A)	1101	1	11101
(B)	0101	0	00101
(C)	0001	1	10001
(D)	1011	1	11011

By comparison, message (D) is in error.

The answer is (D).

9. From Eq. 18.10, the mean squared thermal noise voltage is

$$\overline{v_n^2} = 4kTR(\text{BW})$$
$$v_n = \sqrt{4kTR(\text{BW})}$$

Here, k is Boltzmann's constant, T is the absolute temperature, R is resistance, and BW is the bandwidth. The maximum voltage occurs at the maximum temperature (60°C or 333K), so

$$v = \sqrt{\begin{array}{c}(4)\left(1.38 \times 10^{-23} \; \dfrac{\text{J}}{\text{K}}\right)(333\text{K}) \\ \times (5 \times 10^3 \; \Omega)(6 \times 10^6 \; \text{Hz})\end{array}}$$
$$= \boxed{2.35 \times 10^{-5} \; \text{V}}$$

The answer is (C).

10. From Eq. 18.13, the mean squared shot noise current is given in terms of the change per carrier, q.

$$\overline{i_n^2} = 2qI(\text{BW})$$
$$i_n = \sqrt{2qI(\text{BW})}$$
$$= \sqrt{\begin{array}{c}(2)(1.6022 \times 10^{-19} \; \text{C}) \\ \times (500 \times 10^{-3} \; \text{A})(16 \times 10^3 \; \text{Hz})\end{array}}$$
$$= \boxed{5.06 \times 10^{-8} \; \text{A}}$$

The answer is (D).

19 Acoustic and Transducer Theory

PRACTICE PROBLEMS

1. Explain two differences between sound waves and electromagnetic waves.

2. Define the term "transducer."

3. What is the sound pressure level in air for a signal at the pain sensation threshold of 1 W/m^2?

   (A) -120 dB

   (B) 3.0 dB

   (C) 60 dB

   (D) 120 dB

4. Explain the difference between photovoltaic and photoconductive transduction.

5. The reference pressure level, p_0, in air of $20 \text{ } \mu\text{Pa}$ correlates with the threshold audible frequency of 2000 Hz. What is the wavelength of this sound wave?

   (A) 0.17 m

   (B) 1.0 m

   (C) 6.1 m

   (D) 20 m

SOLUTIONS

1. Sound is a longitudinal or compression wave. Electromagnetic waves are transverse. That is, the electric and magnetic fields are at right angles to the direction of propagation. Also, sound waves require a medium in which to travel, while electromagnetic waves do not.

2. A transducer is a communication interface that provides a usable output in response to a specified measurand (i.e., a physical quantity, property, or condition).

3. From Ex. 19.2, the pressure level corresponding to 1 W/m^2 is 29.3 Pa. The reference pressure (corresponding to 0 dB) is defined as $20 \text{ } \mu\text{Pa}$. Substituting from Eq. 19.8 gives

$$
\begin{aligned}
L_p &= 20 \log \frac{p}{p_0} \\
&= 20 \log \frac{29.3 \text{ Pa}}{20 \times 10^{-6} \text{ Pa}} \\
&= \boxed{123 \text{ dB} \quad (120 \text{ dB})}
\end{aligned}
$$

The "pain" threshold is nominally 120 dB in most references.

The answer is (D).

4. The primary difference between photovoltaic and photoconductive transduction is that a photovoltaic device uses dissimilar metals and a photoconductive device uses a semiconductor *pn* junction.

5. From Table 19.1, the speed of sound in air is approximately 330 m/s. Substituting this and the given frequency into Eq. 19.1 gives

$$
\text{v}_s = f\lambda
$$

$$
\lambda = \frac{\text{v}_s}{f} = \frac{330 \text{ } \frac{\text{m}}{\text{s}}}{2000 \text{ Hz}}
$$

$$
= \boxed{0.165 \text{ m} \quad (0.17 \text{ m})}
$$

The answer is (A).

Topic III: Field Theory

Chapter

Field Theory

20 Electrostatics

PRACTICE PROBLEMS

1. Determine the magnitude of the electric field necessary to place a 1 N force on an electron.

2. A point charge of 4.8×10^{-19} C is located at a point $(3, 1, 0)$. What is the electric field strength at $(3, 0, 1)$?

(A) 1.5×10^{-9} V/m $\left(\dfrac{-\mathbf{y}}{\sqrt{2}} + \dfrac{\mathbf{z}}{\sqrt{2}} \right)$

(B) 2.2×10^{-9} V/m $\left(\dfrac{-\mathbf{y}}{\sqrt{2}} + \dfrac{\mathbf{z}}{\sqrt{2}} \right)$

(C) 3.1×10^{-9} V/m $(-\mathbf{y} + \mathbf{z})$

(D) 4.3×10^{-9} V/m $(-\mathbf{y} + \mathbf{z})$

3. Water molecules in vapor form have a fractional dipole charge of approximately 5.26×10^{-20} C. If the dipole bond length is 1.17 Å, what is the magnitude of the dipole moment?

(A) 6.2×10^{-32} C·m

(B) 6.2×10^{-30} C·m

(C) 5.3×10^{-20} C·m

(D) 6.1×10^{-20} C·m

4. The potential of a point charge, in spherical coordinates and relative to infinity, is

$$V = \frac{Q}{4\pi\epsilon_0 r}$$

Determine the expression for the electric field strength, $\mathbf{E}$.

5. The electric field in an n-type material outside the space charge region is given by

$$\mathbf{E} = 2\mathbf{x} + 2\mathbf{y} + 3\mathbf{z}$$

Determine the charge density, ρ, and explain the result.

SOLUTIONS

1. The magnitude of the force is

$$F = QE$$

Rearranging gives

$$E = \frac{F}{Q}$$
$$= \frac{1 \text{ N}}{1.602 \times 10^{-19} \text{ C}}$$
$$= \boxed{6.2 \times 10^{18} \text{ V/m}}$$

For determining the magnitude of the force and not the direction, the minus sign is not required on the charge.

2.

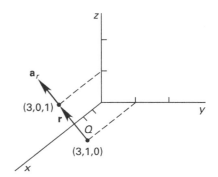

The electric field is

$$\mathbf{E} = \left(\frac{Q}{4\pi\epsilon r^2} \right) \mathbf{a}_r$$

First, determine the value of the unit vector, $\mathbf{a}_r$.

$$a_r = \sqrt{(3-3)^2 + (0-1)^2 + (1-0)^2} = \sqrt{2}$$
$$\mathbf{r} = 0\mathbf{x} - 1\mathbf{y} + 1\mathbf{z}$$

Therefore,

$$\mathbf{a}_r = \frac{\mathbf{r}}{a_r} = \frac{1}{\sqrt{2}} (-\mathbf{y} + \mathbf{z})$$

Substituting this and the given information (free space is assumed) gives

$$\mathbf{E} = \left(\frac{Q}{4\pi\epsilon r^2}\right)\mathbf{a}_r$$

$$= \left(\frac{4.8 \times 10^{-19} \text{ C}}{4\pi\left(8.854 \times 10^{-12}\dfrac{\text{C}^2}{\text{N·m}^2}\right)(\sqrt{2} \text{ m})^2}\right)$$

$$\times \left(\left(\frac{1}{\sqrt{2}}\right)(-\mathbf{y}+\mathbf{z})\right)$$

$$= \boxed{2.16 \times 10^{-9} \text{ N/C}\left(\frac{-\mathbf{y}}{\sqrt{2}} + \frac{\mathbf{z}}{\sqrt{2}}\right)}$$

The answer is (B).

3. The dipole moment vector, given in terms of a single positive charge, is

$$\mathbf{p} + q\mathbf{d}$$

It points from the negative to positive charge. The separation of charges is d.

An angstrom, Å, equals 10^{-10} m. Substituting gives

$$p = qd = (5.26 \times 10^{-20} \text{ C})(1.17 \times 10^{-10} \text{ m})$$

$$= \boxed{6.15 \times 10^{-30} \text{ C·m} \quad (6.2 \times 10^{-30} \text{ C·m})}$$

The answer is (B).

4. The electric field strength is

$$\mathbf{E} = -\boldsymbol{\nabla} V$$

$$= -\boldsymbol{\nabla}\left(\frac{Q}{4\pi\epsilon_0 r}\right)$$

$$= \left(-\frac{Q}{4\pi\epsilon_0}\right)\left(\boldsymbol{\nabla}\frac{1}{r}\right)$$

In spherical coordinates, the del operator can be represented as

$$\boldsymbol{\nabla} = \left(\frac{\partial}{\partial r}\right)\mathbf{r} + \left(\frac{1}{r}\right)\left(\frac{\partial}{\partial \theta}\right)\theta + \left(\frac{1}{r\sin\theta}\right)\left(\frac{\partial}{\partial \phi}\right)\phi$$

For a point charge, the potential and the electric field are constant with respect to θ and ϕ. Thus,

$$\mathbf{E} = \left(-\frac{Q}{4\pi\epsilon_0}\right)\left(\frac{\partial\dfrac{1}{r}}{\partial r}\right)\mathbf{r}$$

$$= \left(\frac{Q}{4\pi\epsilon_0 r^2}\right)\mathbf{r}$$

$$= \boxed{\left(\frac{Q}{4\pi\epsilon_0 r^2}\right)\mathbf{a}_r}$$

The last equation changes the symbol from $\mathbf{r}$ to $\mathbf{a}_r$ only. Many variations exist.

5. The charge density, ρ, can be determined from Gauss' law.

$$\text{div } \mathbf{E} = \boldsymbol{\nabla}\cdot\mathbf{E} = \frac{\rho}{\epsilon}$$

$$\boldsymbol{\nabla}\cdot\mathbf{E} = \boldsymbol{\nabla}\cdot(2\mathbf{x} + 2\mathbf{y} + 3\mathbf{z})$$

$$= \left(\frac{\partial 2}{\partial x}\right)\mathbf{x} + \left(\frac{\partial 2}{\partial y}\right)\mathbf{y} + \left(\frac{\partial 3}{\partial z}\right)\mathbf{z}$$

$$= 0$$

When the electric field is constant, the divergence is zero. This indicates there is no net charge in the region, as is the case for the pn junction not under the influence of an external electric field.

21 Electrostatic Fields

PRACTICE PROBLEMS

1. Determine the electric flux density magnitude if the polarization is 2.0×10^{-6} C/m^2 and the susceptibility is 2.2.

   (A) 1.8×10^{-17} C/m^2

   (B) 1.8×10^{-6} C/m^2

   (C) 2.9×10^{-6} C/m^2

   (D) 4.0×10^{-6} C/m^2

2. Given $\mathbf{D} = (2 \times 10^{-6})\mathbf{a}_r$ C/m^2 and $\epsilon_r = 3.0$, what is the electric field strength?

   (A) $(6.7 \times 10^{-7})\mathbf{a}_r$ N/C

   (B) $(2.0 \times 10^{-4})\mathbf{a}_r$ N/C

   (C) $(7.5 \times 10^{4})\mathbf{a}_r$ N/C

   (D) $(2.3 \times 10^{5})\mathbf{a}_r$ N/C

3. The following values for relative permittivity are taken from a standard reference: paper, 2.0; quartz, 5.0; and marble, 8.3.

The electric field strength is 2.0×10^5 V/m. $\mathbf{D}$ must be greater than or equal to 7×10^{-6} C/m^2. Determine which of the given dielectrics is suitable.

4. Moist soil has a conductivity of approximately 10^{-3} S/m and an ϵ_r of 3.0. A certain electric field strength is given by

$$E = 3.5 \times 10^{-6} \sin(2.5 \times 10^3)t \ \frac{\text{V}}{\text{m}}$$

Determine the displacement current density.

5. In which circuit element is displacement current largest: conductor, inductor, or capacitor? Explain.

SOLUTIONS

1. The polarization for a material with susceptibility χ_e is

$$P = \chi_e \epsilon_0 E$$

This assumes that the electric field $\mathbf{E}$ and the polarization $\mathbf{P}$ are in the same direction; that is, the material is isotropic and linear. Given that, the electic flux density is

$$D = \epsilon_0 E + P$$

Substituting P/χ_e for $\epsilon_0 E$,

$$
\begin{aligned}
D &= \frac{P}{\chi_e} + P \\
&= \frac{2.0 \times 10^{-6} \ \dfrac{\text{C}}{\text{m}^2}}{2.2} + 2.0 \times 10^{-6} \ \frac{\text{C}}{\text{m}^2} \\
&= \boxed{2.9 \times 10^{-6} \ \text{C/m}^2}
\end{aligned}
$$

The answer is (C).

2. The flux density $\mathbf{D}$ in a medium of permittivity ϵ_r is

$$\mathbf{D} = \epsilon_0 \epsilon_r \mathbf{E}$$

Rearranging gives

$$
\begin{aligned}
\mathbf{E} = \frac{\mathbf{D}}{\epsilon_0 \epsilon_r} &= \frac{(2 \times 10^{-6})\mathbf{a}_r \ \dfrac{\text{C}}{\text{m}^2}}{\left(8.854 \times 10^{-12} \ \dfrac{\text{C}^2}{\text{N·m}^2}\right)(3.0)} \\
&= \boxed{(7.5 \times 10^{4})\mathbf{a}_r \ \text{N/C}}
\end{aligned}
$$

The answer is (C).

3. Determine the minimum value of the permittivity, ϵ_r, required.

$$\mathbf{D} = \epsilon_0 \epsilon_r \mathbf{E}$$

Assuming the dielectrics to be isotropic and linear,

$$D = \epsilon_0 \epsilon_r E$$

$$\epsilon_r = \frac{D}{E\epsilon_0} = \frac{7 \times 10^{-6}\,\dfrac{\text{C}}{\text{m}^2}}{\left(2.0 \times 10^5\,\dfrac{\text{V}}{\text{m}}\right)\left(8.854 \times 10^{-12}\,\dfrac{\text{C}^2}{\text{N·m}^2}\right)}$$

$$= 3.95$$

Therefore, the quartz or marble is an acceptable dielectric.

4. Determine the value of the magnitude of the electric flux density, D.

$$\begin{aligned}
D &= \epsilon_0 \epsilon_r E \\
&= \left(8.854 \times 10^{-12}\,\frac{\text{C}^2}{\text{N·m}^2}\right)(3.0) \\
&\quad \times \left(3.5 \times 10^{-6}\sin(2.5 \times 10^3)t\,\frac{\text{V}}{\text{m}}\right) \\
&= 9.30 \times 10^{-17}\sin(2.5 \times 10^3)t \ \text{C/m}^2
\end{aligned}$$

The displacement current is

$$\begin{aligned}
J_d &= \frac{\partial D}{\partial t} \\
&= \frac{\partial}{\partial t}\left((9.30 \times 10^{-17})\sin(2.5 \times 10^3)t\,\frac{\text{C}}{\text{m}^2}\right) \\
&= (9.3 \times 10^{-17})\left((2.5 \times 10^3)\cos(2.5 \times 10^3)t\,\frac{\text{A}}{\text{m}^2}\right) \\
&= \boxed{(2.32 \times 10^{-13})\cos(2.5 \times 10^3)t \ \text{A/m}^2}
\end{aligned}$$

5. The displacement current is largest in a capacitor. The charge stored as a capacitor, of capacitance C, is

$$Q = CV$$

The capacitance is proportional to the area.

$$C \propto A$$

Capacitors, in general, have large areas and store high concentrations of charges. Therefore, the electric field strength is large and, consequently, so is $\mathbf{D}$. At high frequencies this results in a significant displacement current.

22 Magnetostatics

PRACTICE PROBLEMS

1. Flux density is a distance, r, from a current-carrying conductor of length dl. The flux density varies as

(A) $\ln r$

(B) $1/r$

(C) $1/r^2$

(D) constant

2. A wire is wrapped around a cast-iron core with a relative permeability, μ_r, of 400. The wire is rated at 15 A. The core measures 0.10 m × 0.15 m. Determine the magnitude of the core flux density for one turn of the wire.

3. In a straight, infinitely long conductor surrounded by vacuum, how much current does it take to produce a magnetic field that is 1 m away and has a strength of 5×10^{-5} T (i.e., a magnetic field that is equivalent in strength to the earth's)?

4. Consider the following illustration.

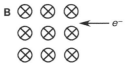

The **B**-field (magnetic field) is uniform and directed into the paper. An electron is injected into the field from the right with some initial velocity. In what direction does the electron initially deflect?

(A) up

(B) down

(C) left

(D) right

5. A galvanometer's moving coil is placed between two permanent magnets with a **B**-field strength of 4.0×10^{-3} T in the directions shown. The radius of the coil is 0.005 m. The coil contains 10 turns, or loops, and has a current rating of 0.001 A. Determine the maximum torque that the meter can generate against the restoring spring.

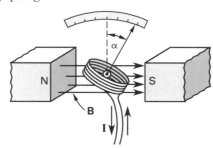

SOLUTIONS

1. The Biot-Savart law gives the magnitude of flux density in differential form as

$$dB = \left(\frac{\mu_0 I dl}{4\pi r^2}\right)\sin\theta$$

The answer is (C).

2. The flux produced by a current, I, that travels N turns over a distance, l, is given by Eq. 22.4.

$$\Phi = \mu N I l = BA$$

Rearranging to solve for the flux density, B, over the area gives

$$B = \frac{\mu N I l}{A} = \frac{\mu_0 \mu_r N I l}{A}$$

$$= \frac{\left(1.2566 \times 10^{-6} \ \dfrac{\text{N}}{\text{A}^2}\right)(400)(1)(15 \text{ A})}{\times(0.10 \text{ m} + 0.15 \text{ m} + 0.10 \text{ m} + 0.15 \text{ m})}$$
$$= \boxed{0.25 \text{ T}}$$

3. For a straight, infinitely long conductor, the magnitude of the magnetic field, B, is given by Eq. 22.16 as

$$B = \frac{\mu I}{2\pi r}$$

Rearranging and substituting gives

$$I = \frac{2\pi r B}{\mu_0}$$

$$= \frac{2\pi \ (1 \text{ m})(5 \times 10^{-5} \text{ T})}{1.2566 \times 10^{-6} \ \dfrac{\text{H}}{\text{m}}}$$

$$= \boxed{250 \text{ A}}$$

4. From Eq. 22.28, the magnetic force on a moving charged particle, Q, moving with velocity in the magnetic field is

$$\mathbf{F} = q\mathbf{v} \times \mathbf{B}$$

Using the right-hand rule to determine the direction of the cross-product, the initial deflection appears to be

down toward the bottom of the paper. But because the electron's charge is negative, the deflection is upward toward the top of the paper.

The answer is (A).

5. From Eq. 22.28, the torque on the magnetic moment, $\mathbf{m}$, in the magnetic field, $\mathbf{B}$, is

$$\mathbf{T} = \mathbf{m} \times \mathbf{B}$$

The strength of the $\mathbf{B}$-field is given. The magnetic moment, m, can be determined from Eq. 22.29.

$$m = IA$$

Because there are N turns of wire, each carrying a current, I, Eq. 22.29 becomes

$$m = NIA = NI\pi r^2$$
$$= (10)(0.001 \text{ A})\pi(0.005 \text{ m})^2$$
$$= 7.85 \times 10^{-7} \text{ A·m}^2$$

The maximum magnetic torque can be determined by substitution.

$$\mathbf{T} = \mathbf{m} \times \mathbf{B}$$
$$T_{\max} = mB\sin\frac{\pi}{2} = mB$$
$$= (7.85 \times 10^{-7} \text{ A·m}^2)(4.0 \times 10^{-3} \text{ T})$$
$$= \boxed{3.1 \times 10^{-9} \text{ N·m}}$$

Field Theory

23 Magnetostatic Fields

PRACTICE PROBLEMS

1. Define magnetization, and state its mathematical formula with appropriate units.

2. Write the formula for the magnetic flux density that is always valid regardless of the linearity of the material. Use the SI form of the equation.

3. High-purity iron is substituted for commercial iron as the core of a transformer. Most nearly, by what factor does the flux density increase?

   (A)   4

   (B)   30

   (C)   6000

   (D) 2×10^5

4. The magnetic field strength of Earth is approximately 40 A/m. Determine the current needed in a single turn of wire located at the equator to generate the same magnetic field at distances far from Earth's surface.

5. What type of magnetic material generates an internal magnetic field that opposes the applied field?

   (A)   diamagnetic

   (B)   paramagnetic

   (C)   ferromagnetic

   (D)   antiferromagnetic

SOLUTIONS

1. Magnetization, $\mathbf{M}$, is the magnetic moment per unit volume and is given by

$$\mathbf{M} = \frac{\mathbf{m}}{v}$$

The units are A/m.

2. The formula gives the magnetic flux density, $\mathbf{B}$, in terms of the magnetic field, $\mathbf{H}$, and the magnetization, $\mathbf{M}$, in the permittivity of free space, μ_0.

$$\boxed{\mathbf{B} = \mu_0 \mathbf{H} + \mu_0 \mathbf{M}}$$

3. The magnitude of the magnetic flux density is

$$B = \mu_0 \mu_r H$$

The only factor that changes when the core is changed is the relative permeability, μ_r. For commercial-grade iron, μ_r is approximately 6000, while for high-purity iron, μ_r is 2×10^5. When the core is changed, the magnetic flux density will increase by a factor of

$$\begin{aligned} f &= \frac{\mu_r \text{ (high purity)}}{\mu_r \text{ (commercial)}} \\ &= \frac{2 \times 10^5}{6000} \\ &= \boxed{33 \quad (30)} \end{aligned}$$

The answer is (B).

4. At distances far from Earth's surface, the magnitude of the magnetic field strength is

$$H = \frac{NI}{l}$$

Earth's radius is approximately 6.4×10^6 m. The length of a single turn of wire around the equator would be

$$\begin{aligned} l = 2\pi r &= 2\pi (6.4 \times 10^6 \text{ m}) \\ &= 4.0 \times 10^7 \text{ m} \end{aligned}$$

The current needed is found by rearranging the equation used to find the magnetic field strength.

$$I = \frac{Hl}{N} = \frac{\left(40 \ \frac{A}{m}\right)(4.0 \times 10^7 \ m)}{1}$$
$$= \boxed{1.6 \times 10^9 \ A}$$

5. Diamagnetic materials tend to reduce the flux by generating an opposing internal magnetic field.

The answer is (A).

Field Theory

24 Electrodynamics

PRACTICE PROBLEMS

1. The following information is obtained from a portable generator's specifications: the velocity is 1450 m/s, the magnetic field is 1.2×10^{-2} T, the loop length is 0.4 m, and the output voltage is 120 V. Approximately how many turns are required to induce the indicated voltage?

(A) 1.0

(B) 17

(C) 18

(D) 120

2. Which law describes induced electromotance?

(A) Faraday's law

(B) Ampère's law

(C) Coulomb's law

(D) Gauss' law

3. Which of the following terms represents the magnetomotive force, or mmf?

(A) $\oint \mathbf{E} \cdot d\mathbf{l}$

(B) $\oint \epsilon \mathbf{D} \cdot d\mathbf{l}$

(C) $\mu_0 \mathbf{H}$

(D) $\oint \mathbf{H} \cdot d\mathbf{l}$

4. In the following illustration, determine the direction of energy flow.

5. What fundamental action must occur to a charge in order to produce an electromagnetic wave capable of transporting energy in free space?

SOLUTIONS

1. The induced voltage resulting from a conductor moving in a uniform magnetic field is

$$v = -NBl\mathbf{v}$$

Rearranging and solving gives

$$N = \left| \frac{v}{Bl\mathbf{v}} \right|$$
$$= \frac{120 \text{ V}}{(1.2 \times 10^{-2} \text{ T})(0.4 \text{ m})\left(1450 \, \dfrac{\text{m}}{\text{s}}\right)}$$
$$= 17.2$$

Rounding down to 17 turns would result in an output voltage lower than 120 V. Therefore, 18 turns are required.

The answer is (C).

2. Faraday's law of magnetic induction is used to determine the amount of induced voltage, also called induced electromotance.

The answer is (A).

3. The mmf is the integral of the magnetic field strength, $\mathbf{H}$, along the path, $d\mathbf{l}$, where the magnetic flux is present. It is often given as NI, the number of turns times the current. The force produced per unit flux set up in the conductor is $\mathbf{H}$, hence the name.

The answer is (D).

4. The direction of energy flow, or power density, is given by Poynting's vector, $\mathbf{S}$.

$$\mathbf{S} = \mathbf{E} \times \mathbf{H}$$

In terms of the $\mathbf{B}$-field, assuming a linear, isotropic medium,

$$\mathbf{S} = \frac{1}{\mu} \mathbf{E} \times \mathbf{B}$$

If the material were not isotropic, the permeability would remain with the **B**-field as part of the cross product. The direction of **S** would then depend on the product of the permeability and magnetic flux density fields, $\mu\mathbf{B}$, as well.

Using the right-hand rule, the direction of the cross product **S** is into the plane of the paper, $\otimes$.

5. The charge must be accelerated or decelerated by some means.

A stationary charge is related to an electric field, a charge in motion with uniform velocity is related to a magnetic field, and an accelerating charge is related to a radiated field.

25 Maxwell's Equations

PRACTICE PROBLEMS

1. Which of the following point forms of Maxwell's equations implies the nonexistence of magnetic monopoles?

(A) $\nabla \cdot \mathbf{D} = \rho$

(B) $\nabla \cdot \mathbf{B} = 0$

(C) $\nabla \times \mathbf{E} = -\dfrac{\partial \mathbf{B}}{\partial t}$

(D) $\nabla \cdot \mathbf{D} = 0$

2. What is missing from the following equation?

$$\nabla \times \mathbf{H} = \mathbf{J}_c + \frac{\partial}{\partial t}$$

(A) $\mathbf{B}$

(B) $\mathbf{D}$

(C) $\mathbf{E}$

(D) $\mathbf{H}$

3. Which of the following equations for magnetic circuits is analogous to the equation shown for an electric circuit?

$$\text{emf} = V = IR$$

(A) $G = \dfrac{1}{R}$

(B) $P_m = \dfrac{\mu A}{l}$

(C) $\mathcal{R} = \dfrac{l}{\mu A}$

(D) $V_m = \phi \mathcal{R}$

4. Which of the following is the continuity equation?

(A) $\nabla \cdot \mathbf{D} = \rho$

(B) $\mathbf{J} = \sigma \mathbf{E}$

(C) $\nabla \cdot \mathbf{J} = -\dfrac{\partial \rho}{\partial t}$

(D) $\mathbf{F} = Q(\mathbf{E} + \mathbf{v} + \mathbf{B})$

5. What is missing in the following equation?

$$\nabla \times \mathbf{E} = -\frac{\partial}{\partial t}$$

(A) $\mathbf{B}$

(B) $\mathbf{D}$

(C) $\mathbf{H}$

(D) $\mathbf{J}$

SOLUTIONS

1. The nonexistence of magnetic monopoles is implied by the divergence of the magnetic flux density, **B**, being equal to zero.

$$\boxed{\boldsymbol{\nabla}\cdot\mathbf{B} = 0}$$

The answer is (B).

2.

$$\boldsymbol{\nabla}\times\mathbf{H} = \mathbf{J}_c + \mathbf{J}_d$$
$$\mathbf{J}_d = \frac{\partial\mathbf{D}}{\partial t}$$

> The electric flux density, **D**, is missing from the displacement current term.

The answer is (B).

3. The magnetomotive force (mmf) is analogous to the electromotive force (emf). The analogous magnetic equation is

$$\boxed{\mathrm{mmf} = V_m = \phi\mathcal{R}}$$

The answer is (D).

4. The continuity equation states that the divergence of the current density equals the negative rate of change of the charge density.

$$\boxed{\boldsymbol{\nabla}\cdot\mathbf{J} = -\frac{\partial\rho}{\partial t}}$$

The answer is (C).

5. The curl of the electric field is equal to the negative rate of change of the magnetic flux density.

$$\boldsymbol{\nabla}\times\mathbf{E} = -\frac{\partial\mathbf{B}}{\partial t}$$

> The magnetic flux density term, **B**, is missing.

The answer is (A).

Topic IV: Circuit Theory

Circuit Theory

26 DC Circuit Fundamentals

PRACTICE PROBLEMS

1. What is the power dissipated in R_3 of the circuit shown?

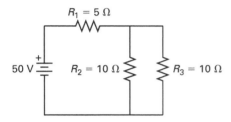

2. For the circuit shown, what resistance will result in the maximum power transfer?

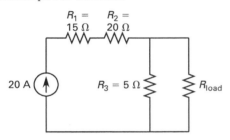

3. For the circuit shown, which method will result in fewer equations: the loop-current method or the node-voltage method?

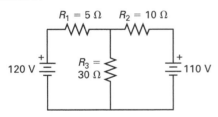

4. Using the loop-current method, solve for the current in R_3 in Prob. 3.

5. Using the node-voltage method, solve for the current in R_3 in Prob. 3.

6. 150 m of wire with a resistivity of 1.72×10^{-8} $\Omega \cdot$m and a diameter of 0.5 mm is used in a project. Most nearly, what is the DC voltage drop across this wire in a 15 A circuit?

(A)   13 V

(B)   15 V

(C)   20 V

(D)   200 V

7. Most nearly, what is the power dissipated in the 2 Ω resistor of the circuit shown?

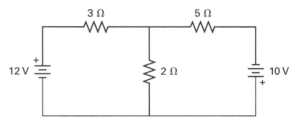

(A)   1 W

(B)   2 W

(C)   4 W

(D)   5 W

8. A power plant tracks power usage based on the decibel level referenced to 1 kW. From Friday to Sunday, the total power usage at peak demand is 12 MW, 13 MW, and 11 MW for each respective day. Most nearly, what is the average decibel level for the weekend?

(A)   1.1 dB

(B)   11 dB

(C)   41 dB

(D)   82 dB

9. A DC generator is monitored at the output terminals. The no-load voltage is 250 V. The voltage regulation is 3%. Most nearly, what is the full-load voltage of the generator?

(A) 8.0 V

(B) 240 V

(C) 250 V

(D) 260 V

10. Voltage sources in the circuit shown are ideal.

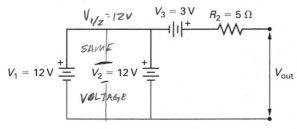

Most nearly, what is the output voltage?

(A) 3.0 V

(B) 9.0 V

(C) 15 V

(D) 27 V

11. The circuit shown is the equivalent of a voltage source for a power supply. The source voltage is 4 V and the battery's internal resistance is 0.75 Ω.

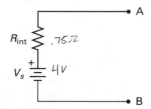

The battery will be combined with a transistor to create a current source.

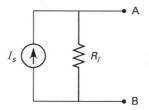

Most nearly, what are the values of the current source, I_s, and the resistor, R_I, respectively?

(A) 5.3 A, 0.75 Ω

(B) 5.3 A, 1.3 Ω

(C) 10 A, 1.5 Ω

(D) 10 A, 2.7 Ω

12. The load requires 2.0 V to operate properly.

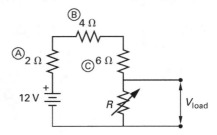

Most nearly, what is the required resistance, R?

(A) 0.40 Ω

(B) 1.2 Ω

(C) 2.4 Ω

(D) 14 Ω

13. Most nearly, what is the resistance, R, for the circuit shown?

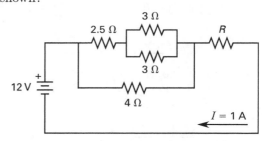

(A) 2 Ω

(B) 4 Ω

(C) 8 Ω

(D) 10 Ω

14. Most nearly, what is the resistance, R_A, in the wye circuit shown?

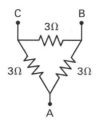

(A) 1.0 Ω

(B) 1.5 Ω

(C) 3.0 Ω

(D) 9.0 Ω

15. Most nearly, what is the current through the center leg of the circuit shown?

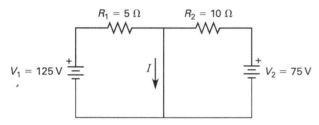

(A) 8.0 A

(B) 16 A

(C) 25 A

(D) 32 A

SOLUTIONS

1. Power is given by

$$P = VI = I^2 R = \frac{V^2}{R}$$

Using the last form of the equation, determine the voltage across R_3. Combine the parallel resistors R_2 and R_3.

$$R_{2/3} = \frac{R_2 R_3}{R_2 + R_3} = \frac{(10\ \Omega)(10\ \Omega)}{10\ \Omega + 10\ \Omega} = 5\ \Omega$$

The circuit is transformed as follows.

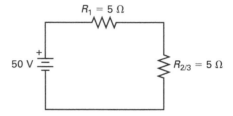

Using the voltage-divider concept, the voltage across $R_{2/3}$ (and R_3) is

$$V_{2/3} = V_3 = (50\ \text{V})\left(\frac{R_{2/3}}{R_1 + R_{2/3}}\right)$$
$$= (50\ \text{V})\left(\frac{5\ \Omega}{5\ \Omega + 5\ \Omega}\right)$$
$$= 25\ \text{V}$$

The power dissipated in R_3 is

$$P_3 = \frac{V_3^2}{R_3} = \frac{(25\ \text{V})^2}{10\ \Omega}$$
$$= \boxed{62\ \text{W}}$$

Circuit Theory

2. Maximum power transfer occurs when the load resistance equals the Thevenin or Norton equivalent resistance.

Change the current source to an open circuit and determine R_{Th}.

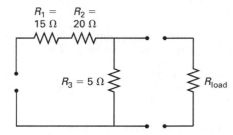

The 15 Ω and 20 Ω resistors are in an open-circuit path.

$$R_{\text{Th}} = R_3 = \boxed{5 \ \Omega}$$

3. The loop-current method using Kirchhoff's voltage law (KVL) uses $n - 1$ equations, where n is the number of loops. There are three loops present. The number of equations required is two.

The node-voltage method using Kirchhoff's current law (KCL) requires $n - 1$ equations, where n is the number of principal nodes. There are two principal nodes. Only one equation is required.

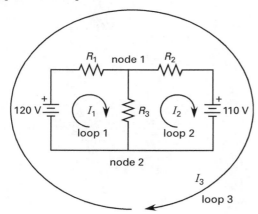

The node-voltage method requires fewer equations.

4. Consider the following circuit.

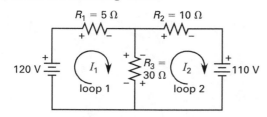

Using the selected loops with polarities for the resistive elements based on the loop-current direction, KVL for loop 1 is

$$120 \ \text{V} - I_1 R_1 - (I_1 - I_2)R_3 = 0$$

Rearranging the equation for KVL for loop 1 gives

$$I_1(R_1 + R_3) - I_2 R_3 = 120 \ \text{V} \qquad \text{[I]}$$

KVL for loop 2 is

$$-(I_2 - I_1)R_3 - I_2 R_2 - 110 \ \text{V} = 0$$

Rearranging the equation for KVL for loop 2 gives

$$I_1 R_3 - I_2(R_3 + R_2) = 110 \ \text{V} \qquad \text{[II]}$$

There are two equations, [I] and [II], with two unknowns, I_1 and I_2. Cramer's rule and matrix algebra are used to solve for I_1 and I_2.

$$\begin{vmatrix} R_1 + R_3 & -R_3 \\ R_3 & -(R_3 + R_2) \end{vmatrix} \begin{vmatrix} I_1 \\ I_2 \end{vmatrix} = \begin{vmatrix} 120 \ \text{V} \\ 110 \ \text{V} \end{vmatrix}$$

$$\begin{vmatrix} 35 \ \Omega & -30 \ \Omega \\ 30 \ \Omega & -40 \ \Omega \end{vmatrix} \begin{vmatrix} I_1 \\ I_2 \end{vmatrix} = \begin{vmatrix} 120 \ \text{V} \\ 110 \ \text{V} \end{vmatrix}$$

$$I_1 = \frac{\begin{vmatrix} 120 \ \text{V} & -30 \ \text{V} \\ 110 \ \text{V} & -40 \ \text{V} \end{vmatrix}}{\begin{vmatrix} 35 \ \Omega & -30 \ \Omega \\ 30 \ \Omega & -40 \ \Omega \end{vmatrix}} = \frac{-1500 \ \text{V}}{-500 \ \Omega} = 3 \ \text{A}$$

$$I_2 = \frac{\begin{vmatrix} 35 \ \text{V} & 120 \ \text{V} \\ 30 \ \text{V} & 110 \ \text{V} \end{vmatrix}}{\begin{vmatrix} 35 \ \Omega & -30 \ \Omega \\ 30 \ \Omega & -40 \ \Omega \end{vmatrix}} = \frac{250 \ \text{V}}{-500 \ \Omega} = -0.5 \ \text{A}$$

Summing to obtain the current through R_3 and noting that the direction of I_2 was incorrect gives

$$I_3 = |I_1| + |I_2| = 3 \ \text{A} + 0.5 \ \text{A}$$
$$= \boxed{3.5 \ \text{A}}$$

The direction of the current I_3 is positive and identical to the direction of I_1.

5. Consider the following circuit.

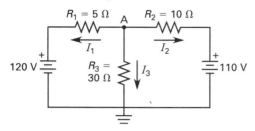

The principal node A is labeled. The node at the lower portion of the circuit is chosen as the reference or ground node. All currents are assumed to be positive.

Writing KCL for node A gives

$$\frac{V_A - 120 \text{ V}}{R_1} + \frac{V_A - 0 \text{ V}}{R_3} + \frac{V_A - 110 \text{ V}}{R_2} = 0$$

Solve for V_A.

$$\frac{V_A}{R_1} - \frac{120 \text{ V}}{R_1} + \frac{V_A}{R_3} + \frac{V_A}{R_2} - \frac{110 \text{ V}}{R_2} = 0$$

$$\frac{V_A}{R_1} + \frac{V_A}{R_3} + \frac{V_A}{R_2} = \frac{120 \text{ V}}{R_1} + \frac{110 \text{ V}}{R_2}$$

$$V_A = \frac{\dfrac{120 \text{ V}}{R_1} + \dfrac{110 \text{ V}}{R_2}}{\dfrac{1}{R_1} + \dfrac{1}{R_3} + \dfrac{1}{R_2}}$$

$$= \frac{\dfrac{120 \text{ V}}{5 \text{ }\Omega} + \dfrac{110 \text{ V}}{10 \text{ }\Omega}}{\dfrac{1}{5 \text{ }\Omega} + \dfrac{1}{30 \text{ }\Omega} + \dfrac{1}{10 \text{ }\Omega}}$$

$$= 105 \text{ V}$$

Use V_A to determine the desired current.

$$I_3 = \frac{V_A}{R_3} = \frac{105 \text{ V}}{30 \text{ }\Omega} = \boxed{3.5 \text{ A}}$$

6. The resistance of the wire is calculated from Eq. 26.1.

$$R = \frac{\rho l}{A} = \frac{\rho l}{\pi \left(\dfrac{d}{2}\right)^2}$$

$$= \frac{(1.72 \times 10^{-8} \text{ }\Omega \cdot \text{m})(150 \text{ m})}{\pi \left(\dfrac{0.5 \times 10^{-3} \text{ m}}{2}\right)^2}$$

$$= 13.14 \text{ }\Omega$$

From Eq. 26.11, Ohm's law, the voltage drop is

$$V = IR$$
$$= (15 \text{ A})(13.14 \text{ }\Omega)$$
$$= \boxed{197.10 \text{ V} \quad (200 \text{ V})}$$

The answer is (D).

7. Apply Kirchhoff's voltage law (KVL) around each loop as shown.

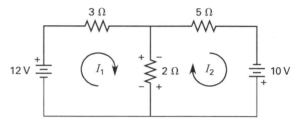

KVL for loop 1 is

$$12 \text{ V} - I_1 R_{3 \text{ }\Omega} - I_1 R_{2 \text{ }\Omega} + I_2 R_{2 \text{ }\Omega} = 0$$

Rearranging the equation for KVL for loop 1 gives

$$12 \text{ V} = I_1(R_{2 \text{ }\Omega} + R_{3 \text{ }\Omega}) + I_2 R_{2 \text{ }\Omega}$$
$$12 \text{ V} = I_1(5 \text{ }\Omega) + I_2(2 \text{ }\Omega)$$

KVL for loop 2 is

$$10 \text{ V} - I_2 R_{2 \text{ }\Omega} + I_1 R_{2 \text{ }\Omega} - I_2 R_{5 \text{ }\Omega} = 0$$

Rearranging the equation for KVL for loop 2 gives

$$10 \text{ V} = I_1 R_{2 \text{ }\Omega} + I_2(R_{2 \text{ }\Omega} + R_{5 \text{ }\Omega})$$
$$10 \text{ V} = I_1(2 \text{ }\Omega) + I_2(7 \text{ }\Omega)$$

For loop 1, from Cramer's rule and matrix algebra, the current is

$$I_1 = \frac{|\mathbf{A_1}|}{|\mathbf{A}|} = \frac{\begin{vmatrix} 12 \text{ V} & 2 \text{ }\Omega \\ 10 \text{ V} & 7 \text{ }\Omega \end{vmatrix}}{\begin{vmatrix} 5 \text{ }\Omega & -2 \text{ }\Omega \\ -2 \text{ }\Omega & 7 \text{ }\Omega \end{vmatrix}}$$

$$= \frac{(12 \text{ V})(7 \text{ }\Omega) - (10 \text{ V})(-2 \text{ }\Omega)}{(5 \text{ }\Omega)(7 \text{ }\Omega) - (-2 \text{ }\Omega)(-2 \text{ }\Omega)}$$

$$= 3.354 \text{ A}$$

Circuit Theory

For loop 2, from Cramer's rule and matrix algebra, the current is

$$I_2 = \frac{|\mathbf{A}_2|}{|\mathbf{A}|} = \frac{\begin{vmatrix} 5\ \Omega & 12\ \text{V} \\ 2\ \Omega & 10\ \text{V} \end{vmatrix}}{\begin{vmatrix} 5\ \Omega & 2\ \Omega \\ 2\ \Omega & 7\ \Omega \end{vmatrix}}$$

$$= \frac{(5\ \Omega)(10\ \text{V}) - (-2\ \Omega)(12\ \text{V})}{(5\ \Omega)(7\ \Omega) - (-2\ \Omega)(-2\ \Omega)}$$

$$= 2.387\ \text{A}$$

The actual current through the 2 Ω resistor is the combination of the two currents, using the directions assumed. The sign of the result determines the correct direction of the current.

$$I_{2\Omega} = I_1 - I_2$$
$$= 2.06\ \text{A} - 0.84\ \text{A}$$
$$= 1.22\ \text{A}$$

From Eq. 26.13, the power dissipated in the 2 Ω resistor is

$$P = I^2 R = I_{2\Omega}^2 R_{2\Omega}$$
$$= (1.22\ \text{A})^2(2\ \Omega)$$
$$= \boxed{2.97\ \text{W} \quad (3\ \text{W})}$$

The answer is (B).

8. The average power usage for the weekend is

$$P = \frac{P_t}{N_{\text{days}}}$$
$$= \frac{12\ \text{MW} + 13\ \text{MW} + 11\ \text{MW}}{3}$$
$$= 12\ \text{MW}$$

From Eq. 26.14, the decibel level is

$$\text{ratio (in dB)} = 10 \log_{10} \frac{P}{P_0}$$
$$= 10 \log_{10} \frac{12 \times 10^6\ \text{W}}{1 \times 10^3\ \text{W}}$$
$$= \boxed{40.79\ \text{W} \quad (41\ \text{dB})}$$

The answer is (C).

9. The voltage regulation is given by Eq. 26.15.

$$\text{VR} = \frac{V_{\text{nl}} - V_{\text{fl}}}{V_{\text{fl}}} \times 100\%$$

Equation 26.15 is rearranged in order to find the generator's full load voltage.

$$V_{\text{fl}} = \frac{V_{\text{nl}}}{\dfrac{\text{VR}}{100\%} + 1} = \frac{250\ \text{V}}{\dfrac{3\%}{100\%} + 1}$$
$$= \boxed{242.7\ \text{V} \quad (240\ \text{V})}$$

The answer is (B).

10. Use Millman's theorem. The two voltage sources in parallel, V_1 and V_2, can be combined into a single voltage source of 12 V. This 12 V source is in series with V_3. The resistor does not have an impact until the current flows (i.e., until a load is added). Sum the voltages to find the total voltage in the circuit.

$$V_t = V_{1,2} + V_3$$
$$= 12\ \text{V} + 3\ \text{V}$$
$$= \boxed{15\ \text{V}}$$

The answer is (C).

11. The two circuits are equivalent if the voltages at the terminals and the resistance as seen from the terminals are identical in each circuit.

Determine the source current by shorting terminals A and B and applying Ohm's law, Eq. 26.11.

$$V_s = I_s R_{\text{int}}$$
$$I_s = \frac{V_s}{R_{\text{int}}}$$
$$= \frac{4\ \text{V}}{0.75\ \Omega}$$
$$= \boxed{5.33\ \text{A} \quad (5.3\ \text{A})}$$

The resistance is found from Ohm's law. The voltage across R_I must be 4 V.

$$V_{\text{AB}} = V_{\text{terminal}} = V_s$$
$$= I_s R_I$$
$$R_I = \frac{V_s}{I_s} = \frac{4\ \text{V}}{5.33\ \text{A}}$$
$$= \boxed{0.75\ \Omega}$$

The answer is (A).

Circuit Theory

12. Resistors A, B, and C are in series. The total resistance for the three resistors is

$$R_{\text{series}} = R_A + R_B + R_C$$
$$= 2\ \Omega + 4\ \Omega + 6\ \Omega$$
$$= 12\ \Omega$$

The equivalent circuit is shown.

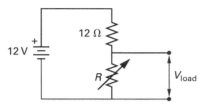

Use the voltage-divider concept.

$$V_{\text{load}} = V_s\left(\frac{R}{R + R_{\text{series}}}\right)$$
$$= V_s\left(\frac{R}{R + 12\ \Omega}\right)$$

Solve for R.

$$R + 12\ \Omega = R\left(\frac{V_s}{V_{\text{load}}}\right)$$
$$1 + \frac{12\ \Omega}{R} = \frac{V_s}{V_{\text{load}}}$$
$$\frac{12\ \Omega}{R} = \frac{V_s}{V_{\text{load}}} - 1$$
$$R = \frac{12\ \Omega}{\dfrac{V_s}{V_{\text{load}}} - 1} = \frac{12\ \Omega}{\dfrac{12\ \text{V}}{2\ \text{V}} - 1} = \frac{12}{5}\ \Omega$$
$$= 2.4\ \Omega$$

The answer is (C).

13. Combine the parallel 3 Ω resistors.

$$R_{\|,1} = \frac{R_{3\Omega}R_{3\Omega}}{R_{3\Omega} + R_{3\Omega}}$$
$$= \frac{(3\ \Omega)(3\ \Omega)}{3\ \Omega + 3\ \Omega}$$
$$= 1.5\ \Omega$$

The circuit can be simplified as shown.

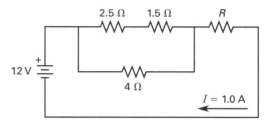

Combine the series resistors.

$$R_{s,1} = R_{2.5\Omega} + R_{1.5\Omega}$$
$$= 2.5\ \Omega + 1.5\ \Omega$$
$$= 4\ \Omega$$

The resulting circuit is simplified as shown.

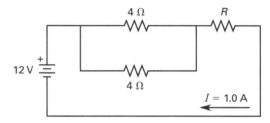

Combine the parallel 4 Ω resistors.

$$R_{\|,2} = \frac{R_{4\Omega}R_{4\Omega}}{R_{4\Omega} + R_{4\Omega}}$$
$$= \frac{(4\ \Omega)(4\ \Omega)}{4\ \Omega + 4\ \Omega}$$
$$= 2\ \Omega$$

The resulting circuit is simplified as shown.

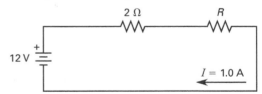

Combine the two remaining series resistors and use Ohm's law, Eq. 26.11, to calculate the resistance of R.

$$V = IR$$
$$V = I(R_{2\Omega} + R)$$
$$R = \frac{V}{I} - R_{2\Omega} = \frac{12\ \text{V}}{1.0\ \text{A}} - 2\ \Omega$$
$$= \boxed{10\ \Omega}$$

The answer is (D).

14. The delta circuit is a balanced three-phase system. The resistance through R_1 in Eq. 26.36 is equivalent to R_A in the wye circuit.

$$R_A = \frac{R_{AB}R_{AC}}{R_{AB} + R_{BC} + R_{CA}}$$
$$= \frac{(3\ \Omega)(3\ \Omega)}{3\ \Omega + 3\ \Omega + 3\ \Omega}$$
$$= \boxed{1.0\ \Omega}$$

The answer is (A).

15. Using the principle of superposition, turn "off" (i.e., short) the 75 V voltage source, and calculate the current through R_1. No current will flow through R_2.

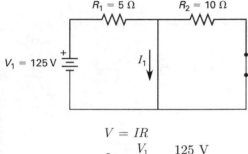

$$V = IR$$
$$I_1 = \frac{V_1}{R_1} = \frac{125\ \text{V}}{5\ \Omega}$$
$$= 25\ \text{A}$$

Short the 125 V voltage source, and calculate the current through R_2. No current will flow through R_1.

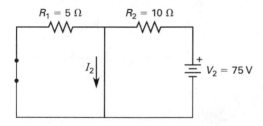

$$V = IR$$
$$I_2 = \frac{V_2}{R_2} = \frac{75\ \text{V}}{10\ \Omega}$$
$$= 7.5\ \text{A}$$

Find the total current through the center leg of the circuit by combining the two currents.

$$I_{\text{total}} = I_1 + I_2$$
$$= 25\ \text{A} + 7.5\ \text{A}$$
$$= \boxed{32.5\ \text{A} \quad (32\ \text{A})}$$

The answer is (D).

27 AC Circuit Fundamentals

PRACTICE PROBLEMS

1. Express the phasor voltage 120 V∠30° as a function of time if the frequency is 60 Hz. The voltage is given in "effective value phasor" notation.

2. Express each of the following in rectangular form.

(a) $15 \angle -180°$

(b) $10 \angle 37°$

(c) $50 \angle 120°$

(d) $21 \angle -90°$

3. Express each of the following in phasor form.

(a) $6 + j7$

(b) $50 - j60$

(c) $-75 + j45$

(d) $90 - j180$

4. Calculate the following, and express each in phasor form.

(a) $(7 + j5)/6$

(b) $(13 + j17)/(15 - j10)$

(c) $(0.020 \angle 90°)/0.034 \angle 56°$

5. A sinusoidal waveform has a peak value at $t = 1$ ms and a period of 10 ms. Express this signal as a cosine and sine function.

6. A voltage of $(10 \text{ V})\cos(100t + 25°)$ is applied to a resistance of 25 Ω and an inductance of 0.50 H in parallel. What are the (a) phase angle, and (b) rms value of the sum of the currents?

7. Draw the power triangle for the circuit in Prob. 6.

8. What is the capacitance, in units of farads, to completely correct the power factor of the circuit in Prob. 6, that is, to make $Q = 0$?

9. An 80 μF capacitor is in series with a 9 mH inductor. (a) What is the total impedance (at 60 Hz)? (b) Is the circuit considered a leading or lagging circuit?

10. If the capacitor and inductor in Prob. 9 are connected in parallel, what is the total impedance, and is the circuit considered leading or lagging?

11. Two 10 μF capacitors are in series with a 20 μF capacitor. What is most nearly the total capacitance?

(A) 0.4 μF

(B) 4 μF

(C) 20 μF

(D) 40 μF

12. What is most nearly the power dissipated in the load impedance of the circuit shown?

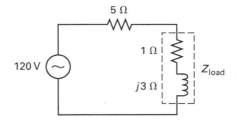

(A) 100 W

(B) 120 W

(C) 320 W

(D) 1900 W

13. What is most nearly the power dissipated by the circuit shown?

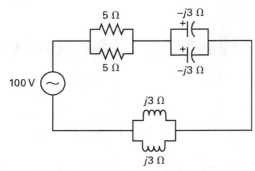

(A)   50 W

(B)   400 W

(C)   800 W

(D)   900 W

14. What is most nearly the root-mean-square (rms) value of the voltage in the signal shown?

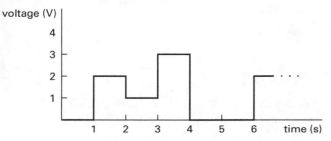

(A)   1.0 V

(B)   1.7 V

(C)   3.0 V

(D)   4.0 V

15. Current in a given circuit increases linearly from 0 A at $t = 0$ s to 15 A at $t = 8$ s. What is most nearly the amount of magnetic energy stored in a 0.5 H inductor at $t = 4$ s?

(A)   2.0 J

(B)   4.0 J

(C)   15 J

(D)   30 J

SOLUTIONS

1. As a function of time, t, with peak voltage V_p and angular frequency ω,

$$v(t) = V_p \sin(\omega t + \theta)$$

The peak voltage is proportional to the effective voltage, V_{eff},

$$\begin{aligned} V_p &= V_{\text{eff}}\sqrt{2} \\ &= 120 \text{ V}\sqrt{2} \\ &= 169.71 \text{ V} \end{aligned}$$

The angular frequency is

$$\begin{aligned} \omega &= 2\pi f = 2\pi(60 \text{ Hz}) \\ &= 376.99 \text{ rad/s} \end{aligned}$$

The given angle of 30° equals $\pi/6$ rad, so

$$v(t) = \boxed{170 \sin\left(377t + \frac{\pi}{6}\right)}$$

2. (a)

$$15\cos(-180°) + j15\sin(-180°)$$
$$= \boxed{-15 + j0}$$

(b)

$$10\cos 37° + j10\sin 37°$$
$$= \boxed{7.99 + j6.02}$$

(c)

$$50\cos 120° + j50\sin 120°$$
$$= \boxed{-25 + j43.3}$$

(d)

$$.21\cos(-90°) + j21\sin(-90°)$$
$$= \boxed{0 - j21}$$

3. (a)

$$\sqrt{(6)^2 + (7)^2} \angle \arctan \frac{7}{6}$$
$$= \boxed{9.22 \angle 49.4°}$$

Circuit Theory

(b)

$$\sqrt{(50)^2 + (-60)^2} \angle \arctan \frac{-60}{50}$$
$$= \boxed{78.10 \angle -50.19°}$$

(c)

$$\sqrt{(-75)^2 + (45)^2} \angle \arctan \frac{45}{-75}$$
$$= \boxed{87.46 \angle 149.04°}$$

(d)

$$\sqrt{(90)^2 + (-180)^2} \angle \arctan \frac{-180}{90}$$
$$= \boxed{201.25 \angle -63.43°}$$

4. (a)

$$\frac{1}{6}\sqrt{(7)^2 + (5)^2} \angle \arctan \frac{5}{7}$$
$$= \boxed{1.4 \angle 35.54°}$$

(b)

$$\frac{\sqrt{(13)^2 + (17)^2} \arctan \frac{17}{13}}{\sqrt{(15)^2 + (-10)^2} \arctan \frac{-10}{15}}$$
$$= \boxed{1.19 \angle 86.28°}$$

(c)

$$\boxed{0.588 \angle 34°}$$

5.

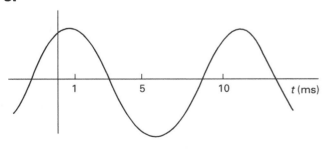

$$\omega = \frac{2\pi}{T} = \frac{2\pi}{0.01} = 200\pi \text{ rad/s}$$
$$10 \text{ ms} = 360°$$
$$1 \text{ ms} = 36°$$

For the sine, the positive zero crossing is at -1.5 ms, or $-54°$.

$$f = \boxed{F_p \sin(200\pi t + 54°) \text{ or } F_p \sin(200\pi t + 0.9)}$$
$$= \boxed{F_p \cos(200\pi t - 36°) \text{ or } F_p \cos(200\pi t - 0.6)}$$

6. Using a cosine reference,

(a)

$$I_R = \frac{10 \text{ V} \angle 25°}{25 \text{ }\Omega} = 0.4 \text{ A} \angle 25°$$
$$= 0.3625 + j0.1690 \text{ A}$$
$$I_L = \frac{10 \text{ V} \angle 25°}{(0.5)(100) \angle 90° \text{ }\Omega}$$
$$= 0.2 \text{ A} \angle -65°$$
$$= 0.0845 - j0.1813 \text{ A}$$
$$I_t = 0.4470 - j0.0123 \text{ A}$$
$$= 0.4472 \text{ A} \angle -1.6°$$
$$\text{phase angle} = \boxed{-1.6°}$$

(b) The calculations were made using peak values. Divide the total peak current by $\sqrt{2}$ to obtain the rms values.

$$I_{\text{rms}} = \frac{0.4472 \text{ A}}{\sqrt{2}} = \boxed{0.316 \text{ A}}$$

7. Use rms values. The magnitude, S, of the complex power vector is the product of the voltage, V, and the magnitude of the complex conjugate of the current, I^*.

$$S = VI^* = \frac{(10 \text{ V} \angle 25°)(0.4472 \text{ A} \angle 1.6°)}{\sqrt{2}\sqrt{2}}$$
$$= 2.236 \angle 26.6° \text{ VA}$$
$$= 2 + j1 \text{ VA}$$

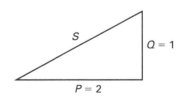

8. The capacitance current must cancel the inductance current, so its reactance must be $-j\omega L = -j50\ \Omega$.

$$X_C = \frac{1}{j\omega C} = \frac{-j}{100\,C} \rightarrow -j50\ \Omega$$

$$C = \frac{1}{(50\ \Omega)\left(100\ \dfrac{\text{rad}}{\text{s}}\right)}$$

$$= \boxed{200 \times 10^{-6}\ \text{F}\quad (200\ \mu\text{F})}$$

9. The impedance, Z_C, for the capacitor is

$$Z_C = \frac{1}{j\omega C} = \frac{1}{j\left(377\ \dfrac{\text{rad}}{\text{s}}\right)(80 \times 10^{-6}\ \text{F})}$$

$$= -j\,(33.16\ \Omega)$$

$$= 33.16\ \Omega\angle - 90°$$

The impedance, Z_L, for the inductor is

$$Z_L = j\omega L = j\left(377\ \dfrac{\text{rad}}{\text{s}}\right)(9 \times 10^{-3}\ \text{H})$$

$$= j3.39\ \Omega$$

$$= 3.39\ \Omega\angle 90°$$

For a series circuit, the impedance is

$$\begin{aligned}
Z_{\text{total}} &= Z_C + Z_L \\
&= -j33.16\ \Omega + j3.39\ \Omega \\
&= -j29.77\ \Omega \\
&= \boxed{29.77\ \Omega\angle - 90°}
\end{aligned}$$

The impedance angle is negative. The circuit is capacitative and therefore leading.

10. For a parallel circuit, the impedance is

$$\begin{aligned}
\frac{1}{Z_{\text{total}}} &= \frac{1}{Z_C} + \frac{1}{Z_L} \\
&= \frac{1}{33.16\ \Omega\angle - 90°} + \frac{1}{3.39\ \Omega\angle 90°} \\
&= 0.0302\ \Omega\angle 90° + 0.2950\ \Omega\angle - 90° \\
&= j0.0302\ \Omega - j0.2950\ \Omega \\
&= -j0.2648\ \Omega
\end{aligned}$$

So,

$$\begin{aligned}
Z_{\text{total}} &= \frac{1}{-j0.2648\ \Omega} = \frac{j}{0.2648\ \Omega} \\
&= \boxed{3.78\ \Omega\angle 90°}
\end{aligned}$$

The impedance angle is positive. The circuit is inductive and, therefore, lagging.

11. Series capacitance is found by summing the reciprocals of the individual capacitors, similar to how parallel resistance is found by summing the reciprocals of the individual resistors.

$$\frac{1}{C_{\text{eq}}} = \frac{1}{C_1} + \frac{1}{C_2} + \frac{1}{C_3}$$

$$\begin{aligned}
C_{\text{eq}} &= \frac{1}{\dfrac{1}{C_1} + \dfrac{1}{C_2} + \dfrac{1}{C_3}} \\
&= \frac{1}{\dfrac{1}{10\ \mu\text{F}} + \dfrac{1}{10\ \mu\text{F}} + \dfrac{1}{20\ \mu\text{F}}} \\
&= \boxed{4\ \mu\text{F}}
\end{aligned}$$

The answer is (B).

12. From Ohm's law for AC circuits, Eq. 27.41, the current through the series is

$$\begin{aligned}
\mathbf{V} &= \mathbf{IZ} \\
\mathbf{I} &= \frac{\mathbf{V}}{\mathbf{Z}} = \frac{120\ \text{V}\angle 0°}{5\ \Omega + (1 + j3\ \Omega)} \\
&= \frac{120\ \text{V}\angle 0°}{6 + j3\ \Omega} \\
&= \frac{120\ \text{V}\angle 0°}{6.7\ \Omega\angle 26.5°} \\
&= 17.9\ \text{A}\angle - 26.5°
\end{aligned}$$

Use the magnitude of the current to determine the power dissipated in the load impedance. (The inductor is ideal given that the resistance component was shown as a separate lumped parameter.)

$$\begin{aligned}
P &= I^2 R \\
&= (17.9\ \text{A})^2(1\ \Omega) \\
&= \boxed{320.4\ \text{W}\quad (320\ \text{W})}
\end{aligned}$$

The answer is (C).

13. Combine the parallel resistors.

$$\begin{aligned}
\mathbf{Z}_R &= \frac{\mathbf{Z}_{5\ \Omega}\mathbf{Z}_{5\ \Omega}}{\mathbf{Z}_{5\ \Omega} + \mathbf{Z}_{5\ \Omega}} \\
&= \frac{(5\ \Omega\angle 0°)(5\ \Omega\angle 0°)}{5\ \Omega\angle 0° + 5\ \Omega\angle 0°} \\
&= 2.5\ \Omega\angle 0°
\end{aligned}$$

Combine the parallel capacitors.

$$\mathbf{Z}_C = \mathbf{Z}_{3\,\Omega} + \mathbf{Z}_{3\,\Omega}$$
$$= -j3\ \Omega - j3\ \Omega$$
$$= -j6\ \Omega$$

Combine the parallel inductors.

$$\mathbf{Z}_L = \frac{\mathbf{Z}_{3\,\Omega}\mathbf{Z}_{3\,\Omega}}{\mathbf{Z}_{3\,\Omega} + \mathbf{Z}_{3\,\Omega}}$$
$$= \frac{(j3\ \Omega)(j3\ \Omega)}{j3\ \Omega + j3\ \Omega}$$
$$= j1.5\ \Omega$$

The circuit can then be reduced to the equivalent circuit shown.

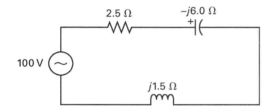

Calculate the total impedance of the circuit.

$$\mathbf{Z}_t = \mathbf{Z}_R + \mathbf{Z}_C + \mathbf{Z}_L$$
$$= 2.5\ \Omega \angle 0\,° - j6\ \Omega + j1.5\ \Omega$$
$$= 2.5\ \Omega \angle 0\,° - j4.5\ \Omega$$
$$= 5.15\ \Omega \angle -61°$$

From Ohm's law, Eq. 27.41, the total current in the circuit is

$$\mathbf{V} = \mathbf{IZ}$$
$$\mathbf{I} = \frac{\mathbf{V}}{\mathbf{Z}}$$
$$= \frac{100\ \text{V} \angle 0°}{5.15\ \Omega \angle -61°}$$
$$= 19.42\ \text{A} \angle 61°$$

Calculate the power dissipated in the resistor only. From Eq. 29.6, the dissipated power is

$$P = I^2 R$$
$$= (19.42\ \text{A})^2 (2.5\ \Omega)$$
$$= \boxed{942.84\ \text{W} \quad (900\ \text{W})}$$

The answer is (D).

14. From Eq. 27.19, the rms value of the voltage is

$$V_{\text{rms}} = \sqrt{\frac{1}{T} \int_0^T v^2(t)\, dt}$$
$$= \sqrt{\left(\frac{1}{5\ \text{s}}\right)\left(\begin{array}{c}(0\ \text{V})^2(1\ \text{s}) + (2\ \text{V})^2(1\ \text{s}) \\ + (1\ \text{V})^2(1\ \text{s}) + (3\ \text{V})^2(1\ \text{s}) \\ + (0\ \text{V})^2(1\ \text{s})\end{array}\right)}$$
$$= \boxed{1.67\ \text{V} \quad (1.7\ \text{V})}$$

The answer is (B).

15. Since the current increases linearly, the current at $t = 4$ s is one-half the current at 8 s, or 7.5 A.

From Eq. 29.20, the stored energy is

$$U = \frac{1}{2} L I^2$$
$$= \left(\frac{1}{2}\right)(0.5\ \text{H})(7.5\ \text{A})^2$$
$$= \boxed{14.06\ \text{J} \quad (15\ \text{J})}$$

The answer is (C).

Circuit Theory

28 Transformers

PRACTICE PROBLEMS

1. Which of the following describes leakage flux?

(A) flux outside the core

(B) flux linking the primary

(C) flux linking the secondary

(D) flux other than the mutual flux

2. For the transformer shown, which of the following statements is true?

(A) $M > 0$

(B) $M < 0$

(C) $X_m > 0$ and $X_p > 0$

(D) Both (A) and (C) are true.

3. An ideal transformer has 90 turns on the primary and 2200 turns on the secondary. The load draws 8 A at 0.8 pf. Most nearly, what is the current in the primary?

(A) $90 \text{ A} \angle 37°$

(B) $196 \text{ A} \angle 37°$

(C) $320 \text{ A} \angle 1°$

(D) $333 \text{ A} \angle 0.2\pi$

4. An ideal transformer with a 120 V, 60 Hz source has a turns ratio of 10. Most nearly, what is the peak flux in the secondary?

(A) 19 mWb

(B) 23 mWb

(C) 37 mWb

(D) 45 mWb

5. A step-down transformer consists of 200 primary turns and 40 secondary turns. The primary voltage is 550 V. The impedance in the secondary load is 4.2 Ω.

What are the (a) secondary voltage, (b) primary current, and (c) secondary current?

6. For the circuit shown, if the amplifier and load impedances are matched, what is the turns ratio?

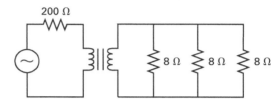

7. A 100 V (rms) source with an internal resistance of 8 Ω is connected to a 200/100-turn step-down transformer. An impedance of $6 \text{ Ω} + j8$ is connected to the secondary. Find the (a) primary current and (b) secondary power.

8. A 480 V transformer has a full-load current rating of 250 A. The breaker supplying power to the transformer has the overcurrent characteristics shown. The pickup current is 200 A.

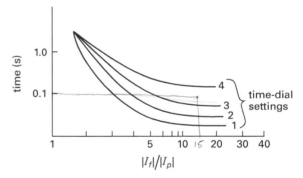

The inrush current is 11–12 times the transformer full-load current. Additionally, the inrush current is typically expected to be over in 0.1 s. Regarding the magnitude of the inrush current, what is the lowest time-dial setting that will avoid tripping the breaker during transformer startup?

(A) 1

(B) 2

(C) 3

(D) 4

9. A transformer is rated for 480 V by arc flash analysis using a software program. The program output indicates a bolted fault current of 1000 A. Most nearly, what is the lowest current at which an arc will be self-sustaining?

(A)   100 A

(B)   380 A

(C)   760 A

(D)   1000 A

10. A 480/208/120 Y three-phase wye transformer powers a secondary side load of $40 + j20$ Ω. What is the turns ratio of the transformer?

(A)   1:1

(B)   2:1

(C)   3:1

(D)   4:1

SOLUTIONS

1. In a two-coil transformer, leakage flux is flux that links only one coil or the other, but not both, and so does not contribute to the transfer of electrical energy. Both the primary and secondary coils can produce leakage flux. The flux linking both coils is the mutual flux.

The answer is (D).

2. Both assumed currents enter at the dotted terminals, so from the clot rules, the mutual inductor is positive. Further, the mutual reactance, X_m, and the self-inductance of the coils, X_p (or X_s), have the same sign as the mutual inductor.

The answer is (D).

3. The turns ratio, a, for a transformer with N_p and N_s turns in the primary and secondary windings, respectively, is

$$a = \frac{N_p}{N_s} = \frac{I_s}{I_p}$$

I_p and I_s are the primary and secondary currents, respectively. The magnitude of the primary current is

$$I_p = \frac{I_s N_s}{N_p} = \frac{(8 \text{ A})(2200)}{90} = 195.6 \text{ A}$$

In an ideal transformer, no phase shift occurs, so the power factor is

$$\text{pf} = \cos \phi$$
$$\phi = \arccos \text{pf} = \arccos 0.8 = 36.87°$$

Therefore,

$$\boxed{\mathbf{I}_p = 196 \text{ A} \angle 37°}$$

The answer is (B).

4. Within a transformer with mutual flux Φ_m, the effective voltage that is induced in the secondary is

$$v_s(t) = -4.44N_sf\Phi_m \cos\omega t$$

For the peak flux only, this is

$$V_s = 4.44N_sf\Phi_m$$

$$\Phi_m = \frac{V_s}{4.44N_sf}$$

Use the turns ratio, a, to find the secondary voltage.

$$a = \frac{N_p}{N_s} = \frac{V_p}{V_s}$$

$$V_s = \frac{V_p}{a} = \frac{120 \text{ V}}{10} = 12 \text{ V}$$

The peak flux in the secondary is

$$\Phi_m = \frac{V_s}{4.44N_sf} = \frac{12 \text{ V}}{(4.44)(1)(60 \text{ Hz})}$$

$$= 4.54 \times 10^{-2} \text{ Wb} \quad (45 \text{ mWb})$$

This is the same flux present in the primary.

The answer is (D).

5. (a) Use the turns ratio to find the secondary voltage.

$$a = \frac{N_p}{N_s} = \frac{V_p}{V_s}$$

$$V_s = \frac{V_pN_s}{N_p} = \frac{(550 \text{ V})(40)}{200}$$

$$= \boxed{110 \text{ V}}$$

(b) Use the turns ratio and the impedance in the secondary load to find the primary current.

$$a = \frac{V_p}{V_s} = \frac{I_s}{I_p}$$

$$I_p = \frac{V_sI_s}{V_p} = \frac{V_s^2}{V_pZ_s} = \frac{(110 \text{ V})^2}{(550 \text{ V})(4.2 \text{ }\Omega)}$$

$$= \boxed{5.24 \text{ A}}$$

(c) The secondary current is

$$I_s = \frac{V_s}{Z_s} = \frac{110 \text{ V}}{4.2 \text{ }\Omega}$$

$$= \boxed{26.2 \text{ A}}$$

6. The turns ratio is

$$a = \sqrt{\frac{Z_p}{Z_s}} = \sqrt{\frac{200 \text{ }\Omega}{\dfrac{8 \text{ }\Omega}{3}}} = \boxed{8.66}$$

7. Consider the following circuit.

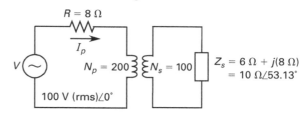

This circuit can be simplified to

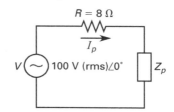

$$\mathbf{Z}_p = \left(\frac{N_p}{N_s}\right)^2\mathbf{Z}_s = \left(\frac{200}{100}\right)^2(10 \text{ }\Omega \angle 53.13°)$$

$$= 40 \text{ }\Omega \angle 53.13°$$

$$= (24 + j32) \text{ }\Omega$$

The circuit can be further simplified to

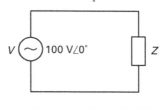

$$Z = \sqrt{\left(\sum R\right)^2 + \left(\sum X_L - \sum X_C\right)^2}$$

$$= \sqrt{\left(R + R_{Z_m}\right)^2 + \left(X_{Z_m}\right)^2}$$

$$= \sqrt{(8 \text{ }\Omega + 24 \text{ }\Omega)^2 + (32 \text{ }\Omega)^2}$$

$$= 45.25 \text{ }\Omega$$

$$\phi\mathbf{z} = \arctan\frac{\sum X_L - \sum X_C}{\sum R}$$

$$= \arctan\frac{X_{Z_m}}{R + R_{Z_m}}$$

$$= \arctan\frac{32 \text{ }\Omega}{8 \text{ }\Omega + 24 \text{ }\Omega}$$

$$= 45°$$

$$\mathbf{Z} = 45.25 \text{ }\Omega \angle 45°$$

(a) The primary current is

$$\mathbf{I}_p = \frac{\mathbf{V}}{\mathbf{Z}} = \frac{100 \text{ V} \angle 0°}{45.25 \ \Omega \angle 45°}$$
$$= \boxed{2.21 \text{ A}_{\text{rms}} \angle -45°}$$

(b) The secondary power is

$$P_s = I_s^2 R_s = I_s^2 Z_s \cos \phi$$
$$= \left(\frac{N_p I_p}{N_s} \right)^2 Z_s \cos \phi$$
$$= \left(\frac{(200)(2.21 \text{ A})}{100} \right)^2 (10 \ \Omega) \cos 53.13°$$
$$= \boxed{117.2 \text{ W}}$$

8. The inrush current is

$$I_{\text{inrush}} = 12 I_{\text{full load}}$$
$$= (12)(250 \text{ A})$$
$$= 3000 \text{ A}$$

Although the inrush current does not represent a fault, the breaker protection curves, or time current characteristic (TCC) curves, can be used to calculate a dial setting that will not trip the breaker during the inrush time frame.

On the graph shown, curves are entered on the y-axis at 0.1 s and move in the direction of the x-axis.

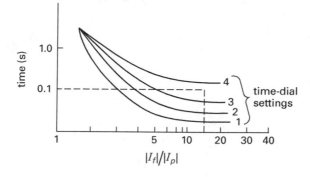

The ratio of the fault current to the pickup current is

$$\frac{|I_f|}{|I_p|} = \frac{3000 \text{ A}}{200 \text{ A}} = 15$$

Time dial setting $\boxed{4}$ is the only dial setting that does not trip in 0.1 s. This can be confirmed by entering curves on the x-axis—as though the inrush were a fault at 15—and moving in the direction of the y-axis.

The answer is (D).

9. Per Annex D of NFPA 70E, *Standard for Electrical Safety in the Workplace*, the industry-accepted minimum level for sustaining an arc fault is 38% of the available bolted fault current.

Calculate the arc flash minimum current.

$$I_{\text{AF}} = 0.38 I_{\text{BF}} = (0.38)(1000 \text{ A})$$
$$= \boxed{380 \text{ A}}$$

The answer is (B).

10. IEEE standards define the turns ratio of a transformer in several different ways, such as the ratio of primary to secondary voltage, primary to secondary current, and primary to secondary power. All calculations result in the same turns ratio as long as the correct values and equations are used.

The only known values for the three-phase transformer are the voltages, so calculate the turns ratio based on voltage. (The load value given can be used to find the currents in the transformer, but this would be unnecessarily time consuming.) The calculation can be based on either line-to-line voltages or line-to-neutral (phase) voltages.

In this transformer, 480 V is the primary voltage value for both line-to-line voltage calculations and phase voltage calculations. Two secondary voltages are given; 208 V is the secondary line-to-line voltage, $V_{L,s}$, and 120 V is the secondary phase voltage, $V_{\phi,s}$.

Find the turns ratio using the phase voltages.

$$a = \frac{V_{\phi,p}}{V_{\phi,s}} = \frac{480 \text{ V}}{120 \text{ V}} = 4 \quad (4:1)$$

The turns ratio is calculated with line-to-line voltages. Phase-to-phase values are used, so include the square root of three in the calculation.

$$a = \frac{V_{\phi,p}}{V_{\phi,s}}$$
$$V_{\phi,p} = a V_{\phi,s} = a \left(\frac{V_{L,s}}{\sqrt{3}} \right)$$
$$a = \sqrt{3} \left(\frac{V_{\phi,p}}{V_{L,s}} \right) = \sqrt{3} \left(\frac{480 \text{ V}}{208 \text{ V}} \right)$$
$$= \boxed{4 \quad (4:1)}$$

The answer is (D).

29 Linear Circuit Analysis

PRACTICE PROBLEMS

1. For which of the following circuit elements is super-position valid?

(A) independent source

(B) dependent source

(C) charged capacitor

(D) inductor with $I_{initial} \neq 0$

2. A 5 Ω length of copper wire is used in a circuit designed for a maximum temperature of 50°C. What is the resistance of the wire at maximum design temperature specification?

3. Two square parallel plates (0.04 m × 0.04 m) are separated by a 0.1 cm thick insulator with a dielectric constant of 3.4. What is the capacitance?

4. Two square parallel plates (0.04 m × 0.04 m) are separated by a 0.1 cm thick insulator with a dielectric constant of 3.4. The plates are connected across 200 V. What charge exists on the plates?

5. An 8 mH inductor carries 15 A of current. What is the average magnetic flux present?

6. For the circuit shown, which resistor(s) is (are) redundant?

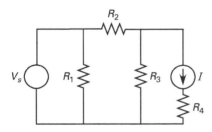

(A) R_1 only

(B) R_1 and R_3

(C) R_2 and R_4

(D) R_1 and R_4

7. For the circuit shown, what current is flowing in the 8 Ω resistor?

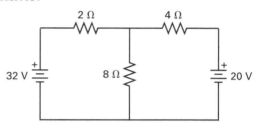

8. For the circuit shown, what current is flowing in the 6 Ω resistor?

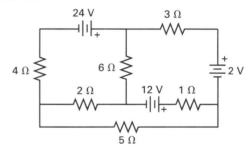

9. At $t = 0$, a switch connects to position A. At $t = 5 \times 10^{-4}$ s, the switch is moved to position B. There is no current anywhere in the circuit prior to $t = 0$.

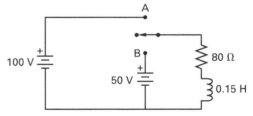

(a) What is the current flowing in the circuit just prior to the switch moving to position B? (b) What is the equation of the current after the switch is moved to position B?

10. Find the open-circuit impedance parameters for the circuit shown.

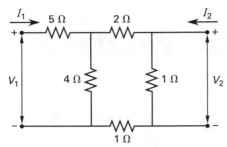

SOLUTIONS

1. Capacitors and inductors are linear elements for which superposition would be valid only if they were initially uncharged or without an initial magnetic field, respectively. Therefore, neither option (C) nor option (D) is the answer. The output of dependent sources depends on electrical parameters elsewhere in the circuit and is not linear. Option (B) is not the answer.

Independent sources are linear, and superposition can be applied.

The answer is (A).

2. Equation 29.4 gives resistance as a function of the resistivity, ρ, the length, l, and the area, A.

$$R = \frac{\rho l}{A}$$

Since the length and cross-sectional area are unknown, the following ratio is used.

$$\frac{R_{50}}{R_{20}} = \frac{\rho_{50}\dfrac{l}{A}}{\rho_{20}\dfrac{l}{A}} = \frac{\rho_{50}}{\rho_{20}}$$

Therefore,

$$R_{50} = R_{20}\frac{\rho_{50}}{\rho_{20}}$$

Using Eq. 29.3 to solve for ρ_{50},

$$
\begin{aligned}
\rho_{50} &= \rho_{20}(1 + \alpha_{20}(T - 20°C)) \\
&= (1.8 \times 10^{-8}\ \Omega\cdot\text{m}) \\
&\quad \times \left(1 + (3.9 \times 10^{-3}\ °C^{-1})(50°C - 20°C)\right) \\
&= 20.11 \times 10^{-9}\ \Omega\cdot\text{m}
\end{aligned}
$$

Substituting the given and calculated information gives

$$
\begin{aligned}
R_{50} = R_{20}\frac{\rho_{50}}{\rho_{20}} &= (5\ \Omega)\left(\frac{20.11 \times 10^{-9}\ \Omega\cdot\text{m}}{1.8 \times 10^{-8}\ \Omega\cdot\text{m}}\right) \\
&= \boxed{5.6\ \Omega}
\end{aligned}
$$

3. The capacitance is a function of the permittivity, $\epsilon_r\epsilon_0$, the area of the plates, A, and the distance, r, between them.

From Eq. 29.11,

$$C = \frac{\epsilon A}{r} = \frac{\epsilon_r\epsilon_0 A}{r}$$

$$= \frac{(3.4)\left(8.854 \times 10^{-12} \dfrac{\text{C}^2}{\text{N}\cdot\text{m}^2}\right)(0.04 \text{ m})^2}{0.001 \text{ m}}$$

$$= \boxed{4.817 \times 10^{-11} \text{ F}}$$

4. The charge on the plates is

$$Q = CV_{\text{plates}} = (4.817 \times 10^{-11} \text{ F})(200 \text{ V})$$

$$= \boxed{9.63 \times 10^{-9} \text{ C}}$$

5. The average energy in the inductor's magnetic field depends on the self-inductance, L, and the effective current, I.

From Eq. 29.20,

$$U = \frac{1}{2}LI^2 = \left(\frac{1}{2}\right)(8 \times 10^{-3} \text{ H})(15 \text{ A})^2$$

$$= 0.90 \text{ J}$$

Using Eq. 29.20, the energy in terms of the magnetic flux, Ψ, is

$$U = \frac{1}{2}\left(\frac{\Psi^2}{L}\right)$$

$$\Psi = \sqrt{2UL} = \sqrt{(2)(0.90 \text{ J})(8 \times 10^{-3} \text{ H})}$$

$$= \boxed{0.120 \text{ Wb}}$$

6. Resistors in parallel with voltage sources, or in series with current sources, do not affect the remainder of the circuit. Thus, R_1 and R_4 are redundant.

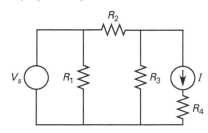

The answer is (D).

7. The circuit can be represented as

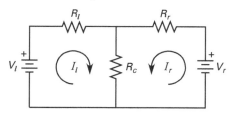

Using the loop-current method,

$$V_l = R_l I_l + R_c(I_l + I_r)$$
$$V_r = R_r I_r + R_c(I_l + I_r)$$
$$I_l + I_r = \frac{R_r V_l + R_l V_r}{R_r R_l + R_r R_c + R_l R_c}$$
$$= \frac{(4 \text{ }\Omega)(32 \text{ V}) + (2 \text{ }\Omega)(20 \text{ V})}{(4 \text{ }\Omega)(2 \text{ }\Omega) + (4 \text{ }\Omega)(8 \text{ }\Omega) + (2 \text{ }\Omega)(8 \text{ }\Omega)}$$
$$= \boxed{3 \text{ A} \quad [\text{down}]}$$

If the first two equations are solved simultaneously,

$$I_l = 4 \text{ A}$$
$$I_r = -1 \text{ A}$$

8. The circuit can be represented as

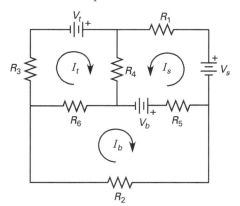

Using the loop-current method,

$$V_t = R_4(I_t + I_s) + R_6(I_t - I_b) + R_3 I_t$$
$$V_s + V_b = R_1 I_s + R_4(I_t + I_s) + R_5(I_s + I_b)$$
$$V_b + R_6(I_t - I_b) = R_5(I_b + I_s) + R_2 I_b$$

After rearranging, substituting in the constants, and evaluating the coefficients, the previous three equations become

$$24 \text{ V} = (12 \text{ }\Omega)I_t + (6 \text{ }\Omega)I_s - (2 \text{ }\Omega)I_b$$
$$14 \text{ V} = (6 \text{ }\Omega)I_t + (10 \text{ }\Omega)I_s + (1 \text{ }\Omega)I_b$$
$$12 \text{ V} = -(2 \text{ }\Omega)I_t + (1 \text{ }\Omega)I_s + (8 \text{ }\Omega)I_b$$

Circuit Theory

Solving simultaneously,

$$I_t = 2.53 \text{ A}$$
$$I_s = -0.336 \text{ A}$$
$$I_t + I_s = \boxed{2.194 \text{ A} \quad [\text{down}]}$$

9. (a) Before the switch is moved to position B,

$$I(t) = \left(\frac{V_{100}}{R}\right)(1 - e^{-Rt/L})$$

$$I(5 \times 10^{-4}) = \left(\frac{100 \text{ V}}{80 \text{ }\Omega}\right)(1 - e^{-(80 \text{ }\Omega)(5 \times 10^{-4} \text{ s})/0.15 \text{ H}})$$

$$= \boxed{0.2926 \text{ A} \quad [\text{clockwise}]}$$

(b) Move the switch to position B. Solve for currents I_1 and I_2, whose sum is I_N, the current flowing in the circuit.

$$I_1(t) = I_0 e^{-Rt/L} \quad [\text{clockwise}]$$

$$I_2(t) = \left(\frac{V_{50}}{R}\right)(1 - e^{-Rt/L}) \quad [\text{clockwise}]$$

$$I_N(t) = I_1(t) + I_2(t)$$

$$= I_0 e^{-Rt/L} + \left(\frac{V_{50}}{R}\right)(1 - e^{-Rt/L})$$

$$= \frac{50 \text{ V}}{80 \text{ }\Omega} + \left(0.2926 \text{ A} - \frac{50 \text{ V}}{80 \text{ }\Omega}\right)e^{-(80 \text{ }\Omega)t/0.15 \text{ H}}$$

$$= \boxed{0.625 - 0.332 e^{-(533 s^{-1})t} \text{ A} \quad [\text{clockwise}]}$$

10. Consider the circuit shown.

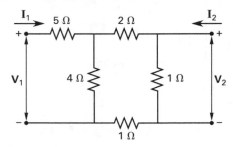

From the deriving equations in Table 29.3, the open-circuit impedance, z_{11}, is

$$z_{11} = \left.\frac{V_1}{I_1}\right|_{I_2 = 0 \text{ A}}$$

Since all the impedances are resistors, the phasor notation is omitted. The circuit, redrawn with $I_2 = 0$ A, is

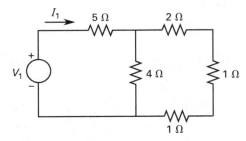

The equivalent impedance, z_e, from the left-hand side of the circuit to the right, into the input terminals, is

$$z_e = R_e = 5 \text{ }\Omega + \frac{(4 \text{ }\Omega)(2 \text{ }\Omega + 1 \text{ }\Omega + 1 \text{ }\Omega)}{4 \text{ }\Omega + 2 \text{ }\Omega + 1 \text{ }\Omega + 1 \text{ }\Omega}$$

$$= 7 \text{ }\Omega$$

Redrawing the circuit again gives

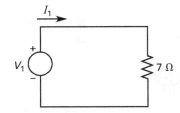

Using Eq. 29.2 (Ohm's law),

$$V_1 = I_1 R = I_1(7 \text{ }\Omega)$$

$$\frac{V_1}{I_1} = 7 \text{ }\Omega$$

Therefore,

$$z_{11} = \boxed{7 \text{ }\Omega}$$

From the deriving equations in Table 29.3, the open-circuit forward transfer impedance, z_{21}, is

$$z_{21} = \left.\frac{V_2}{I_1}\right|_{I_2 = 0 \text{ A}}$$

The redrawn circuit for z_{21} is identical to that of z_{11}, since $I_2 = 0$ A.

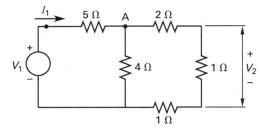

For the current through the branch containing V_2, using the concept of a current divider at node A, as in Eq. 29.38,

$$I_{V_2} = I_1 \left(\frac{4\ \Omega}{4\Omega + (2\ \Omega + 1\ \Omega + 1\ \Omega)} \right)$$
$$= \frac{1}{2} I_1$$

Ohm's law is applied across R, the $1\ \Omega$ resistor associated with V_2, using the calculated current.

$$V_2 = \frac{1}{2} I_1 R$$
$$= \frac{1}{2} I_1 (1\ \Omega)$$
$$= \frac{1}{2} I_1$$
$$\frac{V_2}{I_1} = \frac{1}{2}\ \Omega$$

Therefore,

$$\boxed{z_{21} = \left. \frac{V_2}{I_1} \right|_{I_2 = 0\ \text{A}} = \frac{1}{2}\ \Omega}$$

From the deriving equations in Table 29.3, the open-circuit reverse transfer impedance, z_{12}, is

$$z_{12} = \left. \frac{V_1}{I_2} \right|_{I_1 = 0\ \text{A}}$$

The circuit, redrawn with $I_1 = 0\ \text{A}$, is

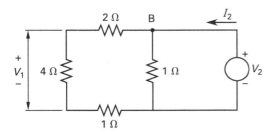

For the branch current flowing through the $4\ \Omega$ resistor associated with V_1, using a current divider at node B, as in Eq. 29.38,

$$I_{V_1} = I_2 \left(\frac{1\ \Omega}{1\ \Omega + (2\ \Omega + 4\ \Omega + 1\ \Omega)} \right)$$
$$= \frac{1}{8} I_2$$

Applying Ohm's law across the $4\ \Omega$ resistor associated with V_1 gives

$$V_1 = \frac{1}{8} I_2 R = \frac{1}{8} I_2 (4\ \Omega) = \frac{1}{2} I_2$$
$$\frac{V_1}{I_2} = \frac{1}{2}\ \Omega$$

Therefore,

$$\boxed{z_{12} = \left. \frac{V_1}{I_2} \right|_{I_1 = 0\ \text{A}} = \frac{1}{2}\ \Omega}$$

From the deriving equations in Table 29.3, the open-circuit output impedance, z_{22}, is

$$z_{22} = \left. \frac{V_2}{I_2} \right|_{I_1 = 0\ \text{A}}$$

The redrawn circuit for z_{22} is identical to that of z_{12}, since $I_1 = 0\ \text{A}$.

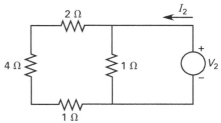

The equivalent impedance, z_e, from the right-hand side of the circuit to the left, into the output terminals, is

$$z_e = R_e = \frac{(1\ \Omega)(2\ \Omega + 4\ \Omega + 1\ \Omega)}{1\ \Omega + (2\ \Omega + 4\ \Omega + 1\ \Omega)} = \frac{7}{8}\ \Omega$$

Redrawing the circuit again gives

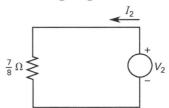

Using Eq. 29.2 (Ohm's law),

$$V_2 = I_2 R = I_2 \left(\frac{7}{8}\ \Omega \right)$$
$$\frac{V_2}{I_2} = \frac{7}{8}\ \Omega$$

Therefore,

$$\boxed{z_{22} = \frac{7}{8}\ \Omega}$$

30 Transient Analysis

PRACTICE PROBLEMS

1. Consider the circuit shown. What is the time constant?

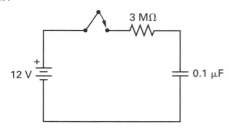

(A) $1.0\ \mu s$

(B) $36\ \mu s$

(C) $0.30\ s$

(D) $3.0\ s$

2. With an initially uncharged capacitor in the circuit in Prob. 1, what is the initial current?

(A) $0.0\ A$

(B) $4.0\ \mu A$

(C) $6.0\ \mu A$

(D) $120\ A$

3. How long does it take the circuit shown in Prob. 1 to reach steady-state response?

4. What is the rise time of the circuit in Prob. 1?

5. Consider the circuit shown. Write the equations for loops 1 and 2.

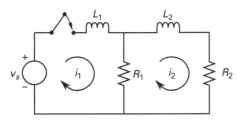

6. A series RC circuit contains a $4\ \Omega$ resistor and an uncharged $300\ \mu F$ capacitor. The switch is closed at $t = 0$ s, connecting the circuit across a 500 V source. For the circuit shown, determine the current as a function of time.

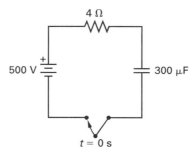

7. The switch in the circuit shown is closed at $t = 0$ s and opened again at $t = 0.05$ s. How much energy is stored in the initially uncharged capacitor?

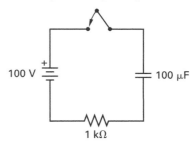

8. Find the current through the $80\ \Omega$ resistor (a) immediately after the switch is closed and (b) 2 s after the switch is closed. The switch has been open for a long time.

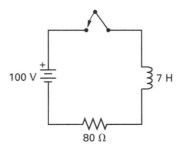

9. For the circuit shown, determine the steady-state inductance current, i_L, (a) with the switch closed and (b) with the switch open.

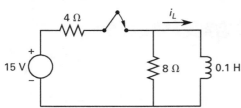

10. For the circuit shown, determine the steady-state capacitance voltage, v_C, (a) with the switch closed and (b) with the switch open.

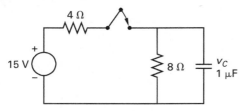

11. For the circuit shown, determine the steady-state inductance current, i_L, (a) with the switch closed and (b) with the switch open.

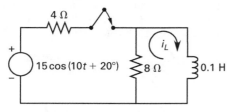

12. The circuit in Prob. 10 has had the switch closed for a long time. At $t = 0$ s, the switch opens. Determine the capacitance voltage, $v_C(t)$, for $t > 0$ s, for a capacitance of 1×10^{-6} F.

13. The circuit in Prob. 10 has had the switch open for a long time. At $t = 0$ s, the switch closes. Determine the capacitance voltage, $v_C(t)$, for $t > 0$ s.

14. A system is described by the equation

$$12 \sin 2t = \frac{d^2 v}{dt^2} + 5 \frac{dv}{dt} + 4v$$

The initial conditions on the voltage are

$$v(0) = -5 \text{ V}$$
$$\frac{dv(0)}{dt} = 2 \text{ V/s}$$

Determine $v(t)$ for $t > 0$ s.

15. For the circuit shown, the current source is

$$i_s(t) = 0.01 \cos 500t$$

Find the transfer function.

$$G(s) = \frac{V_C(s)}{I_s(s)}$$

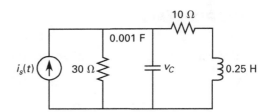

16. For a series RLC circuit of 10 Ω, 0.5 H, and 100 μF, determine the (a) resonant frequency, (b) quality factor, and (c) bandwidth.

17. For a parallel RLC circuit of 10 Ω, 0.5 H, and 100 μF, determine the (a) resonant frequency, (b) quality factor, and (c) bandwidth.

18. The switch of the circuit shown has been open for a long time and is closed at $t = 0$ s. Determine the currents flowing in the two capacitors at the instant the switch closes. Give the magnitudes and directions.

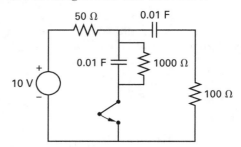

19. A system is described by the following differential equation.

$$20 \sin 4t = \frac{d^2 i}{dt^2} + 4 \frac{di}{dt} + 4i$$

The initial conditions are

$$i(0) = 0 \text{ A}$$
$$\frac{di(0)}{dt} = 4$$

Determine $i(t)$ for $t > 0$ s.

20. The circuit shown has the 1 μF capacitance charged to 100 V and the 2 μF capacitance uncharged before the switch is closed. Determine the (a) energy stored in each capacitance at the instant the switch closes, (b) energy stored in each capacitance a long time after the switch closes, and (c) energy dissipated in the resistance.

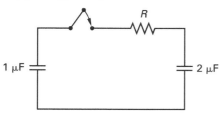

SOLUTIONS

1.

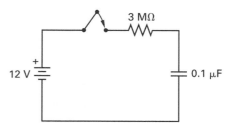

The time constant is

$$\tau = RC = (3 \times 10^6 \ \Omega)(0.1 \times 10^{-6} \ \text{F})$$
$$= \boxed{0.3 \ \text{s} \quad (0.30 \ \text{s})}$$

The answer is (C).

2. The capacitor initially acts as a short. From Ohm's law,

$$V = IR$$
$$I = \frac{V}{R} = \frac{12 \ \text{V}}{3 \times 10^6 \ \Omega}$$
$$= \boxed{4.0 \times 10^{-6} \ \text{A} \quad (4.0 \ \mu\text{A})}$$

The answer is (B).

3. Steady-state response is reached in approximately 5 time constants. Thus,

$$t = 5\tau$$

The time constant is $RC = 0.3$ s. Substituting,

$$t = (5)(0.3 \ \text{s}) = \boxed{1.5 \ \text{s}}$$

4. The rise time is most often determined by measurements in the lab. Nevertheless, the rise time is

$$t_r = 2.2\tau$$

The time constant was found in Prob. 1. Substituting gives

$$t_r = (2.2)(0.3 \ \text{s}) = \boxed{0.660 \ \text{s}}$$

Circuit Theory

5.

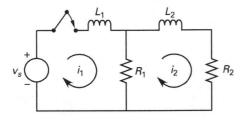

Kirchhoff's voltage law (KVL) for loop 1 gives

$$v_s - L_1\left(\frac{di_1}{dt}\right)$$
$$- (i_1 - i_2)R_1 = 0$$

$$v_s = \boxed{L_1\left(\frac{di_1}{dt}\right) + (i_1 - i_2)R_1}$$

KVL for loop 2 gives

$$-(i_2 - i_1)R_1 - L_2\left(\frac{di_2}{dt}\right) - i_2R_2 = 0$$

$$-i_2R_1 + i_1R_1 - L_2\left(\frac{di_2}{dt}\right) - i_2R_2 = 0$$

$$\boxed{L_2\left(\frac{di_2}{dt}\right) + R_1(i_2 - i_1) + i_2R_2 = 0}$$

6. From Table 30.1, the current response for a series RC circuit is

$$i(t) = \left(\frac{V_{\text{bat}} - V_0}{R}\right)e^{-t/\tau} = \left(\frac{500\ \text{V} - 0\ \text{V}}{4\ \Omega}\right)e^{-t/(1.2 \times 10^{-3})}$$
$$= \boxed{125e^{-833t}\ \text{A}}$$

7.

$$Q(t) \quad = CV_C(t) = CV(1 - e^{-t/RC})$$
$$E_C(t) \quad = \frac{1}{2}CV_C^2(t) = \frac{1}{2}CV^2(1 - e^{-t/RC})^2$$
$$E_C(0.05) = \left(\frac{1}{2}\right)(100 \times 10^{-6}\ \text{F})(100\ \text{V})^2$$
$$\times\left(1 - e^{-0.05\ \text{s}/\left((10^3\ \Omega)(100 \times 10^{-6}\ \text{F})\right)}\right)^2$$
$$= \boxed{0.0774\ \text{J}}$$

8. (a) Since the current in the inductor cannot change instantaneously,

$$I(0) = \boxed{0\ \text{A}}$$

(b) $$I(t) = \left(\frac{V}{R}\right)(1 - e^{-Rt/L})$$

$$I(2) = \left(\frac{100\ \text{V}}{80\ \Omega}\right)(1 - e^{-(80\ \Omega)(2\ \text{s})/7\ \text{H}})$$
$$= \boxed{1.25\ \text{A} \quad [\text{down}]}$$

9. (a) Use KVL around the outer loop. With the switch closed, KVL gives

$$15\ \text{V} = 4i_L + v_L$$
$$v_L = L\left(\frac{di_L}{dt}\right) = 0 \quad [\text{in steady state,ss}]$$
$$i_{L_{\text{ss}}} = i_{\text{ss}} = \frac{15\ \text{V}}{4\ \Omega}$$
$$= 3.75\ \text{A} \quad (4\ \text{A})$$

The current through the 8 Ω resistor is $v_L/8 = 0$ V; therefore, in the steady-state condition, all of i_{ss} flows through the inductance.

(b) With the switch open, the inductor dissipates its energy in the 8 Ω resistor.

$$i_{L_{\text{ss}}} = \boxed{i_{\text{ss}} = 0}$$

10. (a) With the switch closed,

$$i_C = C\left(\frac{dv_C}{dt}\right)$$

In steady-state condition,

$$\frac{dv_C}{dt} = 0$$

Therefore, $i_C \to 0$. By voltage division,

$$v_C = \left(\frac{8\ \Omega}{12\ \Omega}\right)(15\ \text{V}) = \boxed{10\ \text{V}}$$

With the switch open, the capacitor dissipates its energy in the 8 Ω resistor.

(b)

$$\boxed{v_C = 0}$$

11. (a) A Thevenin equivalent circuit, seen from the inductance, is obtained.

$$v_{oc} = v_{Th} = \left(\frac{8\ \Omega}{12\ \Omega}\right)v_s = 10\cos(10t + 20°)\ \text{V}$$

$$i_{sc} = \frac{15\cos(10t + 20°)\ \text{V}}{4\ \Omega} = 3.75\cos(10t + 20°)\ \text{A}$$

$$R_{Th} = \frac{v_{oc}}{i_{sc}} = \frac{10\ \text{V}}{3.75\ \text{A}} = 8/3\ \Omega$$

Taking $\cos(10t + 20°)$ as the phasor reference and using the maximum value, for convenience, instead of the rms value, gives

$$V_{Th} = 10\ \text{V} \angle 0°$$

$$Z_L = j\omega L = \left(j10\ \frac{\text{rad}}{\text{s}}\right)(0.1\ \text{H}) = j1\ \Omega$$

$$i_L = \frac{V_{Th}}{R_{Th} + j1\ \Omega}$$

$$= \left(\frac{10\ \text{V}}{\frac{8}{3}\ \Omega + j1\ \Omega}\right)\left|\frac{\frac{8}{3}\ \Omega - j1\ \Omega}{\frac{8}{3}\ \Omega - j1\ \Omega}\right|$$

$$= \left(\frac{90}{73}\ \text{A}\right)\left(\frac{8}{3} - j1\right) = \boxed{3.51\ \text{A} \angle -20.6°}$$

As a time function,

$$i_L(\infty) = 3.51\cos(10t + 20° - 20.6°)$$
$$= 3.51\cos(10t - 0.6°)\ \text{A}$$

(b) With the switch open,

$$i_L(\infty) = \boxed{0}$$

12. With the switch closed for a long time, the capacitance voltage is 10 V (see Prob. 10). The initial voltage when the switch opens is therefore 10 V.

With the switch open, KVL is

$$v_C = 8i$$

$$i_C = 10^{-6}\left(\frac{dv_C}{dt}\right)$$

$$i_C = -i$$

$$v_C + (8 \times 10^{-6})\left(\frac{dv_C}{dt}\right) = 0$$

$$\frac{dv_C}{dt} + (1.25 \times 10^5)v_C = 0$$

Any appropriate mathematical method may now be used to solve the differential equation.

Let $\alpha = 1.25 \times 10^5$. The solution is of the form

$$v_C = V(t_1)e^{-\alpha(t - t_1)}$$

Taking $t_1 = 0$ s, $v_C(0) = 10$ V.

$$\frac{dv_C}{dt} = -\alpha v_C$$

$$\alpha = 1.25 \times 10^5$$

$$v_C(t) = \boxed{10e^{-(1.25 \times 10^5)t}\ \text{V}}$$

13. The initial capacitance voltage is zero. Kirchhoff's current law (KCL) at the upper node, with the lower node as a reference, is

$$\frac{15\ \text{V} - v_C}{4\ \Omega} = \frac{v_C}{8\ \Omega} + (10^{-6}\ \text{F})\left(\frac{dv_C}{dt}\right)$$

This is manipulated to

$$\left(\frac{8}{3} \times 10^{-6}\right)\left(\frac{dv_C}{dt}\right) + v_C = 10\ \text{V}$$

$$V_{ss} = 10\ \text{V}$$

$$\tau = \frac{8}{3} \times 10^{-6}\ \text{s} \quad (2.67 \times 10^{-6}\ \text{s})$$

$$V_0 = 0$$

$$v_C(t) = V_{ss} + (V_0 - V_{ss})e^{-t/\tau}$$

$$= \boxed{(10)\left(1 - e^{-(3.75 \times 10^5)t}\right)u(t)\ \text{V}}$$

14. The homogeneous differential equation is

$$\frac{d^2v}{dt^2} + 5\frac{dv}{dt} + 4v = 0$$

This has solutions e^{st},

$$\frac{d^2v}{dt^2} \rightarrow s^2V$$

$$\frac{dv}{dt} \rightarrow sV$$

$$s^2 + 5s + 4 = 0$$

$$(s + 4)(s + 1) = 0$$

Transient solutions are of the form

$$Ae^{-4t} + Be^{-t}$$

The steady-state sinusoidal solution must be found before A and B can be evaluated.

The phasor solution is

$$12 \sin 2t \Leftrightarrow 12 \text{ V} \angle 0°$$
$$\omega = 2$$
$$\frac{d}{dt} \Rightarrow j\omega = j2$$
$$\frac{d^2}{dt^2} \Rightarrow -\omega^2 = -4$$

The phasor equation is

$$12 \text{ V} = -4 V_{ss} + (5)(j2) V_{ss} + 4 V_{ss}$$
$$= j10 V_{ss}$$
$$V_{ss} = \frac{1.2}{j} = 1.2 \text{ V} \angle -90°$$
$$v_{ss} = 1.2 \text{ V} \sin(2t - 90°)$$
$$= -1.2 \text{ V} \cos 2t$$
$$v(t) = A e^{-4t} + B e^{-t} - 1.2 \cos 2t$$
$$v(0) = A + B - 1.2 \text{ V} = -5 \text{ V}$$
$$A + B = -3.8 \text{ V}$$
$$\frac{dv}{dt} = -4A e^{-4t} - B e^{-t} + 2.4 \sin 2t$$
$$\frac{dv}{dt}(0) = -4A - B = 2$$
$$A = 0.6 \text{ V}$$
$$B = -4.4 \text{ V}$$
$$\boxed{v(t) = (0.6 e^{-4t} - 4.4 e^{-t} - 1.2 \cos 2t) u(t) \text{ V}}$$

15. The impedance of the inductance branch is

$$Z_{LR} = 10 + 0.25s \ \Omega$$

KCL applied to the upper node is

$$I_s = \frac{V_C}{30 \ \Omega} + (0.001 \text{ F}) s V_C + \frac{V_C}{10 \ \Omega + (0.25 \text{ H}) s}$$

Solving,

$$G = \frac{V_C}{I_s} = \left(\frac{3 \times 10^4 + 750s}{0.75s^2 + 55s + 4000} \right) \Omega$$
$$= \boxed{(1000) \left(\frac{s + 40}{s^2 + 73.33s + 5333} \right) \Omega}$$

16. (a)

$$Z = 10 \ \Omega + 0.5s \ \Omega + \frac{10^4}{s} \ \Omega$$
$$= \left(\frac{1}{2} \right) \left(\frac{s^2 + 20s + 2 \times 10^4}{s} \right) \Omega$$
$$= \left(\frac{1}{2s} \right) \left(s^2 + \frac{\omega_0}{Q} s + \omega_0^2 \right) \Omega$$

The resonant frequency is

$$\omega_0 = \boxed{\sqrt{2} \times 10^2 \text{ rad/s}}$$

(b) The quality factor is

$$Q = \frac{\omega_0}{20} = \boxed{5\sqrt{2}}$$

(c) The bandwidth is

$$\text{BW} = \frac{\omega_0}{Q} = \boxed{20 \text{ rad/s}}$$

17. (a)

$$Y = 0.1 + \frac{1}{0.5s} + 10^{-4}s$$
$$= \left(\frac{10^{-4}}{s} \right) (s^2 + 10^3 s + 2 \times 10^4)$$
$$= \left(\frac{10^{-4}}{s} \right) \left(s^2 + \frac{\omega_0}{Q} s + \omega_o^2 \right)$$

The resonant frequency is

$$\omega_0 = \sqrt{2 \times 10^4} = \boxed{100\sqrt{2} \text{ rad/s}}$$

(b) The quality factor is

$$Q = \frac{\omega_0}{\text{BW}} = \frac{\sqrt{2}}{10} = \boxed{0.1\sqrt{2}}$$

(c) The bandwidth is

$$\text{BW} = \frac{\omega_0}{Q} = \boxed{10^3 \text{ rad/s}}$$

18. With the switch open for a long time, the upper-right capacitor is charged to 10 V and the lower (center) capacitor is discharged to 0 V. At the instant of switching, the capacitors can be modeled as ideal voltage sources with their initial voltages. The resulting equivalent circuit is shown.

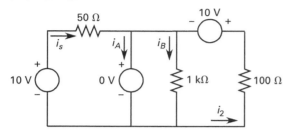

KVL on the left loop is

$$10 \text{ V} - (50 \text{ }\Omega)i_s + 0.0 = 0$$
$$i_s = 1/5 \text{ A}$$

KVL on the right loop is

$$0.0 + 10 \text{ V} - (100 \text{ }\Omega)i_2 = 0$$
$$i_2 = 1/10 \text{ A}$$

The current in the right (upper) capacitor is i_2, with 0.1 A flowing left.

As the voltage across the 1 kΩ resistor is zero, $i_B = 0$ A. By KCL, the other capacitor current is

$$i_A = i_s + i_2 = \tfrac{1}{5}\text{ A} + \tfrac{1}{10}\text{ A} = \boxed{0.3 \text{ A}}$$

19. The homogeneous equation is

$$s^2 + 4s + 4 = 0$$

The roots are $(s+2)^2$. The homogeneous solution is

$$Ae^{-2t} + Bte^{-2t}$$

The steady-state solution is

$$s \rightarrow j\omega$$
$$\omega = 4$$
$$\frac{d}{dt} \rightarrow j\omega$$
$$\frac{d^2}{dt^2} \rightarrow -\omega^2 = -16$$

The sine reference is used.

$$20 \text{ V} = (-16 \text{ }\Omega + 16j \text{ }\Omega + 4 \text{ }\Omega)I_{ss}$$
$$I_{ss} = \frac{20}{-12 + j16} = -0.6 + j(-0.8) \text{ A}$$
$$i_{ss} = -0.6 \sin 4t - 0.8 \sin(4t + 90°) \text{ A}$$
$$= -0.6 \sin 4t - 0.8 \cos 4t \text{ A}$$
$$i = i_{ss} + Ae^{-2t} + Bte^{-2t}$$
$$i(0) = -0.8 \text{ A} + A = 0$$
$$A = 0.8 \text{ A}$$
$$\frac{di}{dt} = -2.4 \cos 4t + 3.2 \sin 4t$$
$$\qquad -2Ae^{-2t} + Be^{-2t} - 2Bte^{-2t}$$
$$\frac{di}{dt}(0) = -2.4 - 2A + B = 4 \text{ A}$$
$$B = 2.4 + 2A + 4 = 8 \text{ A}$$
$$i(t) = 0.8e^{-2t} + 8te^{-2t} - 0.6 \sin 4t - 0.8 \cos 4t \text{ A}$$
$$= \boxed{0.8e^{-2t} + 8te^{-2t} - \sin(4t + 53.1°) \text{ A}}$$

20. The initial charge on the 1 μF capacitor is

$$Q_{1,\text{initial}} = CV_0$$
$$= (10^{-6} \text{ F})(100 \text{ V})$$
$$= 10^{-4} \text{ C}$$

The initial energy, $W_{1,\text{initial}}$, is

(a)

$$W_{1,\text{initial}} = \tfrac{1}{2}C_1 V_0^2$$
$$= \left(\frac{1}{2}\right)(10^{-6} \text{ F})(100 \text{ V})^2$$
$$= 0.005 \text{ J} \quad (5 \text{ mJ})$$

In the final condition, the voltage across the two capacitors is the same.

$$W_{2,\text{initial}} = 0$$
$$V_f = \frac{Q_1}{C_1} = \frac{Q_2}{C_2}$$
$$Q_2 = 2Q_1$$

As $Q_1 + Q_2 = Q_0$,

$$Q_1 = \frac{1}{3} \times 10^{-4} \text{ C}$$

$$Q_2 = \frac{2}{3} \times 10^{-4} \text{ C}$$

$$
\begin{aligned}
W_{\text{final}} &= \frac{1}{2}\left(\frac{Q_1^2}{C_1}\right) + \frac{1}{2}\left(\frac{Q_2^2}{C_2}\right) \\
&= \left(\frac{1}{2}\right)\left(\frac{\left(\frac{1}{3} \times 10^{-4} \text{ C}\right)^2}{1 \times 10^{-6} \text{ F}}\right) + \left(\frac{1}{2}\right)\left(\frac{\left(\frac{2}{3} \times 10^{-4} \text{ C}\right)^2}{2 \times 10^{-6} \text{ F}}\right) \\
&= 0.001667 \text{ J} \quad (1.667 \text{ mJ}) \\
W_{\text{loss}} &= W_{\text{initial}} - W_{\text{final}} \\
&= 5 \text{ mJ} - 1.667 \text{ mJ} \\
&= \boxed{3.333 \text{ mJ}}
\end{aligned}
$$

(b) $\quad W_{\text{final}} = \frac{3}{18} \times 10^{-2} \text{ J} = \boxed{0.00167 \text{ J}}$

(c) $\quad W_{\text{loss}} = 0.005 \text{ J} - 0.00167 \text{ J} = \boxed{0.00333 \text{ J}}$

Circuit Theory

31 Time Response

PRACTICE PROBLEMS

1. Which of the following is in the form of the solution to a first-order circuit?

(A) $\kappa + Ae^{-t/\tau}$

(B) $\kappa + Ae^{s_1 t} + Be^{s_2 t}$

(C) $\kappa + Ae^{st} + Bte^{st}$

(D) $\kappa + e^{\alpha t}(A\cos\beta t + B\sin\beta t)$

2. In the circuit shown, how long will it take for the capacitor voltage to reach 98% of its final value, assuming the capacitor is initially uncharged?

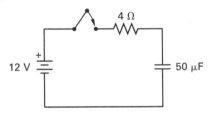

3. Determine the time-domain equation for the circuit shown.

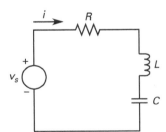

4. A series RLC circuit is shown.

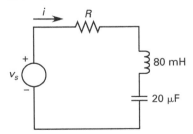

Most nearly, what is the minimum value of the resistance required for the circuit to be overdamped?

(A) 25 Ω

(B) 50 Ω

(C) 100 Ω

(D) 150 Ω

5. Complete the following table for a sine reference in the appropriate units.

function	$A_p \angle \phi$	s
5		
$5e^{-200t}$		
$10\sin 377t$		
$10\sin(500t + 30°)$		
$20\cos(1000t + 60°)$		
$20e^{-200t}\sin 1000t$		

6. The time response of an AC generator is controlled by a proportional-integral-derivative (PID) controller. The illustration shows the overall gain effect at values from 0.01 to 0.5.

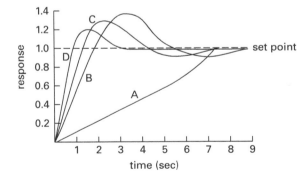

Which of the curves shows the greatest overall gain for the system?

(A) curve A

(B) curve B

(C) curve C

(D) curve D

7. The time response for a voltage transient beginning at $t = 1$ is shown. The response time required for the system to stabilize is greater than 5 sec.

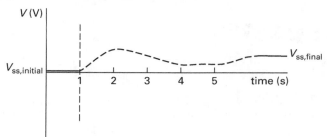

What gain adjustment should be made to shorten the transient time to stabilize?

(A)   The derivative gain should be decreased.

(B)   The derivative gain should be increased.

(C)   The proportional gain should be decreased.

(D)   The proportional gain should be increased.

8. The time response for a voltage transient beginning at $t = 0$ is shown. The system exhibits ringing.

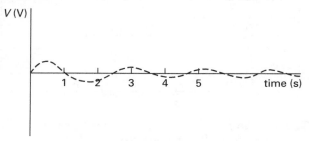

What gain constant adjustment should be made to eliminate the ringing?

(A)   The derivative gain constant should be decreased.

(B)   The derivative gain constant should be increased.

(C)   The integral gain constant should be decreased.

(D)   The integral gain constant should be increased.

9. The time response for a voltage transient beginning at $t = 1$ is shown. The system exhibits too much voltage overshoot.

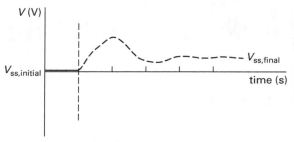

What gain constant adjustment should be made to reduce the overshoot?

(A)   The derivative gain should be decreased.

(B)   The derivative gain should be increased.

(C)   The proportional gain should be increased.

(D)   Either the derivative gain should be decreased, or the proportional time constant should be increased.

SOLUTIONS

1. The form of a first-order circuit solution is

$$x_t = \boxed{\kappa + Ae^{-t/\tau}}$$

The answer is (A).

2. The time constant, τ, is

$$\tau = RC = (4\ \Omega)(50 \times 10^{-6}\ \text{F})$$
$$= 200 \times 10^{-6}\ \text{s} \quad (200\ \mu\text{s})$$

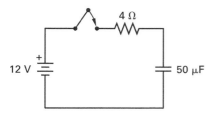

The build-up factor after a step change takes the form

$$1 - e^{-t/\tau}$$

Therefore, the time to 98% of the final value, regardless of the actual value, is

$$0.98 = 1 - e^{-t/\tau}$$

For the given circuit, $\tau = 200\ \mu\text{s}$. Substituting and solving for the time gives

$$0.98 = 1 - e^{-t/200 \times 10^{-6}\ \text{s}}$$
$$-0.02 = -e^{-t/200 \times 10^{-6}\ \text{s}}$$
$$\ln 0.02 = -\frac{t}{200 \times 10^{-6}\ \text{s}}$$
$$t = -\ln 0.02(200 \times 10^{-6}\ \text{s})$$
$$= \boxed{782.4 \times 10^{-6}\ \text{s} \quad (782\ \mu\text{s} \quad [\approx 4\tau])}$$

3.

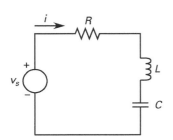

Using Kirchhoff's voltage law (KVL) around the circuit gives

$$v_s - iR - v_L - v_C = 0\ \text{V}$$
$$v_s = iR + v_L + v_C$$
$$= iR + L\left(\frac{di}{dt}\right) + \frac{1}{C}\int i\,dt$$

Differentiating gives

$$\frac{dv_s}{dt} = \left(\frac{di}{dt}\right)R + L\left(\frac{d^2i}{dt^2}\right) + \frac{i}{C}$$
$$= \boxed{L\left(\frac{d^2i}{dt^2}\right) + R\left(\frac{di}{dt}\right) + \frac{1}{C}i}$$

4. The overdamped condition is determined by the coefficients of the differential equation. Specifically, for the overdamped condition, $b^2 > 4ac$.

For a series second-order circuit,

$$\frac{dv_s}{dt} = L\left(\frac{d^2i}{dt^2}\right) + R\left(\frac{di}{dt}\right) + \frac{1}{C}i$$

This indicates that $a = L$, $b = R$, and $c = 1/c$. Substituting these values into $b^2 > 4ac$ gives

$$R^2 > 4L\left(\frac{1}{C}\right)$$
$$R < \sqrt{(4)(80 \times 10^{-3}\ \text{H})\left(\frac{1}{20 \times 10^{-6}\ \text{F}}\right)} = \boxed{126.5\ \Omega}$$

The answer is (D).

5.

function	$A_p \angle \phi$	s
5	$5\angle 0°$	$0 + j0$
$5e^{-200t}$	$5\angle 0°$	$-200 + j0$ Np/s
$10 \sin 377t$	$10\angle 0°$	$0 \pm j377$ rad/s
$10 \sin(500t + 30°)$	$10\angle 30°$	$0 \pm j500$ rad/s
$20 \cos(1000t + 60°)$	$20\angle 150°$	$0 \pm j1000$ rad/s
$20e^{-200t} \sin 1000t$	$20\angle 0°$	$-200 \pm j1000$ s^{-1}

6. The time response of an AC generator is controlled by its excitation control system, which is commonly called a voltage regulator. Excitation control systems, however, do more than their name suggests. Such systems are usually digital and are adjusted by setting an overall gain, K_g. Individual time response issues are adjusted by setting the terms of the gain equation.

For a PID controller,

$$K_g = K_P + K_I + K_D$$

As the overall gain increases, the response of the system is faster for any given transient with less overshoot and undershoot. Curve D has the fastest response and thus the greatest overall gain.

The answer is (D).

7. The derivative gain controls the magnitude of the overshoot, while the proportional gain controls the response time. When the response time of a control system is too great, the proportional gain is increased to decrease the response time.

The answer is (D).

8. The derivative gain constant controls the magnitude of the overshoot, while the integral gain constant controls the number of oscillations over time. Ringing is usually a result of the integral gain being set too high, so decreasing the integral gain is the most likely way to eliminate the ringing.

The answer is (C).

9. The proportional gain controls the response time. When too much voltage overshoot occurs in a control system, the derivative gain is increased to reduce the overshoot.

The answer is (B).

Circuit Theory

32 Frequency Response

PRACTICE PROBLEMS

1. For the circuit shown, determine the transfer function with $V(s)$ as the input and $I(s)$ as the output.

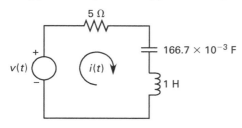

2. Plot the zeros and poles on the s domain for the circuit in Prob. 1.

3. Plot the Bode magnitude plot for the circuit in Prob. 1.

4. For the circuit shown, plot the idealized Bode plot of magnitude and phase.

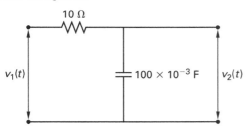

5. In the Bode plot for the circuit in Prob. 4, what type of filter is represented?

(A)   low-pass filter

(B)   high-pass filter

(C)   bandpass filter

(D)   notch filter

6. For the circuit shown, determine the (a) voltage transfer function, (b) frequency response, and (c) unit step voltage response.

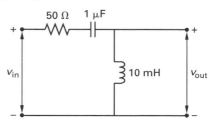

7. For the circuit shown, determine the (a) voltage transfer function, (b) frequency response, and (c) unit step voltage response.

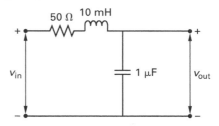

8. The graph shows the frequency response of a circuit.

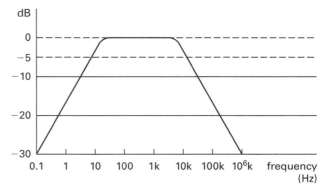

What type of filter is represented and what is the bandwidth?

(A)   bandpass, 10 Hz–10 kHz

(B)   bandpass, 0.1 Hz–10 MHz

(C)   high-pass, 10 kHz–10 MHz

(D)   notch, 0.1 Hz–10 Hz and 10 kHz–10 MHz

9. In the simple filter shown, the resistance is 25 Ω, and the capacitance is 5×10^{-9} F.

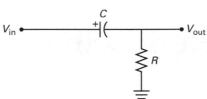

What type of filter is shown, and what is most nearly the cutoff frequency of the filter?

(A) high-pass, 0.008 GHz

(B) high-pass, 125 GHz

(C) low-pass, 0.008 GHz

(D) low-pass, 125 GHz

10. The filter shown is used to adjust frequency response so that only audible frequencies are sent to downstream amplifier circuits. The resistance is 5 Ω, and the inductance is variable.

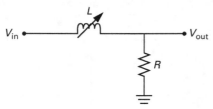

What is most nearly the inductive reactance required to establish a cutoff frequency of 20 kHz (i.e., the upper end of the audible range)?

(A) 5 Ω

(B) 20 Ω

(C) 200 Ω

(D) 250 Ω

SOLUTIONS

1. The transfer function is

$$T_{\text{net}}(s) = \frac{I(s)}{V(s)} = \frac{1}{Z(s)}$$

First, transfer the circuit into the s domain.

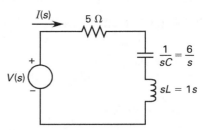

The impedance for this series circuit is found in terms of the complex frequency, s.

$$Z(s) = R + Z_C + Z_L = 5 + \frac{6}{s} + 1s$$

$$= \frac{5s + 6 + s^2}{s}$$

$$= \frac{s^2 + 5s + 6}{s}$$

The transfer function is

$$\boxed{T_{\text{net}} = \frac{1}{Z(s)} = \frac{s}{s^2 + 5s + 6}}$$

2. Factor the transfer function into standard form.

$$T_{\text{net}}(s) = \frac{s}{s^2 + 5s + 6}$$

$$= \frac{s}{(s+2)(s+3)}$$

One zero at the origin is given by the numerator term. Two poles, one at -2 and the other at -3, exist. Plotting gives

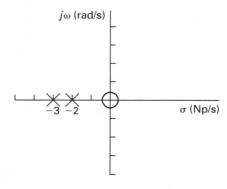

3. The transfer function was determined in Prob. 2 to be

$$T_{\text{net}}(s) = \frac{s}{(s+2)(s+3)}$$

The zero break frequency is $\omega = 1$. The break frequency for the poles is $\omega = 2$ and $\omega = 3$. These points are plotted on the following diagram.

For the zero, a $+20$ dB/decade line through the origin is plotted. For the poles, the low-frequency asymptotes are plotted on the frequency axis extending to lower frequencies from $\omega = 2$ and $\omega = 3$. The high-frequency asymptotes break down at -20 dB/decade from these points. Summing the asymptotic lines gives the net Bode magnitude plot as shown.

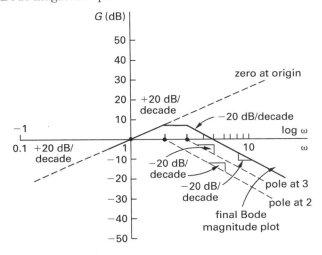

The value of the constant, K, is determined from

$$T_{\text{net}}(j\omega) = K\left(\frac{j\omega}{\left(1 + \dfrac{j\omega}{2}\right)\left(1 + \dfrac{j\omega}{3}\right)}\right)$$

$$K = A\left(\frac{\displaystyle\prod_1^n z_n}{\displaystyle\prod_1^d p_d}\right)$$

The transfer $T_{\text{net}}(s)$ gives the value of A as 1. For a zero at the origin in the s domain, the break frequency is ω equals 1. Therefore, z_n equals 1.

$$K = (1)\left(\frac{1}{(2)(3)}\right) = 1/6$$

This makes $20\log K$ equal to -15.563 dB. Either the Bode plot should be shifted downward by -15.563 or the zero dB point should be changed to -15.563 to obtain the final solution.

4. Because the calculations will take place in the s domain, transform the circuit.

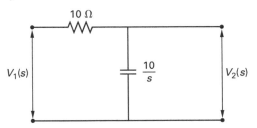

The transfer function, using the voltage divider concept, is as shown.

$$T_{\text{net}}(s) = \frac{V_2(s)}{V_1(s)} = \frac{\dfrac{10}{s}}{\dfrac{10}{s} + 10} = \frac{\dfrac{10}{s}}{\dfrac{10 + s10}{s}}$$

$$= \frac{10}{10 + s10}$$

$$= \frac{1}{s+1}$$

The break frequency, or pole, occurs at $\omega = 1$. At this point the phase is $-45°$. Using the standard method, the resulting plot is as shown.

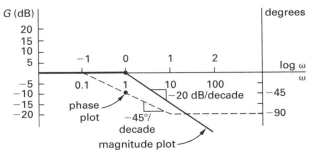

5. The Bode plot of Prob. 4 indicates that low frequencies are passed with a gain of 0 dB. At high frequencies, the output is attenuated and the gain steadily drops. The circuit is called a $\boxed{\text{low-pass filter.}}$

The answer is (A).

6. (a) By voltage division,

$$\left(\frac{V_{\text{out}}}{V_{\text{in}}}\right)s = \frac{0.01s}{0.01s + 50 + \dfrac{10^6}{s}}$$

$$= \boxed{\frac{s^2}{s^2 + 5000s + 10^8}}$$

(b) The denominator needs an underdamped response, so it is put into standard form to take advantage of the Laplace transform pairs table.

$$\left(\frac{V_{\text{out}}}{V_{\text{in}}}\right)s = \frac{s^2}{(s+2500)^2 + (9682)^2}$$

For the frequency response $s \to j\omega$,

$$\left(\frac{V_{\text{out}}}{V_{\text{in}}}\right)j\omega = \frac{-\omega^2}{10^8 - \omega^2 + j5000\omega}$$

At low frequency,

$$\frac{V_{\text{out}}}{V_{\text{in}}} \to \frac{-\omega^2}{10^8}$$

This has a slope of 40 dB per decade and a phase of 180° on a Bode diagram.

At high frequency,

$$\frac{V_{\text{out}}}{V_{\text{in}}} \to 1$$

At $\omega = 10^4$ rad/s,

$$\frac{V_{\text{out}}}{V_{\text{in}}} = 2\angle 90°$$

The frequency response value is $\boxed{6 \text{ dB.}}$

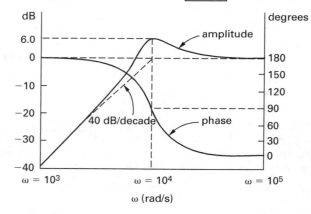

(c) The step response with $V_{\text{in}} = 1/s$ is

$$V_{\text{out}} = \frac{s}{(s+2500)^2 + (9682)^2}$$

$$= \frac{s+2500}{(s+2500)^2 + (9682)^2}$$

$$- \left(\frac{2500}{9682}\right)\left(\frac{9682}{(s+2500)^2 + (9682)^2}\right)$$

$$v_{\text{out}}(t) = e^{-2500t}(\cos 9682t - 0.258\sin 9682t) \text{ V}$$

$$= \boxed{1.03 e^{-2500t}\cos(9682t + 14.5°) \text{ V}}$$

7. (a) By voltage division,

$$\frac{V_{\text{out}}}{V_{\text{in}}} = \frac{\dfrac{10^6}{s}}{10^{-2}s + 50 + \dfrac{10^6}{s}}$$

$$= \boxed{\frac{10^8}{s^2 + 5000s + 10^8}}$$

$$\left(\frac{V_{\text{out}}}{V_{\text{in}}}\right)j\omega = \frac{10^8}{10^8 - \omega^2 + j5000\omega}$$

(b) At low frequency,

$$\frac{V_{\text{out}}}{V_{\text{in}}} \to 1\angle 0°$$

At high frequency,

$$\frac{V_{\text{out}}}{V_{\text{in}}} \to \frac{10^8}{-\omega^2}$$

On the Bode diagram, this has a slope of −40 dB per decade and an angle of −180°. At $\omega = 10^4$ rad/s,

$$\frac{V_{\text{out}}}{V_{\text{in}}} = 2\angle -90°$$

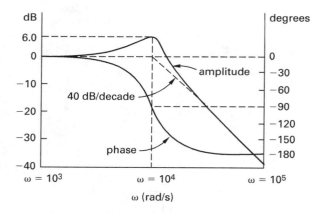

(c) The step response with $V_{\text{in}} = 1/s$ is

$$V_{\text{out}} = \frac{10^8}{s(s^2 + 5000s + 10^8)}$$

This is expanded in partial fractions to

$$V_{\text{out}} = \frac{1}{s}\text{ V} - \frac{s + 5000}{(s+2500)^2 + (9682)^2}\text{ V}$$

This is recast to fit the Laplace transform pairs table.

$$V_{\text{out}} = \frac{1}{s} \text{ V} - \frac{s + 2500}{(s + 2500)^2 + (9682)^2} \text{ V}$$

$$- \frac{\left(\dfrac{2500}{9682}\right)(9682)}{(s + 2500)^2 + (9682)^2} \text{ V}$$

$$v_{\text{out}}(t) = 1 - e^{-2500t}(\cos 9682t + 0.258 \sin 9682t) \text{ V}$$

$$\boxed{= 1 - 1.033 e^{-2500t} \cos(9682t - 14.5°) \text{ V}}$$

8. The filter is a $\boxed{\text{bandpass}}$ filter since it limits the low and high frequencies. The bandwidth is determined by the 3 dB downpoints.

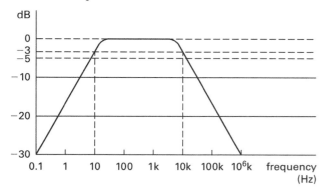

The bandwidth of the filter is from $\boxed{\text{10 Hz to 10 kHz}}$.

The answer is (A).

9. To determine the filter type, consider the effects of the circuit at both low frequency ($\omega = 0$) and high frequency ($\omega = \infty$).

At low frequency, with a DC signal applied at V_{in}, the circuit is as shown.

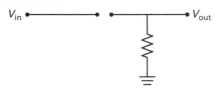

The output voltage is zero, since none of the input signal is passed to the output.

At high frequency, with a high-frequency AC signal applied at V_{out}, the impedance of the capacitor is essentially zero, and the capacitor acts as a short circuit as shown.

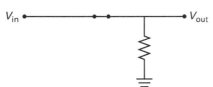

The output voltage is equal to the input voltage, since there is an equal short circuit between the two.

Little or no signal passes at low frequency, and the entire signal passes at high frequency. The filter is a $\boxed{\text{high-pass}}$ filter.

The cutoff frequency is the point at which the resistance of the circuit is equal to the absolute value of the capacitive reactance of the circuit. The resistance of the circuit is known to be 25 Ω, so calculate the cutoff frequency based on an absolute capacitive reactance value of 25 Ω.

$$X_C = \frac{1}{\omega_0 C}$$

$$\omega_0 = \frac{1}{X_C C}$$

$$= \frac{1}{(25 \ \Omega)(5 \times 10^{-9} \ \text{F})}$$

$$\boxed{= 8 \times 10^6 \ \text{Hz} \quad (8 \ \text{MHz})}$$

The answer is (A).

10. The cutoff frequency is the point at which the resistance is equal to the absolute value of the inductive reactance.

$$R = |X_L|$$

Since the resistance is given as 5 Ω, the inductive reactance must be $\boxed{5 \ \Omega.}$

For completeness, consider the determination of the inductor value corresponding to 5 Ω.

The inductance that results in the desired cutoff frequency of 20 kHz is calculated from the inductive reactance.

$$X_L = \omega_0 L$$

$$L = \frac{X_L}{\omega_0}$$

$$= \frac{5 \ \Omega}{20 \times 10^3 \ \text{Hz}}$$

$$= 0.25 \ \text{mH}$$

The answer is (A).

Topic V: Generation

Generation

33 Generation Systems

PRACTICE PROBLEMS

1. What term describes the process of minimizing irreversibilities by raising the temperature of the water returning to a steam generator?

(A)   superheat   ↑ T of steam ⟹ turbine

(B)   reheat   ⟹ turbine

(C)   regeneration

(D)   both (A) and (C)

2. What fission rate (fissions/s) is required for a nuclear plant to have a 1000 MW thermal output?

3. What type of first stage, or control stage, is normally used on large turbines that are used for electric power generation?

(A)   impulse

(B)   Rateau

(C)   Curtis

(D)   Parsons

4. In the load sharing diagram shown, the speed droop for both generators is 0.6 Hz/1000 kW. If generator A is to carry all the load, what no-load frequency setpoint is required on generator A while maintaining the system frequency at 60 Hz?

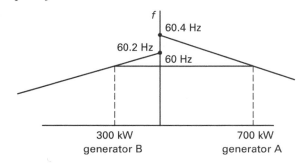

5. Draw the load sharing diagram for Prob. 4 with generator A carrying all the load. Label all significant points.

6. The field of a DC generator is normally located on the

(A)   stator

(B)   rotor

(C)   armature

(D)   round rotor

7. What is the maximum speed for 60 Hz AC generator operation?

8. What power quality term describes a transient of short duration and high magnitude?

(A)   overvoltage

(B)   harmonic

(C)   sag

(D)   surge

SOLUTIONS

1. Superheat raises the temperature of the steam supplied to the turbine. Reheat raises the temperature of the steam at the input of the low-temperature stages of a turbine. Regeneration raises the temperature of the water returning to the steam generator.

The answer is (C).

2. The fission rate, fr, is in units of fissions/s. The average energy released per fission is 200 MeV. Unit analysis gives the power output, P_s.

$$P_s = \left(\frac{\text{fissions}}{\text{s}}\right)\left(\frac{\text{energy}}{\text{fission}}\right)$$

$$= \text{fr}\left(200 \ \frac{\text{MeV}}{\text{fission}}\right)$$

$$\text{fr} = \frac{P_s}{200 \ \frac{\text{MeV}}{\text{fission}}}$$

$$= \frac{1000 \times 10^6 \,\text{W}}{\left(200 \times 10^6 \ \frac{\text{eV}}{\text{fission}}\right)\left(\frac{1.602 \times 10^{-19}\text{J}}{1 \ \text{eV}}\right)}$$

$$= \boxed{3.12 \times 10^{19} \ \text{fission/s}}$$

3. Large turbines use impulse control stages because no pressure drop occurs across such a stage, thereby allowing partial arc admission of the steam without significant forces being exerted on the turbine rotor.

The answer is (A).

4.

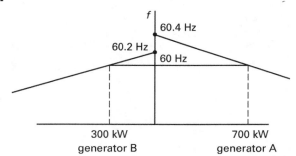

The power carried by an AC generator is

$$P = \frac{f_{\text{n1}} - f_{\text{sys}}}{f_{\text{droop}}}$$

Solving for the no-load frequency gives

$$f_{\text{n1}} = P f_{\text{droop}} + f_{\text{sys}}$$

Solving for the no-load frequency increase necessary to carry an additional 300 kW gives

$$\Delta f_{\text{n1}} = \Delta P f_{\text{droop}}$$

$$= (300 \ \text{kW})\left(\frac{0.6 \ \text{Hz}}{1000 \ \text{kW}}\right)$$

$$= 0.18 \ \text{Hz}$$

The required no-load frequency is

$$f_{\text{n1}} = f_{\text{n1,current}} + \Delta f_{\text{n1}}$$

$$= 60.4 \ \text{Hz} + 0.18 \ \text{Hz}$$

$$= \boxed{60.58 \ \text{Hz} \quad (60.6 \ \text{Hz})}$$

Generator B's frequency will have to be lowered by 0.2 Hz to maintain the system frequency at 60 Hz.

5. The real load sharing diagram with generator A carrying all the load is

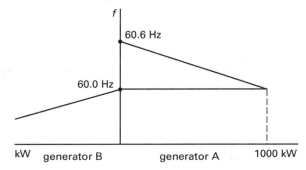

6. The field of a DC machine is normally on the stationary portion, or stator. This allows commutation to be accomplished on the rotor.

The answer is (A).

7. Synchronous speed is derived by

$$n_s = \frac{120 f}{p}$$

The minimum number of poles determines the maximum speed. Two is the minimum number.

$$n_s = \frac{(120)(60 \ \text{Hz})}{2} = \boxed{3600 \ \text{rpm}}$$

8. A transient condition of short duration and high magnitude is called a surge.

The answer is (D).

34 Three-Phase Electricity and Power

PRACTICE PROBLEMS

1. A three-phase 208 V (rms) system supplies heating elements connected in a wye configuration. What is the resistance of each element if the total balanced load is 3 kW?

2. A three-phase system has a balanced delta load of 5000 kW at 84% power factor. If the line voltage is 4160 V (rms), what is the line current?

3. A 120 V (per phase, rms) three-phase system has a balanced load consisting of three 10 Ω resistances. What total power is dissipated if the connection is (a) a wye configuration and (b) a delta configuration?

4. A balanced delta load consists of three 20 Ω∠25° impedances. The 60 Hz line voltage is 208 V (rms). Find the (a) phase current, (b) line current, (c) phase voltage, (d) power consumed by each phase, and (e) total power.

5. A balanced wye load consists of three $3 + j4\Omega$ impedances. The system voltage is 110 V (rms). Find the (a) phase voltage, (b) line current, (c) phase current, and (d) total power.

6. Consider the one-line diagram of a three-phase distribution system. What is the generator base voltage?

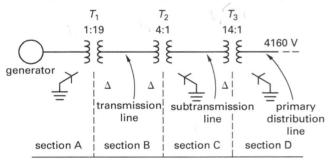

7. A three-phase transformer is rated at 225 kVA with a primary-side voltage of 480 V. The per-unit current in the transformer is 0.4 pu. What is most nearly the actual current?

   (A) 90 A

   (B) 110 A

   (C) 190 A

   (D) 270 A

8. A three-phase, 480 V, 200 hp motor has an efficiency of 96% and a power factor of 0.85. What is most nearly the full-power line current drawn from the source?

   (A) 30 A

   (B) 210 A

   (C) 220 A

   (D) 380 A

9. A transformer on the high-voltage side has a voltage of 480 V, an apparent power of 1500 VA, and an impedance of $j50$ Ω on the primary side. The turns ratio is 4. What is most nearly the per-unit impedance on the secondary side?

   (A) 0.05 pu

   (B) 0.3 pu

   (C) 0.5 pu

   (D) 50 pu

10. The real power drawn by a large three-phase motor is 150 kW. The reactive power drawn is 90 kVAR. What is most nearly the power factor, and is it leading or lagging?

   (A) 0.60, lagging

   (B) 0.60, leading

   (C) 0.86, lagging

   (D) 0.86, leading

SOLUTIONS

1.

$$R = \frac{V_p^2}{P_p} = \frac{\left(\frac{V_l}{\sqrt{3}}\right)^2}{\frac{P_t}{3}} = \frac{V_l^2}{P}$$

$$= \frac{(208 \text{ V})^2}{3 \times 10^3 \text{ W}}$$

$$= \boxed{14.42 \ \Omega}$$

2.

$$I_l = \left(\frac{1}{\sqrt{3}}\right)\left(\frac{P_\Delta}{V_l \cos\phi}\right)$$

$$= \left(\frac{1}{\sqrt{3}}\right)\left(\frac{5000 \times 10^3 \text{ W}}{(4160 \text{ V})(0.84)}\right)$$

$$= \boxed{826 \text{ A}}$$

3. (a)

$$P_p = \frac{V_p^2}{R} = \frac{(120 \text{ V})^2}{10 \ \Omega} = 1440 \text{ W}$$

$$P_t = 3P_p = (3)(1440 \text{ W}) = \boxed{4320 \text{ W}}$$

(b)

$$P_t = 3\left(\frac{V_l^2}{R}\right) = (3)\left(\frac{(120 \text{ V})^2}{10 \ \Omega}\right)$$

$$= \boxed{4320 \text{ W}}$$

4. (a)

$$\mathbf{I}_p = \frac{\mathbf{V}_l}{\mathbf{Z}} = \frac{208 \text{ V} \angle 0^\circ}{20 \ \Omega \angle 25^\circ}$$

$$= \boxed{10.4 \text{ A} \angle -25^\circ}$$

(b)

$$\mathbf{I}_l = \sqrt{3} \ \angle -30^\circ > \mathbf{I}_p$$

$$= (\sqrt{3} \ \angle -30^\circ)(10.4 \text{ A} \angle -25^\circ)$$

$$= \boxed{18 \text{ A} \angle -55^\circ}$$

(c)

$$\mathbf{V}_p = \mathbf{V}_l$$

$$= \boxed{208 \text{ V} \angle 0^\circ}$$

(d)

$$P_p = V_p I_p \cos\phi$$

$$= (208 \text{ V})(10.4 \text{ A})\cos 25^\circ$$

$$= \boxed{1960.5 \text{ W}}$$

(e)

$$P_\Delta = 3P_p = (3)(1960.5 \text{ W})$$

$$= \boxed{5882 \text{ W}}$$

5. (a)

$$V_p = \frac{110 \text{ V}}{\sqrt{3}} = \boxed{63.51 \text{ V}}$$

The impedance is given as $3 + j4\Omega$ in rectangular coordinates. The magnitude of the impedance, Z, in complex algebra, is

(b)

$$Z = \sqrt{(3 \ \Omega)^2 + (4 \ \Omega)^2} = 5 \ \Omega$$

$$\phi = \arctan\frac{4 \ \Omega}{3 \ \Omega} = 53.13^\circ$$

$$\mathbf{I}_l = \frac{\mathbf{V}_p}{\mathbf{Z}} = \frac{63.51 \text{ V} \angle 0^\circ}{5 \ \Omega \angle 53.13^\circ}$$

$$= \boxed{12.70 \text{ A} \angle -53.13^\circ}$$

(c)

$$\mathbf{I}_p = \mathbf{I}_l = \boxed{12.70 \text{ A} \angle -53.13^\circ}$$

(d)

$$P_t = \sqrt{3} \ VI \cos\phi$$

$$= \sqrt{3} \,(110 \text{ V})(12.7 \text{ A})\cos 53.13^\circ$$

$$= \boxed{1451 \text{ W}}$$

Generation

6. For a three-phase system, the line voltage is the base. The base in the primary distribution line is given as 4160 V, or $V = 1.0$ pu $\angle 0°$.

Because the transformers scale the base values in each section, the base voltage value of 4160 V will be determined at the generator. The $\sqrt{3}$ is used to account for the wye connection line-to-phase and phase-to-line conversions.

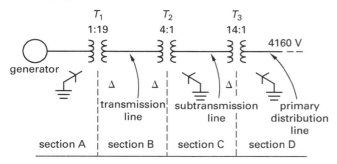

$$V_{\text{gen,base}} = V_{\text{gen,line}}$$
$$= \left(\frac{V_{\text{line,D}}}{\sqrt{3}}\right)\left(\frac{N_{3,\text{pri}}}{N_{3,\text{sec}}}\right)\left(\frac{1}{\sqrt{3}}\right)$$
$$\times \left(\frac{N_{2,\text{pri}}}{N_{2,\text{sec}}}\right)\left(\frac{N_{1,\text{pri}}}{N_{1,\text{sec}}}\right)\sqrt{3}$$
$$= \left(\frac{4160 \text{ V}}{\sqrt{3}}\right)\left(\frac{14}{1}\right)\left(\frac{1}{\sqrt{3}}\right)\left(\frac{4}{1}\right)\left(\frac{1}{19}\right)\sqrt{3}$$
$$= \boxed{7078.9 \text{ V} \quad (7070 \text{ V})}$$

This generator line voltage of 7070 V is the 1.0 pu $\angle 0°$ value in section A of the distribution system.

7. From Eq. 34.43, the base current is

$$I_{\text{base}} = \left(\frac{S}{\sqrt{3} \, V}\right)_{3\phi}$$
$$= \frac{225 \times 10^3 \text{ VA}}{\sqrt{3}\,(480 \text{ V})}$$
$$= 270.63 \text{ A}$$

From Eq. 34.47, the actual current is

$$I_{\text{pu}} = \frac{I_{\text{actual}}}{I_{\text{base}}}$$
$$I_{\text{actual}} = I_{\text{pu}}I_{\text{base}}$$
$$= (0.4 \text{ pu})(270.63 \text{ A})$$
$$= \boxed{108.25 \text{ A} \quad (110 \text{ A})}$$

The answer is (B).

8. For line quantities, it does not matter whether the motor is delta- or wye-wound. From the equation for total power, the line current is

$$P_t = \eta\sqrt{3} \, I_L V_L(\text{pf})$$
$$I_L = \frac{P_t}{\eta\sqrt{3} \, V_L \,(\text{pf})}$$
$$= \frac{(200 \text{ hp})\left(745.7 \, \dfrac{\text{W}}{\text{hp}}\right)}{(0.96)\sqrt{3}\,(480 \text{ V})(0.85)}$$
$$= \boxed{219.84 \text{ A} \quad (220 \text{ A})}$$

The answer is (C).

9. The per-unit impedance is identical on both sides of a transformer. Therefore, from Eq. 34.44 and using the high-voltage side parameters as the base, the base impedance is

$$Z_{\text{base}} = \frac{V_{\text{base}}^2}{S_{\text{base}}}$$
$$= \frac{(480 \text{ V})^2}{1500 \text{ VA}}$$
$$= 153.6 \text{ } \Omega$$

From Eq. 34.49, the per-unit impedance is

$$Z_{\text{pu}} = \frac{Z_{\text{out}}}{Z_{\text{base}}}$$
$$= \frac{j50 \, \Omega}{153.6 \, \Omega}$$
$$= \boxed{j0.33 \text{ pu} \quad (0.3 \text{ pu})}$$

The answer is (B).

10. From Eq. 40.5, the power factor is

$$\text{pf} = \frac{P}{S}$$

From Eq. 34.60,

$$S^2 = P^2 + Q^2$$
$$S = \sqrt{P^2 + Q^2}$$

Combining Eq. 40.5 and Eq. 34.60, the power factor is

$$
\begin{aligned}
\mathrm{pf} &= \frac{P}{S} = \frac{P}{\sqrt{P^2 + Q^2}} \\
&= \frac{150 \times 10^3\,\mathrm{W}}{\sqrt{\left(150 \times 10^3\,\mathrm{W}\right)^2 + \left(90 \times 10^3\,\mathrm{VAR}\right)^2}} \\
&= \boxed{0.857 \quad (0.86)}
\end{aligned}
$$

Since the reactive power is positive, the power factor is lagging (i.e., inductive) as expected for a motor with wound coils of wire.

The answer is (C).

35 Batteries, Fuel Cells, and Power Supplies

PRACTICE PROBLEMS

1. A cell that undergoes a chemical reaction in such a manner that it cannot be recharged is called a

- (A) dry cell
- (B) primary cell
- (C) reserve cell
- (D) secondary cell

2. Within a battery, conventional current flows from

- (A) anion to anode
- (B) anion to cation
- (C) anode to cathode
- (D) positive electrode to negative electrode

3. A series of fuel cells is connected to produce a no-load voltage of 320 V. At a rated voltage of 300 V the cells provide 2.0 MW to a DC-AC converter. Assuming the cell can be modeled as a standard battery, what is the total internal resistance of the fuel cells?

- (A) 0.0001 Ω
- (B) 0.003 Ω
- (C) 0.05 Ω
- (D) 3 Ω

4. A wristwatch mercury cell has a capacity of 200×10^{-3} A·h. How many months can the watch function without replacement if it draws 15×10^{-6} A?

- (A) 3 mo
- (B) 6 mo
- (C) 12 mo
- (D) 18 mo

5. Increasing the current withdrawn from a battery has what effect on the battery capacity and why?

- (A) increased capacity, which is caused by less internal heat loss
- (B) increased capacity, which is caused by lower internal resistance
- (C) decreased capacity, which is caused by higher load voltage drop
- (D) decreased capacity, which is caused by increased polarization

6. The specific conductivity of a sodium hydroxide electrolyte solution is 2.21 S/m. The electrode is 0.15 m^2 in size and the distance between the electrodes is 1.3 cm. Most nearly, what is the resistance of the solution?

- (A) 40 mΩ
- (B) 2000 mΩ
- (C) 4000 mΩ
- (D) 40 000 mΩ

7. The anode reaction in a lead-acid storage battery is shown.

$$Pb + HSO_4^- + H_2O \rightarrow PbSO_4 + H_3O^+ + ne^-$$

What is the number of electrons, n, released during this reaction?

- (A) 1
- (B) 2
- (C) 3
- (D) 4

8. The standard cell emf, E^0, is given by

$$E^0 = e_1^0 - e_2^0$$

e_1^0 and e_2^0 are the standard electrode potentials for reversible half-cell reactions.

When a zinc electrode and a copper electrode are placed in solution, the reaction shown takes place.

$$Zn + Cu^{++} \rightarrow Zn^{++} + Cu$$

The standard electrode potential is -0.76 V for zinc (Zn^{++}/Zn) and $+0.34$ V for copper (Cu^{++}/Cu). What is most nearly the standard cell emf for these two electrodes?

(A) -1.1 V

(B) 0.42 V

(C) 1.1 V

(D) 2.2 V

9. What is most nearly the standard electrode potential for hydrogen?

(A) 0 V

(B) 0.62 V

(C) 1.0 V

(D) 1.2 V

10. Battery charge for an aqueous electrolyte is measured using the specific gravity of the electrolyte. When a battery discharges, the specific gravity

(A)   decreases

(B)   increases

(C)   does not change

(D)   cannot be determined

SOLUTIONS

1. A cell in which the electrodes are consumed and in general cannot be recharged is called a primary cell.

The answer is (B).

2. Within a battery, the conventional current flows from anode to cathode. (The anode is labeled as the positive terminal. Externally, the anode is the terminal into which positive current flows.)

The answer is (C).

3. The circuit is

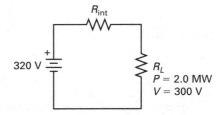

The voltage across the internal resistance is given (20 V). To find the resistance, the current must be determined. From the rated conditions, the load resistance is

$$P = \frac{V^2}{R_L}$$

$$R_L = \frac{V^2}{P} = \frac{(300 \text{ V})^2}{2.0 \times 10^6 \text{ W}} = 0.045 \ \Omega$$

The rated current is

$$P = I^2 R_L$$

$$I = \sqrt{\frac{P}{R_L}} = \sqrt{\frac{2.0 \times 10^6 \text{ W}}{0.045 \ \Omega}} = 6.67 \times 10^3 \text{ A}$$

Using Ohm's law on the internal resistor gives

$$V = I R_{int}$$

$$R_{int} = \frac{V}{I} = \frac{320 \text{ V} - 300 \text{ V}}{6.67 \times 10^3 \text{ A}}$$

$$= \boxed{0.003 \ \Omega}$$

The answer is (B).

4. Amp-hour capacity is

$$Ah = It$$

Therefore,

$$t = \frac{Ah}{I} = \frac{200 \times 10^{-3} \text{ A·h}}{15 \times 10^{-6} \text{ A}} = 1.33 \times 10^4 \text{ h}$$

Assuming 30-day months, the capacity is

$$\text{total months} = \frac{1.33 \times 10^4 \text{ h}}{\left(30 \dfrac{\text{d}}{\text{mo}}\right)\left(24 \dfrac{\text{h}}{\text{d}}\right)}$$
$$= \boxed{18.5 \text{ mo}}$$

The answer is (D).

5. Battery capacity drops as the discharge current increases due to increased losses on the internal resistance and polarization of electrodes; that is, hydrogen bubble insulation of the electrodes.

The answer is (D).

6. The equation for the conductivity of a solution is similar to that for the conductivity of a metal. Use the equation for conductivity to find the resistance of the solution.

$$\sigma = \frac{D_{\text{plates}}}{A_{\text{electrode}} R_{\text{solution}}}$$

$$R_{\text{solution}} = \frac{D_{\text{plates}}}{\sigma A_{\text{electrode}}}$$

$$= \frac{1.3 \times 10^{-2} \text{ m}}{\left(2.21 \dfrac{\text{S}}{\text{m}}\right)(0.15 \text{ m}^2)}$$

$$= \boxed{0.039 \ \Omega \quad (40 \text{ m}\Omega)}$$

The answer is (A).

7. Charge must be conserved. The left side of the reaction shows a total charge of -1, so the right side must also show a final charge of -1. The hydronium ion has a charge of $+1$. (If water is not shown in the reaction, the hydronium ion is shown as H+.) Therefore, $\boxed{\text{two}}$ electrons must be released to make the charge on the right side of the reaction equal -1 (i.e., $+1 - 2 = -1$).

The answer is (B).

8. The standard electrode potentials indicate the direction of the reaction. For example, when Zn^{++} changes to Zn, the potential is -0.76 V. When Zn changes to Zn^{++}, the potential is $+0.76$ V. With this information, the signs can be correctly assigned.

$$E^0 = e_1^0 - e_2^0$$
$$= 0.34 \text{ V} - (-0.76 \text{ V})$$
$$\boxed{= 1.1 \text{ V}}$$

The answer is (C).

9. The standard electrode potential for hydrogen is set at 0 V. The standard reaction is based on equilibrium between hydrogen gas at $\boxed{1 \text{ atm}}$ of pressure.

$$2H^+ + 2e^- \rightarrow H_2$$

The answer is (A).

10. When a battery discharges, the electrolyte components involved in the reaction move out of solution and onto the plates of the battery. They change from ions in solution to solid material on the plates. This results in an increased concentration of water in the solution, which is the reference density (pure water) to which the electrolyte is compared.

$$SG = \frac{\rho_{\text{electrolyte}}}{\rho_{H_2O}}$$

As the battery discharges, the electrolyte density approaches the density of pure water (that is, moves closer to a value of 1). While the precise value is affected by the temperature, the general effect of the discharge on the specific gravity will be the same at any temperature at which the electrolyte solution remains liquid.

The answer is (A).

Generation

Topic VI: Distribution

Distribution

36 Power Distribution

PRACTICE PROBLEMS

1. Which of the following voltages is NOT commonly used in the subtransmission portion of the power distribution system?

(A) 25 kV

(B) 69 kV

(C) 115 kV

(D) 138 kV

2. A single-conductor cable with an outside diameter of 0.5 cm is to be used in an underground installation. What is the optimal conductor radius?

(A) 0.05 cm

(B) 0.1 cm

(C) 0.2 cm

(D) 0.3 cm

3. A single-conductor capacitance-graded cable with an outside diameter of 10 cm can withstand a maximum electric field of 700 kV/m before insulation breakdown occurs. What is the maximum operating voltage of the conductor?

4. What is the total capacitance of 100 m of single-conductor cable with the optimal radii ratio and poly-ethylene insulation ($\epsilon_r = 2.25$)?

5. A three-phase 11.5 kV generator drives a 500 kW, 0.866 lagging power factor load. Determine the (a) line current and (b) necessary generator rating.

6. A transmission line with 8% reactance on a 200 MW base connects two substations. At the first substation the voltage is 1.03 pu∠5°, and at the second the voltage is 0.98 pu∠−2.5°. Find the (a) power and (b) volt-amperes reactive flowing out of the _first_ substation.

7. In the diagram shown, relay 51 is an inverse time overcurrent relay per _IEEE Standard Electrical Power System Device Function Numbers, Acronyms, and Contact Designations_ (IEEE Std C37.2).

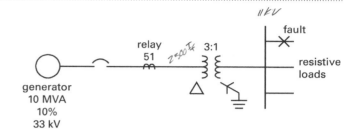

The relay has the characteristics shown on the graph with a minimum pickup current of 400 A.

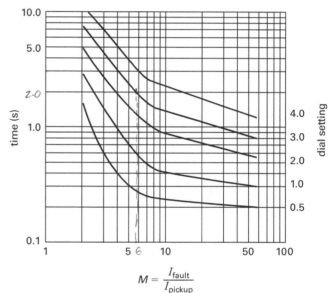

$$M = \frac{I_{fault}}{I_{pickup}}$$

Analysis of a three-phase fault short circuit at the fault location on the 11 kV bus shown in the diagram results in a current on the 33 kV bus of 2300 A. The relay must operate at 2.0 s. What is the required protective dial setting?

(A) 2

(B) 3

(C) 4

(D) 5

8. A partial power plant installation is placed in service before the remaining portion is complete. A one-line diagram for the installation is shown.

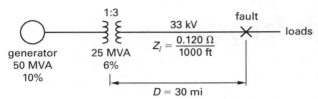

A fault occurs at the location indicated. What is most nearly the short-circuit current?

(A) 200 A

(B) 820 A

(C) 880 A

(D) 1030 A

9. The 60 Hz generator shown provides a line current of 120 A to a three-phase, four-wire, wye-connected, nonlinear load. The load causes 5% current harmonic distortion at the third harmonic.

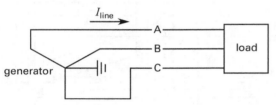

What is most nearly the rms value of the neutral current?

(A) 0 A

(B) 6 A

(C) 18 A

(D) $\dfrac{120}{\sqrt{3}}$ A

10. A 138 kV line-to-line, three-phase, 60 Hz system provides 150 MVA to a balanced delta load with a power factor of 0.8 lagging. Most nearly, what capacitance value, connected in parallel, is necessary to correct the power factor to unity?

(A) 4×10^{-6} F

(B) 6×10^{-6} F

(C) 30×10^{-6} F

(D) 600×10^{-6} F

11. Consider a typical home electrical distribution system consisting of a single-phase 208 Y/120 V setup. The total current measured on the 120 V system is 25 A $\angle -36°$. What is most nearly the capacitive reactance required to correct the power factor to unity?

(A) $-25\ \Omega$

(B) $-20\ \Omega$

(C) $-15\ \Omega$

(D) $-8\ \Omega$

12. The manufacturer's data for two power distribution generators are shown.

generator 1	generator 2
15 MVA	20 MVA
12.5 kV	12.5 kV
pf = 0.8 lagging	pf = 0.8 lagging
$Z_{pu} = 10\%$	$Z_{pu} = 12\%$

Using generator 2 as the base, what is most nearly the per-unit impedance of generator 1?

(A) 8%

(B) 11%

(C) 12%

(D) 13%

13. A distribution system power plant has fuel on hand that contains an estimated 2 quads. The average demand on the power system is 8 GW/d. The plant efficiency is 15%. Most nearly, how many days will the fuel last?

(A) 100 d

(B) 250 d

(C) 350 d

(D) 450 d

14. The percent values for a given section of a distribution system are listed as 80%, 20%, 85%, and 125% for voltage, current, impedance, and apparent power, respectively. The base voltage is 12.5 kV. The base current is 50 A. The base impedance is 30 Ω. The base apparent power is 5 MVA. What is most nearly the actual current?

(A) 10 A

(B) 50 A

(C) 200 A

(D) 400 A

15. Which device is NOT designed to interrupt a fault current?

(A) circuit breaker

(B) current-limiting fuse

(C) electronic solid insulation recloser

(D) automatic single-phase sectionalizer

SOLUTIONS

1. The trend in transmission voltages is toward higher voltages. The most common are 115 kV, 69 kV, 138 kV, and 238 kV. 25 kV is not common.

The answer is (A).

2. The optimal ratio is

$$\frac{r_2}{r_1} = 2.718$$

The outside diameter is given as 0.5 cm. The radius, r_2, is 0.25 cm. Substituting gives

$$r_1 = \frac{r_2}{2.718} = \frac{0.25 \text{ cm}}{2.718} = \boxed{0.0920 \text{ cm} \quad (0.1 \text{ cm})}$$

The answer is (B).

3. The operating voltage for a capacitance-graded cable is

$$V = E_{\max}\left(r_1 \ln \frac{r_2}{r_1} + r_2 \ln \frac{r_3}{r_2}\right)$$
$$= E_{\max}(r_1 \ln \xi + r_2 \ln \xi)$$

The values of r_1 and r_2 are the only unknowns. They are determined from the ratio of radii, ξ.

$$\xi = \frac{r_3}{r_2}$$

$$r_2 = \frac{r_3}{\xi} = \frac{\dfrac{10 \text{ cm}}{2}}{2.718} = 1.84 \text{ cm}$$

$$\xi = \frac{r_2}{r_1}$$

$$r_1 = \frac{r_2}{\xi} = \frac{1.84 \text{ cm}}{2.718} = 0.677 \text{ cm}$$

Substituting known and calculated values gives

$$V = E_{\max}(r_1 \ln \xi + r_2 \ln \xi)$$
$$= \left(700 \; \frac{\text{kV}}{\text{m}}\right)\left(\begin{array}{c}(0.677 \text{ cm})\ln 2.718 \\ +(1.84 \text{ cm})\ln 2.718\end{array}\right)\left(\frac{1 \text{ m}}{100 \text{ cm}}\right)$$
$$= \boxed{17.6 \text{ kV}}$$

Distribution

4. The capacitance per unit length is

$$C_l = \frac{Q}{V} = \frac{2\pi\epsilon}{\ln\xi} = \frac{2\pi\epsilon_r\epsilon_0}{\ln e}$$

$$= \frac{2\pi(2.25)\left(8.854 \times 10^{-12}\ \dfrac{\text{F}}{\text{m}}\right)}{1}$$

$$= 125 \times 10^{-12}\ \text{F/m}$$

The total capacitance is

$$C = C_l L$$

$$= \left(125 \times 10^{-12}\ \frac{\text{F}}{\text{m}}\right)(100\ \text{m})$$

$$= \boxed{125 \times 10^{-10}\ \text{F}}$$

5. (a)

$$P_{\text{phase}} = I_p V_p(\text{pf})$$

$$I_p = \frac{P_{\text{phase}}}{V_p(\text{pf})}$$

$$I_l = \left(\frac{P_p}{V_l(\text{pf})}\right)\sqrt{3}$$

$$= \left(\frac{\dfrac{P_{\text{total}}}{3\ \text{phases}}}{V_l(\text{pf})}\right)\sqrt{3}$$

$$= \frac{\dfrac{500\ \text{kW}}{3\ \text{phases}}}{\left(\dfrac{11.5\ \text{kV}}{\sqrt{3}}\right)(0.866)}$$

$$= \boxed{29\ \text{A}}$$

(b)

$$P = S(\text{pf})$$

$$S = \frac{P}{\text{pf}}$$

$$= \frac{500\ \text{kW}}{0.866}$$

$$= \boxed{577\ \text{kVA}}$$

6.

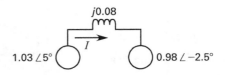

(a)

$$I_{\text{pu}} = \frac{V_{\text{pu}}}{Z_{\text{pu}}} = \frac{1.03\ \text{V}\angle 5° - 0.98\ \text{V}\angle -2.5°}{j0.08}$$

$$= 1.758\ \text{pu}\angle -19.5°$$

$$S_{\text{pu}} = V_{\text{pu}}I_{\text{pu}}^*$$

$$= (1.03\ \text{pu}\angle 5°)(1.758\ \text{pu}\angle 19.5°)$$

$$= 1.811\ \text{pu}\angle 24.5°$$

$$= 1.648 + j0.751\ \text{pu}$$

$$P = P_{\text{pu}}P_{\text{base}} = (1.648)(200\ \text{MW})$$

$$= \boxed{330\ \text{MW}}$$

(b)

$$Q = Q_{\text{pu}}Q_{\text{base}} = (0.751)(200\ \text{MVAR})$$

$$= \boxed{150\ \text{MVAR}}$$

7. Find the multiple of the pickup current, M.

$$M = \frac{I_f}{I_p} = \frac{2300\ \text{A}}{400\ \text{A}} = 5.75$$

Use the pickup current value to find the correct dial setting on the graph for 2.0 s.

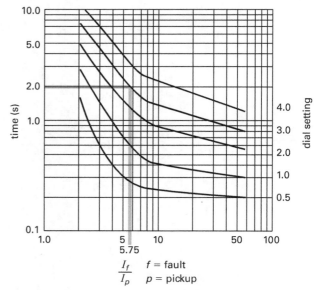

The required dial setting is $\boxed{3.}$

The answer is (B).

8. The power to the fault is limited by the size of the transformer, not the generator. Calculate the short-circuit current.

$$S_{sc} = \sqrt{3}\, I_{sc} V_{base}$$

$$I_{sc} = \frac{S_{sc}}{\sqrt{3}\, V_{base}} = \frac{\dfrac{S_{base}}{Z_{pu,total}}}{\sqrt{3}\, V_{base}} \quad [I]$$

Let $S_{base} = 100$ MVA, which is arbitrary. The base voltage is $V_{base} = 33$ kV because this is the voltage at the location of the fault. The only unknown in Eq. I is the per-unit impedance, $Z_{pu,total}$.

$$Z_{pu,total} = Z_{pu,trans} + Z_{pu,line}$$

$$= Z_{pu,trans}\left(\frac{S_{base,new}}{S_{base,old}}\right) + Z_{line,actual}\left(\frac{S_{base}}{V_{base}^2}\right)$$

$$= Z_{pu,trans}\left(\frac{S_{base,new}}{S_{base,old}}\right) + Z_l D\left(\frac{S_{base}}{V_{base}^2}\right)$$

$$= (j0.06 \text{ pu})\left(\frac{100 \text{ MVA}}{25 \text{ MVA}}\right) + \left(\frac{0.120 \text{ }\Omega}{1000 \text{ ft}}\right)$$

$$\times \left(5280 \text{ }\frac{ft}{mi}\right)(30 \text{ mi})\left(\frac{100 \text{ MVA}}{(33 \text{ kV})^2}\right)$$

$$= j0.24 \text{ pu} + 1.7 \text{ pu}$$

$$= 1.71 \text{ pu } \angle\, 8.04°$$

$$|Z_{pu,total}| = 1.71 \text{ pu}$$

Substitute the per-unit impedance into Eq. I.

$$|I_{sc}| = \frac{\dfrac{S_{base}}{|Z_{pu,total}|}}{\sqrt{3}\, V_{base}} = \frac{\dfrac{100 \text{ MVA}}{1.71 \text{ pu}}}{\sqrt{3}\,(33 \text{ kV})}$$

$$= \frac{\dfrac{100 \times 10^6 \text{ VA}}{1.71 \text{ pu}}}{\sqrt{3}\,(33 \times 10^3 \text{ V})}$$

$$= \boxed{1023 \text{ A} \quad (1030 \text{ A})}$$

The answer is (D).

9. Calculate the third-harmonic current.

$$I_{3rd} = I_{total}(HD_{3rd})$$

$$= (120 \text{ A})(0.05)$$

$$= 6 \text{ A}$$

The harmonic current flows in all three phases, has the same magnitude and phase in each line, and combines at the neutral, so three times the line harmonic current flows in the neutral.

$$I_{neutral} = 3I_{line,3rd} = 3I_{phase,3rd}$$

$$= (3)(6 \text{ A})$$

$$= \boxed{18 \text{ A}}$$

The answer is (C).

10. From the data given in the problem statement,

$$V_{11} = V_{phase} = 138 \times 10^3 \text{ V}$$

$$S_{total} = 150 \times 10^6 \text{ VA}$$

$$pf = 0.8 \text{ lagging}$$

$$\phi = \arccos 0.8 = 36.86°$$

Power factor correction occurs on a phase basis. Therefore, all calculations should be performed on a per-phase basis. Calculate the reactive power that must be compensated.

$$S_{phase} = \frac{S_{total}}{3}$$

$$= \frac{150 \times 10^6 \text{ VA}}{3}$$

$$= 50 \times 10^6 \text{ VA}$$

$$Q_{phase} = S_{phase}\sin\phi$$

$$= (50 \times 10^6 \text{ VA})(\sin 36.86°)$$

$$= 30 \times 10^6 \text{ VAR}$$

Series compensation would use $I^2 \times X$. However, for parallel compensation, the following equation is used.

$$Q_{phase} = \frac{V^2}{X_L}$$

$$X_L = \frac{V^2}{Q_{phase}}$$

$$= \frac{(138 \times 10^3 \text{ V})^2}{30 \times 10^6 \text{ VAR}}$$

$$= 634.8 \text{ }\Omega$$

Distribution

To compensate for the inductive reactance of 634.8 Ω, add the capacitive reactance of -634.8 Ω. Calculate the capacitance.

$$X_C = -\frac{1}{\omega C}$$

$$C = -\frac{1}{\omega X} = -\frac{1}{2\pi f X}$$

$$= -\frac{1}{2\pi (60 \text{ Hz})(-634.8 \text{ } \Omega)}$$

$$= \boxed{4.2 \times 10^{-6} \text{ F} \quad (4 \times 10^{-6} \text{ F})}$$

The answer is (A).

11. Power factor correction (PFC) is performed on a phase basis. This is a single-phase system with a total current of 25 A $\angle -36°$. Only the quadrature of the current (i.e., the current out of phase by 90°) needs to be compensated, since this represents the reactive portion of the current. Determine the quadrature of the current.

$$I_{\text{phase}} = 25 \text{ A} \angle -36°$$

$$= 25 \text{ A} \left(\cos(-36°) + j\sin(-36°)\right)$$

$$= 20.2 - j14.7 \text{ A}$$

The quadrature of the current is $-j14.7$ A. To compensate, add $j14.7$ A. From Ohm's Law, calculate the capacitive reactance associated with the current.

$$\mathbf{V} = \mathbf{IZ} = \mathbf{I}jX_c$$

Solve for the capacitive reactance.

$$jX_C = \frac{\mathbf{V}}{\mathbf{I}}$$

$$= \frac{120 \text{ V} \angle 0°}{14.7 \text{ A} \angle 90°}$$

$$= 8.16 \text{ } \Omega \angle -90°$$

Interpret the angle in terms of j to solve directly for X_c.

$$jX_C = 8.16 \text{ } \Omega \angle -90°$$

$$= -j8.16 \text{ } \Omega$$

$$X_C = \frac{-j8.16 \text{ } \Omega}{j}$$

$$= \boxed{-8.16 \text{ } \Omega \quad (-8 \text{ } \Omega)}$$

The answer is (D).

12. Calculate the per-unit impedance for generator 1 on a base of 20 MVA and 12.5 kV.

$$Z_{\text{pu,new}} = Z_{\text{pu,old}} \left(\frac{V_{\text{base,old}}}{V_{\text{base,new}}}\right)^2 \left(\frac{S_{\text{base,new}}}{S_{\text{base,old}}}\right)$$

$$= (0.10) \left(\frac{12.5 \text{ kV}}{12.5 \text{ kV}}\right)^2 \left(\frac{20 \text{ MVA}}{15 \text{ MVA}}\right)$$

$$= \boxed{0.13 \text{ pu} \quad (13\%)}$$

The answer is (D).

13. The quad is equal to a quadrillion (10^{15}) Btu. Convert the quads to Btus and then to joules.

$$E_{\text{total}} = (2 \text{ quads})\left(10^{15} \frac{\text{Btu}}{\text{quad}}\right)$$

$$= (2 \times 10^{15} \text{ Btu})\left(1.055 \times 10^3 \frac{\text{J}}{\text{Btu}}\right)$$

$$= 2.11 \times 10^{18} \text{ J}$$

Convert the 8 GW to joules.

$$E_{\text{use,daily}} = P_{\text{use,daily}} t$$

$$= (8 \times 10^9 \text{ W})\left(1 \frac{\frac{\text{J}}{\text{s}}}{\text{W}}\right)\left(86\,400 \frac{\text{s}}{\text{d}}\right)$$

$$= 6.91 \times 10^{14} \frac{\text{J}}{\text{d}}$$

Calculate the number of days the fuel will last.

$$t = \eta \frac{E_{\text{total}}}{E_{\text{use,daily}}}$$

$$= (0.15)\left(\frac{2.11 \times 10^{18} \text{ J}}{6.91 \times 10^{14} \frac{\text{J}}{\text{d}}}\right)$$

$$= \boxed{458.03 \text{ d} \quad (450 \text{ d})}$$

The answer is (D).

14. Calculate the actual current from the per-unit current.

$$I_{pu} = \frac{I_{actual}}{I_{base}}$$

$$\begin{aligned} I_{actual} &= I_{pu}I_{base} \\ &= (0.20)(50 \text{ A}) \\ &= \boxed{10 \text{ A}} \end{aligned}$$

The answer is (A).

15. A circuit breaker, a current-limiting fuse, and a recloser are all designed to interrupt a fault current. An automatic single-phase sectionalizer counts the number of times a recloser has operated and, if the fault persists, keeps the recloser open. If the fault clears, the sectionalizer allows the recloser to close.

The answer is (D).

37 Power Transformers

PRACTICE PROBLEMS

1. Which electrical element in the exact (real) transformer model accounts for the eddy current and hysteresis losses?

(A) B_c

(B) R_c

(C) R_p

(D) R_s

2. Which electrical element in the exact (real) transformer model accounts for magnetic flux leakage?

(A) B_c

(B) X_p alone

(C) X_s alone

(D) X_p and X_s

3. A 500 kVA rated, 200 kg iron-core transformer has a coupling coefficient of 4×10^{-4} in a 1.4 T peak magnetic field. What is the eddy current power loss?

(A) 3.0 W

(B) 9.0 W

(C) 400 W

(D) 570 W

4. A transformer is rated for 34.5 kV. The measured secondary voltage at no load is 35.02 kV. What is the percent voltage regulation?

(A) 0.50%

(B) 1.4%

(C) 1.5%

(D) 2.0%

5. An open-circuit test is conducted on a 120 V transformer winding. The results are $V_{1oc} = 120$ V, $V_{2oc} = 240$ V, $I_{1oc} = 0.25$ A, and $P_{oc} = 20$ W. What apparent power is necessary to supply core losses in this transformer?

(A) 20 VA

(B) 30 VA

(C) 60 VA

(D) 120 VA

6. A 13.8 kV single-phase transformer is subjected to an open-circuit test with the following results: $P_{in} = 900$ W, $I_{in} = 0.2$ A, and $V_{out} = 460$ V (secondary). Determine the values of the equivalent circuit parameters that can be found from this test.

7. A 32 kVA single-phase transformer is rated at 15 kV on the primary. A 60 Hz short-circuit test on the primary indicates $P_{in} = 1$ kW, $I_{in} = 20$ A, $I_s = 100$ A, and $V_p = 80$ V. Determine the (a) winding resistances and (b) leakage inductances.

8. A single-phase transformer is rated at 115 kV at 500 kVA. A short-circuit test on the high-voltage side at rated current indicates $P_{sc} = 435$ W and $V_{sc} = 2.5$ kV. Determine the per-unit impedance of the transformer.

SOLUTIONS

1. Core losses include those caused by eddy currents and hysteresis. In the exact transformer model, these are accounted for by $G_c = 1/R_c$.

The answer is (B).

2. Magnetic flux leakage is a term used to describe any flux that is not mutual. The primary and secondary self-inductances represented by X_p and X_s are not mutual.

The answer is (D).

3. The eddy current loss is found from

$$P_e = k_e B_m^2 f^2 m$$

Assume 60 Hz is the frequency. Substitute the given values.

$$P_e = (4 \times 10^{-4})(1.4 \text{ T})^2(60 \text{ Hz})^2(200 \text{ kg})$$
$$= \boxed{565 \text{ W} \quad (570 \text{ W})}$$

The answer is (D).

4. The voltage regulation is

$$\text{VR} = \frac{V_{\text{nl}} - V_{\text{fl}}}{V_{\text{fl}}}$$

$$= \frac{\dfrac{V_p}{a} - V_{s,\text{rated}}}{V_{s,\text{rated}}}$$

Because the voltage was measured at the output of the secondary, it is equivalent to V_p/a. Therefore, the voltage regulation is

$$\text{VR} = \frac{35.02 \text{ kV} - 34.5 \text{ kV}}{34.5 \text{ kV}} \times 100\%$$
$$= \boxed{1.5\%}$$

The answer is (C).

5. The open-circuit test determines the parameters associated with core losses. The apparent power is

$$S_{\text{oc}} = I_{1\text{oc}} V_{1\text{oc}}$$
$$= (0.25 \text{ A})(120 \text{ V})$$
$$= \boxed{30 \text{ VA}}$$

The answer is (B).

6. The open-circuit test obtains the core parameters G and B.

$$G = \frac{P_{\text{oc}}}{V_{\text{oc}}^2} = \frac{900 \text{ W}}{(13,800 \text{ V})^2} = \boxed{4.73 \times 10^{-6} \text{ S}}$$

$$Q^2 = S^2 - P^2 = ((13,800 \text{ VA})(0.2 \text{ A}))^2 - (900 \text{ W})^2$$

$$Q = 2609 \text{ VAR}$$

$$B = \frac{Q}{V^2} = \frac{2609 \text{ VAR}}{(13,800 \text{ V})^2} = \boxed{13.7 \times 10^{-6} \text{ S}}$$

The turns ratio is

$$\frac{V_p}{N_p} = \frac{V_s}{N_s}$$

$$\frac{N_p}{N_s} = \frac{V_p}{V_s} = \frac{13,800 \text{ V}}{460 \text{ V}}$$

$$= \boxed{30}$$

7. (a)

$$N_p I_p = N_s I_s$$

$$a = \frac{N_p}{N_s} = \frac{I_s}{I_p}$$

$$= \frac{100 \text{ A}}{20 \text{ A}}$$

$$= 5$$

$$P_{\text{sc}} = I_p^2 R_p + I_s^2 R_s = 1000 \text{ W}$$

$$I_p^2(R_p + a^2 R_s) = 1000 \text{ W}$$

$$R_p + a^2 R_s = \frac{1000 \text{ W}}{(20 \text{ A})^2} = 2.5 \text{ } \Omega$$

Assume $R_p = a^2 R_s = \boxed{1.25 \text{ } \Omega.}$? WTF

$$R_s = \frac{1.25 \text{ } \Omega}{(5)^2} = \boxed{0.05 \text{ } \Omega}$$

(b)

$$S = V_p I_p = (80 \text{ V})(20 \text{ A}) = 1600 \text{ VA}$$

$$= P + jQ$$

$$Q^2 = S^2 - P^2$$

$$Q = 1249 \text{ VAR}$$

$$= I_1^2(X_p + a^2 X_s)$$

$$X_p + a^2 X_s = \frac{Q}{I_1^2} = \frac{1249 \text{ VAR}}{(20 \text{ A})^2} = 3.12 \text{ } \Omega$$

Assume $X_p = a^2 X_s$, then

$$2X_p = 3.12$$

$$X_p = \frac{3.12}{2} = 1.56 \ \Omega = \omega L_p = 377 L_p$$

$$L_p = \frac{1.56 \ \Omega}{377 \ \frac{\text{rad}}{\text{s}}} = \boxed{4.14 \times 10^{-3} \ \text{H} \quad (4.14 \ \text{mH})}$$

$$X_s = \frac{X_p}{a^2} = \frac{1.56}{a^2} = 0.0624 \ \Omega = \omega L_s = 377 L_s$$

$$L_s = \frac{0.0624 \ \Omega}{377 \ \frac{\text{rad}}{\text{s}}} = \boxed{0.166 \times 10^{-3} \ \text{H} \quad (0.166 \ \text{mH})}$$

8. The per-unit impedance is found from

$$Z_{\text{pu}} = \frac{Z_{\text{actual}}}{Z_{\text{base}}} = \frac{R + jX}{Z_{\text{base}}} \qquad \text{[I]}$$

Find the unknown quantities.

$$Z_{\text{base}} = \frac{V_{\text{base}}}{I_{\text{base}}} \qquad \text{[II]}$$

The short-circuit test is performed at rated current, which is I_{base}.

$$I_{\text{rated}} = I_{\text{base}} = \frac{S_{\text{rated}}}{V_{\text{rated}}}$$

$$= \frac{500 \ \text{kVA}}{115 \ \text{kV}}$$

$$= 4.35 \ \text{A}$$

Substitute the rated current into Eq. II.

$$Z_{\text{base}} = \left(\frac{115 \ \text{kV}}{4.35 \ \text{A}} \right) \left(1000 \ \frac{\text{V}}{\text{kV}} \right) = 26{,}440 \ \Omega$$

Find R from the short circuit (sc) parameters.

$$P_{\text{sc}} = I_{\text{sc}}^2 R = I_{\text{rated}}^2 R$$

$$R = \frac{P_{\text{sc}}}{I_{\text{rated}}^2} = \frac{435 \ \text{W}}{(4.35 \ \text{A})^2} = 23 \ \Omega$$

Find X from

$$Q_{\text{sc}} = I_{\text{rated}}^2 X$$

$$X = \frac{Q}{I_{\text{rated}}^2} \qquad \text{[III]}$$

The reactive power is found using

$$S_{\text{sc}}^2 = P_{\text{sc}}^2 + Q_{\text{sc}}^2$$

$$Q_{\text{sc}} = \sqrt{S_{\text{sc}}^2 - P_{\text{sc}}^2}$$

$$= \sqrt{\left(I_{\text{rated}} V_{\text{sc}} \right)^2 - (435 \ \text{W})^2}$$

$$= \sqrt{\left((4.35 \ \text{A})(2.5 \times 10^3 \ \text{V}) \right)^2 - (435 \ \text{W})^2}$$

$$= 10{,}870 \ \text{VAR}$$

Substitute into Eq. III, giving

$$X = \frac{10{,}870 \ \text{VAR}}{(4.35 \ \text{A})^2} = 574 \ \Omega$$

Substitute the calculated values for R, X, and Z_{base} into Eq. I.

$$Z_{\text{pu}} = \frac{23 \ \Omega + j574 \ \Omega}{26{,}440 \ \Omega} = 0.0009 + j0.022 \ \text{pu}$$

$$= \boxed{0.001 + j0.022 \ \text{pu}}$$

Distribution

38 Power Transmission Lines

PRACTICE PROBLEMS

1. As frequency increases, which of the following parameters also increases due to the skin effect?

- (A) AC resistance
- (B) DC resistance
- (C) internal inductance
- (D) external inductance

2. What distributed parameter is not normally significant and is usually ignored in calculations?

- (A) line resistance
- (B) line inductance
- (C) shunt resistance
- (D) shunt capacitance

3. Most nearly, what is the velocity factor in a single waxwing conductor at 60 Hz?

- (A) 0.29
- (B) 0.61
- (C) 0.90
- (D) 0.97

4. What is parameter A of a 220 km falcon line with $Z_l = 0.85 \ \Omega/\text{mi}\angle 70°$ and $Y_l = 7 \times 10^{-6} \ \text{S/mi}\angle 90°$?

- (A) $4.0 \times 10^{-4}\angle 160°$
- (B) $0.948\angle 1.2°$
- (C) $0.996\angle 0.008°$
- (D) $1.0\angle 20°$

5. A three-phase, 215 kV transmission line supplies a 100 MW wye-connected load at 0.8 pf lagging. The ABCD parameters are $A = 0.9\angle 2.0°$, $B = 150\angle 80°$, $C = 0.001\angle 90°$, and $D = 0.9\angle 2°$. What is the sending-end voltage?

6. A transmission line with a characteristic impedance of 50 Ω and a termination resistance of 70 Ω has an electrical angle of $\beta l = \pi/2$. Determine the (a) reflection coefficient, (b) input impedance, and (c) VSWR.

SOLUTIONS

1. As frequency increases, the skin effect causes AC resistance to increase and internal inductance to decrease.

The answer is (A).

2. The shunt resistance is normally very large and has a minimal effect on current and voltage.

The answer is (C).

3. The velocity of wave propagation is

$$v_w = \frac{1}{\sqrt{L_l C_l}}$$

From App. 38.A, a waxwing conductor has an inductive reactance of

$$X_{L,l} = 0.476 \ \Omega/\text{mi}$$

Solving for the inductance per unit length gives

$$X_{L,l} = \omega L = 2\pi f L$$

$$L = \frac{X_{L,l}}{2\pi f} = \frac{0.476 \ \dfrac{\Omega}{\text{mi}}}{2\pi (60 \ \text{Hz})}$$

$$= 1.26 \times 10^{-3} \ \text{H/mi}$$

The capacitive reactance, from App. 38.A, is

$$X_C = 0.1090 \ \text{M}\Omega\text{-mi}$$

Solving for C_l gives

$$|X_C| = \frac{1}{\omega C_l} = \frac{1}{2\pi f C_l}$$

$$C_l = \frac{1}{2\pi f |X_C|} = \frac{1}{2\pi (60 \ \text{Hz})(0.1090 \times 10^6 \ \Omega\text{-mi})}$$

$$= 2.43 \times 10^{-8} \ \text{F/mi}$$

Substituting the calculated values gives

$$v_w = \frac{1}{\sqrt{L_l C_l}}$$

$$= \frac{1}{\sqrt{\left(1.26 \times 10^{-3}\ \dfrac{H}{mi}\right)\left(2.43 \times 10^{-8}\ \dfrac{F}{mi}\right)}}$$

$$= \left(1.81 \times 10^5\ \frac{mi}{s}\right)\left(1.609\ \frac{km}{mi}\right)\left(1000\ \frac{m}{km}\right)$$

$$= 2.91 \times 10^8\ m/s$$

The velocity factor is the ratio of the velocity of propagation to the speed of light.

$$v_{factor} = \frac{v_w}{C}$$

$$= \frac{2.91 \times 10^8\ \dfrac{m}{s}}{3.00 \times 10^8\ \dfrac{m}{s}}$$

$$= \boxed{0.97}$$

The answer is (D).

4. This is a medium-length transmission line. Regardless of the model chosen,

$$A = 1 + \tfrac{1}{2}YZ = 1 + \tfrac{1}{2}Y_l Z_l$$

$$= 1 + \frac{1}{2}\left(7 \times 10^{-6}\ \frac{\Omega^{-1}}{mi}\ \angle 90°\right)\left(0.85\ \frac{\Omega}{mi}\ \angle 70°\right)$$

$$= 1 + (2.98 \times 10^{-6}\ mi^{-2}\ \angle 160°)(220\ km)^2$$

$$\times \left(\frac{1\ mi}{1.609\ km}\right)^2$$

$$= 1\angle 0° + 5.56 \times 10^{-2}\ \angle 160°$$

$$= (1 + 0j) + (-0.0523 + j0.0190)$$

$$= 0.9477 + j0.0190$$

$$= \boxed{0.948\angle 1.2°}$$

The answer is (B).

5. The sending-end voltage is given by

$$V_S = AV_R + BI_R$$

To find V_S, V_R and I_R must be determined. Because the two-part model represents a single phase, the receiving-end phase voltage for a wye-connected load is

$$V_R = \frac{215\ kV}{\sqrt{3}} = 124\ kV$$

Because the power factor is given for the load, let V_R be the reference voltage.

$$V_R = 124\ kV\ \angle 0°$$

The receiving-end current is found from

$$P = 3I_R V_R(\text{pf})$$

$$I_R = \frac{P}{3V_R(\text{pf})} = \frac{100 \times 10^6\ W}{(3)(124 \times 10^3\ V)(0.8)}$$

$$= 336\ A$$

Because the power factor is 0.8 lagging, the current angle is $-36.8°$.

$$I_R = 336\ A\angle - 36.8°$$

Substituting gives

$$V_S = AV_R + BI_R$$

$$= (0.9\angle 2.0°)(124 \times 10^3\ V\angle 0°)$$

$$\quad + (150\angle 80°)(336\ A\angle - 36.8°)$$

$$= 1.116 \times 10^5\ V\angle 2.0° + 5.04 \times 10^4\ V\angle 43.2°$$

$$= \boxed{153\ kV\angle 14.5°}$$

6. (a)

$$\Gamma_{load} = \frac{Z_L - Z_0}{Z_L + Z_0} = \frac{70\ \Omega - 50\ \Omega}{70\ \Omega + 50\ \Omega}$$

$$= \boxed{1/6}$$

(b)

$$Z_{in} = Z_0\left(\frac{Z_L \cos\beta l + jZ_0 \sin\beta l}{Z_0 \cos\beta l + jZ_L \sin\beta l}\right)$$

$$= (50\ \Omega)\left(\frac{70\cos\frac{\pi}{2} + j50\sin\frac{\pi}{2}}{50\cos\frac{\pi}{2} + j70\sin\frac{\pi}{2}}\right)$$

$$= \boxed{35.71\ \Omega}$$

(c) $\text{VSWR} = \dfrac{1 + |\Gamma|}{1 - |\Gamma|} = \dfrac{1 + \dfrac{1}{6}}{1 - \dfrac{1}{6}} = \boxed{1.40}$

Distribution

39 The Smith Chart

PRACTICE PROBLEMS

1. A transmission line with a characteristic impedance of 50 Ω is terminated with a load impedance of $25 - j25\ \Omega$. (a) Find the electrical angle (βl) where a compensating capacitor can be inserted in the line to cause the impedance seen at that point to be 50 Ω. (b) Determine the reactance of the capacitor.

2. Given a coaxial cable with a characteristic impedance of 50 Ω, determine the length of cable (in wavelengths) necessary to produce (a) a reactance of $-25\ \Omega$, and (b) a reactance of $+50\ \Omega$. (The stubs may be either open or shorted.)

3. A transmission line with a characteristic impedance of 72 Ω and a phase constant of 0.5 rad/m is 7.5 m long. It is terminated with a pure 50 Ω resistance. Determine the impedance seen from the source.

4. A transmission line with a characteristic impedance of 100 Ω is terminated with a load of $25 + j25\ \Omega$. This is to be compensated with a 72 Ω line section. Determine the length of the compensating line section, and specify its placement from the load on the 100 Ω line for (a) series compensation and (b) parallel compensation.

SOLUTIONS

1. (a) $\dfrac{Z_L}{Z_0} = \dfrac{25 - j25\ \Omega}{50\ \Omega} = \dfrac{1}{2} - j\dfrac{1}{2}$ pu

The load point is shown on the Smith chart. The impedance locus is shown as a dashed circle.

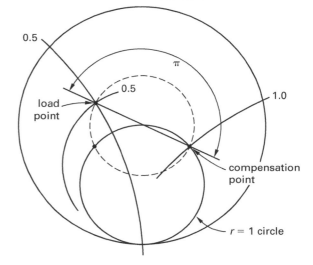

The compensation point for series capacitance is where the impedance locus intersects the $r = 1$ circle on the right-hand side of the chart.

For this particular load, the compensation point is 180° (on the chart) from the load point, at

$$\frac{Z_{\text{in}}}{Z_0} = 1 + j1 \text{ pu}$$

180° on the Smith chart is $\lambda/4$, or $\beta l = \boxed{\pi/2 \text{ rad}}$

(b)
$$\frac{Z_{\text{in}}}{Z_0} = \frac{\dfrac{Z_L}{Z_0}\cos\beta l + j\sin\beta l}{\cos\beta l + j\left(\dfrac{Z_L}{Z_0}\right)\sin\beta l}$$

$$= \frac{\tfrac{1}{2}\cos\beta l + j(\sin\beta l - \tfrac{1}{2}\cos\beta l)}{\cos\beta l + \tfrac{1}{2}\sin\beta l + j\tfrac{1}{2}\sin\beta l}$$

At $\beta l = \pi/2$,

$$\sin \beta l = 1$$
$$\cos \beta l = 0$$
$$\frac{Z_{\text{in}}}{Z_0} = \frac{j}{\frac{1}{2} + j\frac{1}{2}} = \frac{2}{\frac{1}{j} + 1} = 1 + j1$$

The compensating reactance must be $-j1$, or

$$X_C = -Z_0 = \boxed{-50 \ \Omega}$$

2. (a) $\dfrac{Z}{Z_0} = \dfrac{-j25 \ \Omega}{50 \ \Omega} = -0.5j$ pu

The bottom of the chart is open.

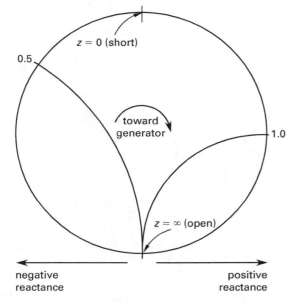

The shortest length to obtain $-j0.5 \times 50$ is to start with an open line, going clockwise $126.5°$ to the intersection with $x = 0.5$ on the left half of the Smith chart. This corresponds to

$$\beta l = \left(\frac{126.5°}{2}\right)\left(\frac{\pi}{180°}\right) \text{rad} = 1.1 \text{ rad}$$
$$= (1.1 \text{ rad})\left(\frac{0.5 \ \lambda}{\pi \text{ rad}}\right)$$
$$= 0.175 \ \lambda$$

(b) $\dfrac{Z}{Z_0} = \dfrac{j50}{50} = j1$ pu

The shortest length is found for a shorted line (short

point at the top of the Smith chart) going clockwise to intersect the $x = 1$ circle (on the right). This is $90°$ on the chart, corresponding to

$$\beta l = (45°)\left(\frac{\pi \text{ rad}}{180°}\right)\left(\frac{0.5 \ \lambda}{\pi \text{ rad}}\right)$$
$$= \boxed{1/8 \ \lambda}$$

3.

$$Z_0 = 72 \ \Omega$$
$$Z_L = 50 \ \Omega$$
$$\beta = 0.5$$
$$l = 7.5 \text{ m}$$
$$\frac{Z_L}{Z_0} = z_L = \frac{50 \ \Omega}{72 \ \Omega} = 0.6944 \text{ pu}$$
$$\beta l = (3.75 \text{ rad})\left(\frac{360°}{2\pi \text{ rad}}\right) = 214.86°$$
$$2\beta l = (2)(214.86°)$$
$$= 429.72° \quad (360° + 69.72°)$$

The load point and source point are shown.

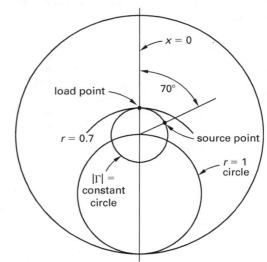

Reading from an enlarged Smith chart, the value is

$$\frac{Z_{\text{in}}}{Z_0} \approx 0.84 + j0.29 \text{ pu}$$
$$Z_{\text{in}} \approx (0.84 + j0.29 \text{ pu})(72 \ \Omega)$$
$$= \boxed{60 + j21}$$

4.

$$Z_0 = 100 \ \Omega$$
$$Z_L = 25 + j25 \ \Omega$$
$$z_L = 0.25 + j0.25$$
$$y_L = 2 - j2$$

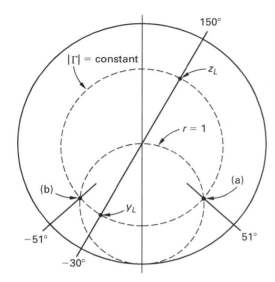

(a) The z_L point, $0.25 + j0.25$, is at $150°$. Following a constant $|\Gamma|$ circle toward the generator, the compensation point is at $51°$, where $z_{\text{in}} = 1 + j1.6$. A series compensation of $-j1.6$ can be added. The line length from the load is

$$\frac{150° - 51°}{720°} \ \lambda = \boxed{0.1375 \ \lambda}$$
$$Z_C = (100 \ \Omega)(-j1.6) = -j160 \ \Omega$$

Using a 72 Ω line for the compensator, its reactance is

$$x_C = -\frac{j160 \ \Omega}{72 \ \Omega} = -2.22 \text{ pu}$$

Starting from the open position (bottom of the Smith chart), the location of $-j2.22$ is about $-48°$ for a compensator length of

$$l_{\text{cp}} = \frac{48°}{720°} \ \lambda \approx \boxed{1/15 \ \lambda}$$

(b) The y_L point is $180°$ from the z_L point on the same $|\Gamma| = $ constant circle.

The parallel compensation point is at $-51°$, so the distance from the load for compensation is at

$$\frac{51° - 30°}{720°} \ \lambda = \boxed{0.0292 \ \lambda}$$

At that point,

$$y_L = 1 - j1.6 \text{ pu}$$
$$y = \frac{Z_0}{Z} = Z_0 Y$$
$$Y_{\text{cp}} = \frac{y}{Z_0} = j\left(\frac{1.6}{100 \ \Omega}\right) \text{S}$$

For the 72 Ω line,

$$y_{\text{cp}} = j\left(\frac{1.6}{100 \ \Omega}\right)(72 \ \Omega) = j1.152 \text{ pu}$$

1.152 is on the right side of the chart at $82°$. The shortest line is from the top of the chart ($y = 0$, open) at $180°$.

$$l = \frac{180° - 82°}{720°} \ \lambda = \boxed{0.136 \ \lambda}$$

Topic VII: System Analysis

Chapter

System Analysis

40 Power System Analysis

PRACTICE PROBLEMS

1. Consider the following generation system with five nodes and five admittances. In order to use a power software analysis tool to model the distribution system attached to the generator, the equations for each node must be known and then placed in the bus admittance matrix. What is the nodal equation for node 2? How many nodal equations are required to determine all branch currents?

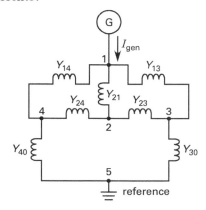

(A) $(V_1 - V_2) Y_{21} + (V_2 - V_3) Y_{23} + (V_4 - V_2) Y_{24} = 0$;
4 equations

(B) $(V_1 - V_2) Y_{21} + (V_2 - V_3) Y_{23} + (V_2 - V_4) Y_{24} = 0$;
5 equations

(C) $V_2(-Y_{21} + Y_{23} + Y_{24}) - V_1 Y_{12} - V_3 Y_{23} - V_4 Y_{24} = 0$;
5 equations

(D) $V_2(Y_{21} + Y_{23} + Y_{24}) - V_1 Y_{21} - V_3 Y_{23} - V_4 Y_{24} = 0$;
4 equations

2. What is most nearly the value of the apparent power, **S**, for a single phase circuit for which the voltage is $480\,\text{V}\angle 0°$ and the current is $30\,\text{A}\angle -37°$?

(A) $0.60\,\text{kVA}\angle 37°$

(B) $14\,\text{kVA}\angle 37°$

(C) $16\,\text{kVA}\angle -37°$

(D) $83\,\text{kVA}\angle -37°$

3. The line-to-line voltages of a balanced three-phase system are 120° apart but differ in magnitude and are 30° ahead of the line-to-neutral voltages. If the line-to-

neutral voltages are considered to be of unit magnitude, what is the approximate magnitude of the line-to-line voltages, which are also 120° apart?

(A) 1.00 V

(B) 1.41 V

(C) 1.73 V

(D) 2.00 V

4. Which of the following statements regarding symmetrical components is NOT true?

(A) A balanced system has phasor components of equal magnitude.

(B) The method of symmetrical components allows the calculation of unbalanced currents during fault conditions. The calculation uses three separate systems of components that are balanced and equal in magnitude.

(C) In a three-phase system, no zero-sequence line voltages exist except during imbalance or fault conditions.

(D) In delta-connected circuits, zero-sequence currents flow around the delta, and therefore do not affect the line currents.

5. Consider the three-phase transmission system shown. The transmission line current is 62.75 A. The load current from the generator is most nearly what value?

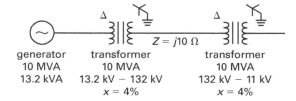

generator	transformer	transformer
10 MVA	10 MVA	10 MVA
13.2 kVA	13.2 kV – 132 kV	132 kV – 11 kV
	$x = 4\%$	$x = 4\%$

(A) 630 A

(B) 1100 A

(C) 6300 A

(D) 7200 A

6. A system's electrical parameters are

$$\mathbf{V} = 480 \text{ V}\angle 0°$$
$$\mathbf{I} = 240 \text{ A}\angle -30°$$
$$\text{THD} = 2.5\%$$

What is most nearly the apparent power, $\mathbf{S}$?

(A) $1.15 \times 10^3 \text{ W}\angle 30°$

(B) $2.50 \times 10^3 \text{ W}\angle -30°$

(C) $1.15 \times 10^5 \text{ W}\angle -30°$

(D) $1.15 \times 10^5 \text{ W}\angle 30°$

7. A balanced wye load has a measured voltage of 173.2 V$\angle 0°$. The phase sequence is A-B-C. What is most nearly the voltage-to-neutral of phase B?

(A) $-150 + j86.6 \text{ V}$

(B) $-86.6 - j50 \text{ V}$

(C) $86.6 + j50 \text{ V}$

(D) $150 - j86.6 \text{ V}$

8. During phasor analysis using the operator a, the term $1 - a^2$ occurs. What is most nearly the value of this term in polar form?

(A) $1 \angle -120°$

(B) $1 \angle 0°$

(C) $1 \angle 120°$

(D) $\sqrt{3} \angle 30°$

9. Three wye-connected generators supply a three-phase distribution system, with the majority of the loads delta-connected. The line voltage is 11 kV on all three generators. From the monitoring system, the data on the generators are

generator 1

   1212 kVA, pf = 0.75 leading

generator 2

   865 kVA, pf = 0.80 lagging

generator 3

   1000 kVA, pf = 0.80 lagging

What is most nearly the total power supplied by the generators?

(A) 2.4 MW

(B) 3.1 MW

(C) 4.0 MW

(D) 6.4 MW

10. The terminal voltage at a wye-connected load is 12.5 kV line-to-line. The load is balanced with an impedance of 15 $\Omega \angle 30°$. The distribution line impedance is 1.2 $\Omega \angle 80°$. What is most nearly the line-to-line voltage of the substation that supplies this load through the distribution line?

(A) 7200 V$\angle 0°$

(B) 7600 V$\angle 3°$

(C) 7600 V$\angle 80°$

(D) 13 200 V$\angle 3°$

SOLUTIONS

1. Write KCL for node 2. Assume all currents flowing out of the node are positive. (Either positive or negative is acceptable, but consistency must be retained for all the nodal equations.)

$$(V_2 - V_1)Y_{21} + (V_2 - V_3)Y_{23} + (V_2 - V_4)Y_{24} = 0$$

Rearrange the equation, expanding the parentheses, and combining like voltage terms.

$$V_2(Y_{21} + Y_{23} + Y_{24}) - V_1 Y_{21} - V_3 Y_{23} - V_4 Y_{24} = 0$$

The first term of this equation is known as the self-admittance of the node. The remaining terms are referred to as the mutual admittances of the node.

The number of nodal equations required to determine all branch currents is one less than the total number of nodes. Because there are five nodes, four node equations are required.

The answer is (D).

2. The apparent power is given by the following equation.

$$\mathbf{S} = VI^*$$

Substitute the given values. Note that the complex conjugate changes the sign of the current angle.

$$\mathbf{S} = VI^* = (480 \text{ V} \angle 0°)(30 \text{ A} \angle 37°)$$
$$= \boxed{14{,}000 \text{ VA} \angle 37° \quad (14\text{k VA} \angle 37°)}$$

The answer is (B).

3. Any two phase-to-neutral phasors (an, bn, or cn) of magnitude 1 with an angle 120° can be combined to obtain the line-to-line quantities (AB, BC, CA). Assume the phasors are located at 0° and 240°. (Consider phase A and phase B in a three-phase system, but with unit magnitude.)

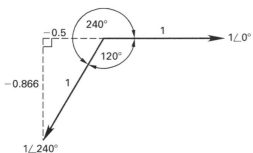

Using KVL around the loop from phase A to neutral, neutral to B, and from the line B to line A, gives the following. (The lower case letters represent phase quantities; the upper case letters represent the desired line quantity. The notation shown for the result uses the *a* operator.)

$$V_{\text{an}} + V_{\text{nb}} + V_{\text{BA}} = 0$$
$$-V_{\text{BA}} = V_{\text{an}} + V_{\text{nb}}$$
$$V_{\text{AB}} = V_{\text{an}} - V_{\text{bn}}$$

Now substitute the values for the phase voltage at phases A and B and solve for the line quantity, or substitute the value of the phase A voltage and the operator *a*.

$$V_{\text{AB}} = V_{\text{an}} - V_{\text{bn}}$$
$$= 1 \text{ V} \angle 0° - 1 \text{ V} \angle 240°$$
$$= 1 \text{ V} \angle 0° + 1 \text{ V} \angle 60°$$
$$= (1 + j0) \text{ V} + (0.5 + j0.866) \text{ V}$$
$$= 1.5 \text{ V} + j0.866 \text{ V}$$
$$= \sqrt{3} \text{ V} \angle 30°$$
$$= \boxed{1.73 \angle 30° \text{ V} \quad (1.73 \text{ V})}$$

Alternatively,

$$V_{\text{AB}} = V_{\text{an}} - aV_{\text{an}}$$
$$= 1 \text{ V} \angle 0° - (-0.5 - j0.866)(1 \text{ V} \angle 0°)$$
$$= 1 \text{ V} \angle 0° + (0.5 + j0.866)(1 + 0j) \text{ V}$$
$$= 1 + (0.5 + j0.866) \text{ V}$$
$$= 1.5 \text{ V} + j0.866 \text{ V}$$
$$= \boxed{1.73 \text{ V} \angle 30° \quad (1.73 \text{ V})}$$

In graphical terms, the result is as shown.

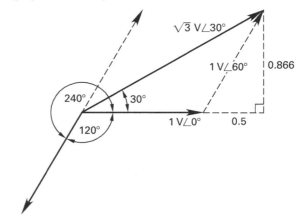

The answer is (C).

4. A balanced system includes components of equal magnitude. Option A is true.

The method of symmetrical components was developed to simplify fault calculations. Fortescue's theorem proves three unbalanced phasors can be resolved in a system of three balanced phasors. Therefore, option B is true.

Because the voltages of a three-phase system always sum to zero, no zero-sequence voltages exist regardless of the imbalance. Option C is not true.

In delta-connected circuits, zero-sequence currents flow around the delta and do not impact the line currents. Therefore, option D is true.

The answer is (C).

5. The transmission line is downstream of the first transformer whose turns ratio is given by

$$n = \frac{13.2 \text{ kV}}{132 \text{ kV}} = 0.1$$

The relationship between the voltages and currents is

$$\frac{V_{\text{gen}}}{V_{\text{trans}}} = \frac{I_{\text{trans}}}{I_{\text{gen}}}$$

Solve for the current on the generator side.

$$I_{\text{gen}} = \left(\frac{V_{\text{trans}}}{V_{\text{gen}}}\right) I_{\text{trans}} = \frac{I_{\text{trans}}}{a}$$

$$I_{\text{gen}} = \frac{I_{\text{trans}}}{n} = \frac{62.75 \text{ A}}{0.1}$$

$$= \boxed{627.5 \text{ A} \quad (630 \text{ A})}$$

The answer is (A).

6. The apparent power, **S**, is given by

$$\mathbf{S} = \mathbf{V}\mathbf{I}^*$$

The current is

$$\mathbf{I} = 240 \text{ A}\angle{-30°}$$
$$= 208 - j120 \text{ A}$$

The complex conjugate of the current is

$$\mathbf{I}^* = 208 + j120 \text{ A}$$
$$= 240 \text{ A}\angle{30°}$$

The apparent power, **S**, is calculated as

$$\mathbf{S} = \mathbf{V}\mathbf{I}^*$$
$$= (480 \text{ V}\angle{0°})(240 \text{ A}\angle{30°})$$
$$= \boxed{1.15 \times 10^5 \text{ W}\angle{30°}}$$

The answer is (D).

7. Balanced wye voltage relationships are shown.

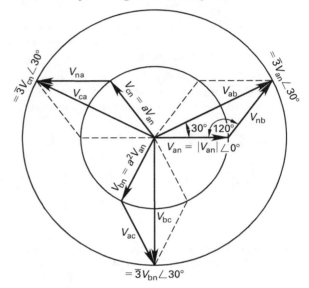

(a) line-to-line versus line-to-neutral voltages

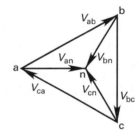

(b) alternative line-to-line versus line-to-neutral voltages

Calculate the voltage-to-neutral of phase A.

$$V_{\text{ab}} = \sqrt{3} \, V_{\text{an}}\angle{30°}$$
$$V_{\text{an}} = \frac{V_{\text{ab}}}{\sqrt{3}}\angle{-30°}$$
$$= \frac{173.2 \text{ V}}{\sqrt{3}}\angle{-30°}$$
$$= 100 \text{ V}\angle{-30°}$$

Use the voltage-to-neutral of phase A to calculate the voltage-to-neutral of phase B.

$$V_{bn} = a^2 V_{an}$$
$$= (1 \angle 240°)(100 \text{ V}\angle -30°)$$
$$= 100 \text{ V} \angle 210°$$

Convert to rectangular form.

$$V_{bn} = 100 \text{ V}\angle 210°$$
$$= \boxed{-86.6 - j50 \text{ V}}$$

The answer is (B).

8. The operator a represents $1 \angle 120°$. It is used for voltage and current analysis of three-phase phasors. Calculate the term $1 - a^2$.

$$1 - a^2 = 1 - 1 \angle 240°$$
$$= 1 - (-0.5 + j0.866)$$
$$= 1.5 + j0.866$$
$$= 1.73 \angle 30°$$
$$= \boxed{\sqrt{3} \angle 30°}$$

The answer is (D).

9. The wye or delta connections determine the phase quantities but do not impact the total power. In addition, the real power is independent of the leading or lagging value of the power factor.

$$\text{pf} = \frac{P}{S}$$
$$P = S(\text{pf})$$

Add the power from all three generators to find the total power supplied.

$$P_{\text{total}} = S_1\text{pf}_1 + S_2\text{pf}_2 + S_3\text{pf}_3$$
$$= (1212 \text{ kVA})(0.75) + (865 \text{ kVA})(0.80)$$
$$\quad + (1000 \text{ kVA})(0.80)$$
$$= \boxed{2.40 \times 10^3 \text{ kW} \quad (2.4 \text{ MW})}$$

The answer is (A).

10. The line-to-line voltage of the substation decreases because of the line losses and the load losses. Use Ohm's law to calculate the substation voltage while taking into account these losses.

Change the terminal voltage into a phase voltage at the load. Make V_{an} the zero reference voltage.

$$V_{an} = \frac{V_{\text{terminal}}}{\sqrt{3}}$$
$$= \frac{(12.5 \text{ kV})\left(1000 \dfrac{\text{V}}{\text{kV}}\right)}{\sqrt{3}}$$
$$= 7217 \text{ V} \angle 0°$$

Calculate the phase current.

$$I_{an} = \frac{V_{an}}{Z_{\text{load}}}$$
$$= \frac{7217 \text{ V}\angle 0°}{15 \, \Omega \angle 30°}$$
$$= 481.13 \text{ A}\angle -30°$$

The phase voltage at the supplying substation is

$$V_{\text{phase,ss}} = V_{an} + I_{an}Z_{\text{line}}$$
$$= 7217 \text{ V}\angle 0° + (481.13 \text{ A}\angle -30°)(1.2\,\Omega \angle 80°)$$
$$= 7217 \text{ V}\angle 0° + 577.36 \text{ V}\angle 50°$$
$$= 7601 \text{ V}\angle 3.34°$$

The substation was assumed to have a wye connection for this calculation and all equations are calculated on a per-phase basis. This assumption is valid and the final results of any such calculation can be then adjusted for the actual connection. Given that the substation connection type is unknown, the line-to-line voltage should be calculated from the above phase result.

$$V_{\text{line,ss}} = \sqrt{3} \; V_{\text{phase,ss}}$$
$$= \sqrt{3} \, (7601 \text{ V}\angle 3.34°)$$
$$= \boxed{13\,165 \text{ V}\angle 3.34° \quad (13\,200 \text{ V}\angle 3°)}$$

The answer is (D).

41 Analysis of Control Systems

PRACTICE PROBLEMS

1. Consider a homogeneous second-order linear differential equation with constant coefficients. The damping ratio is 0.3, and the natural frequency is 10 rad/s. Estimate the amplitude at $t = 0.5$ s graphically.

2. Determine the transfer functions for the following systems.

(a)

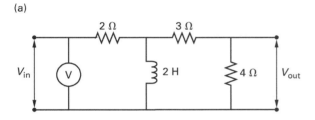

(b)

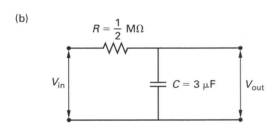

3. An amplifier consists of several stages with gains as shown. A voltage divider provides feedback. What is the overall gain with and without feedback?

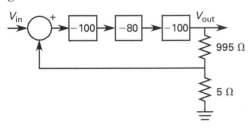

4. A system is composed of a forward path and a unity feedback loop. The gain in the forward loop is $1000 \pm 10\%$. What is the uncertainty of the output signal?

5. (a) Simplify the following block diagrams. (b) Determine the overall system gain.

(a)

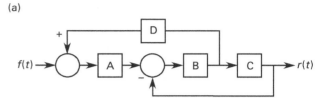

(b)

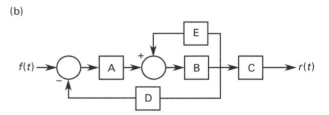

6. (a) Simplify the following block diagram. (b) Determine the overall system gain. (c) What is the system sensitivity if $G_1 = -5$, $G_2 = 2$, $G_3 = 4$, and $G_4 = 3$?

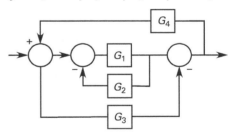

7. What is the steady-state response to an impulse $\delta(t)$ of the system represented by the following differential equation? Assume all initial conditions are zero.

$$r'(t) + 3r'(t) + r(t) = \delta'(t) + \delta(t)$$

8. Find the steady-state response to a step input for a system in which the transfer function is

$$T(s) = \frac{b_0 s^p + b_1 s^{p-1} + \cdots + b_p}{s^n + a_1 s^{n-1} + \cdots + a_n} \quad [a_n \neq 0]$$

9. Consider the following transfer function.

$$T(s) = \frac{1+s}{1+2s+2s^2}$$

What is the (a) amplitude and (b) phase of the steady-state response to a sinusoidal input of $\sin \omega t$?

10. Consider a system with the following transfer function.

$$T(s) = \frac{1}{(s+a)(s+b)}$$

What is the steady-state response to a unit step (a) in the time domain and (b) in the frequency domain?

11. Draw the pole-zero diagram for the following transfer function.

$$T(s) = \frac{(s^2+4)(s-2)}{s^2(s^2+4s+5)(s+1)}$$

12. What is the steady-state response of a system to a sinusoidal input with angular frequency ω_0 if the system's transfer function has a zero at $s = j\omega_0$?

13. Find the (a) bandwidth, (b) peak frequency, (c) half-power points, and (d) quality factor for

$$T(s) = \frac{3s+18}{s^2+12s+3200}$$

14. Is a system with the following transfer function stable?

$$T(s) = \frac{(s-2)(s+3)}{(s^2+s-2)(s+1)}$$

SOLUTIONS

1. At $t = 0.5$ s,

$$\omega t = \left(10 \ \frac{\text{rad}}{\text{s}}\right)(0.5 \text{ s}) = 5 \text{ rad}$$

For a normalized curve of natural response corresponding to $\zeta = 0.3$, see Fig. 41.1.

$$\frac{x(t)}{\omega} \approx -0.22 \text{ s/rad} \quad [\text{at } t = 0.5]$$

$$x(t = 0.5 \text{ s}) = \left(-0.22 \ \frac{\text{s}}{\text{rad}}\right)\left(10 \ \frac{\text{rad}}{\text{s}}\right)$$

$$= \boxed{-2.2}$$

2. (a)

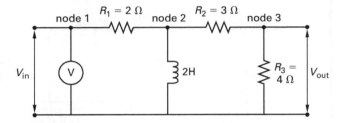

At node 2,

$$0 = \frac{V_2 - V_1}{R_1} + \frac{V_2 - V_3}{R_2} + \frac{1}{L}\int V_2 dt$$

$$= \frac{V_2 - V_1}{2} + \frac{V_2 - V_3}{3} + \frac{1}{2}\int V_2 dt$$

$$= \frac{V_2 - V_1}{2} + \frac{V_2 - V_3}{3} + \frac{V_2}{2s} \quad [\text{Eq. I}]$$

At node 3,

$$0 = \frac{V_3 - V_2}{R_2} + \frac{V_3}{R_3}$$

$$= \frac{V_3 - V_2}{3} + \frac{V_3}{4} \quad [\text{Eq. II}]$$

From Eq. I,

$$V_2 = \frac{3sV_1 + 2sV_3}{5s+3}$$

From Eq. II,

$$V_2 = \frac{7V_3}{4}$$

$$\frac{3sV_1 + 2sV_3}{5s + 3} = \frac{7V_3}{4}$$

$$T(s) = \frac{V_{out}}{V_{in}} = \frac{V_3}{V_1} = \frac{12s}{27s + 21}$$

$$= \boxed{4s/(9s + 7)}$$

The black box representation of this system is

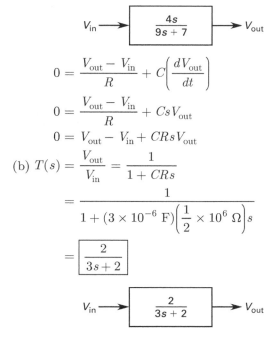

$$0 = \frac{V_{out} - V_{in}}{R} + C\left(\frac{dV_{out}}{dt}\right)$$

$$0 = \frac{V_{out} - V_{in}}{R} + CsV_{out}$$

$$0 = V_{out} - V_{in} + CRsV_{out}$$

(b) $T(s) = \dfrac{V_{out}}{V_{in}} = \dfrac{1}{1 + CRs}$

$$= \frac{1}{1 + (3 \times 10^{-6} \text{ F})\left(\frac{1}{2} \times 10^6 \ \Omega\right)s}$$

$$= \boxed{\dfrac{2}{3s + 2}}$$

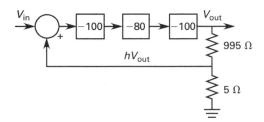

3.

The overall gain without feedback is

$$K = (-100)(-80)(-100) = \boxed{-800{,}000}$$

$$\frac{V_{out} - hV_{out}}{995 \ \Omega} = \frac{hV_{out}}{5 \ \Omega}$$

$$h = \frac{5 \ \Omega}{995 \ \Omega + 5 \ \Omega} = 0.005$$

$K < 0$, but the feedback adds, so the feedback is negative.

$$K_{loop} = \frac{K}{1 - Kh}$$

$$= \frac{-800{,}000}{1 - (-800{,}000)(0.005)}$$

$$= \boxed{-200}$$

4.

$$v_{out} = G_{loop}v_i$$

$$\frac{\Delta v_{out}}{v_{out}} = \frac{\Delta G_{loop}}{G_{loop}}$$

(Assume the input signal has no uncertainty.)

$$\left(\frac{\Delta G_{loop}}{G_{loop}}\right)\left(\frac{G}{\Delta G}\right) = \frac{1}{1 + G}$$

$$\frac{\Delta v_{out}}{v_{out}} = \frac{\Delta G_{loop}}{G_{loop}} = \left(\frac{\Delta G}{G}\right)\left(\frac{1}{1 + G}\right)$$

$$\frac{\Delta G}{G} = 0.1$$

$$\frac{\Delta v_{out}}{v_{out}} = (0.1)\left(\frac{1}{1 + 1000}\right)$$

$$= \boxed{9.99 \times 10^{-5} \quad (0.01\%)}$$

5. (a)

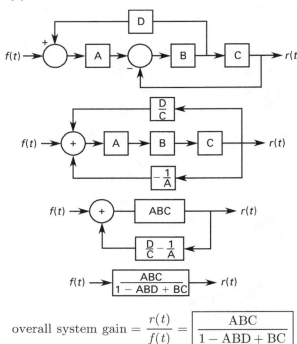

overall system gain $= \dfrac{r(t)}{f(t)} = \boxed{\dfrac{\text{ABC}}{1 - \text{ABD} + \text{BC}}}$

(b)

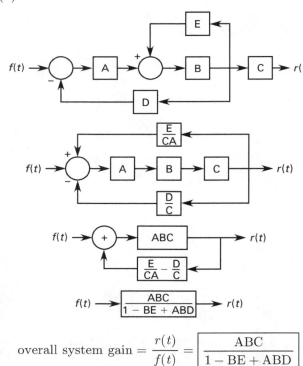

overall system gain $= \dfrac{r(t)}{f(t)} = \boxed{\dfrac{\text{ABC}}{1 - \text{BE} + \text{ABD}}}$

6. (a)

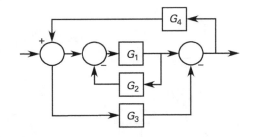

step 1:

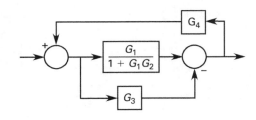

step 2:

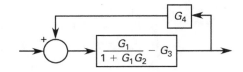

step 3:

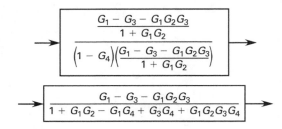

(b) The overall system gain is

$$\boxed{\dfrac{G_1 - G_3 - G_1 G_2 G_3}{1 + G_1 G_2 - G_1 G_4 + G_3 G_4 + G_1 G_2 G_3 G_4}}$$

From step 2,

$$
\begin{aligned}
G(s) &= \frac{G_1}{1 + G_1 G_2} - G_3 \\
&= \frac{-5}{1 + (-5)(2)} - 4 \\
&= -31/9 \\
H(s) &= G_4 = 3 \\
G(s)H(s) &= \left(-\frac{31}{9}\right)(3) = -\frac{31}{3} < 0
\end{aligned}
$$

(c) The system has negative feedback overall. Therefore,

$$\text{sensitivity } S = \frac{1}{1 + G(s)H(s)}$$

$$= \frac{1}{1 + \left(-\dfrac{31}{3}\right)}$$

$$= \boxed{-3/28}$$

7. By definition, the system transfer function is

$$T(s) = \mathcal{L}\left(\frac{r_{\text{sys}}(t)}{f_{\text{sys}}(t)}\right) = \frac{R_{\text{sys}}(s)}{F_{\text{sys}}(s)}$$

$$= \frac{\text{output signal}}{\text{input signal}}$$

The output signal is

$$r_{\text{sys}}(t) = \delta'(t) + \delta(t)$$

$$R_{\text{sys}}(s) = s + 1$$

The input signal is

$$f_{\text{sys}}(t) = r'(t) + 3r'(t) + r(t)$$

$$F_{\text{sys}}(s) = s^2 + 3s + 1$$

$$T(s) = \frac{R_{\text{sys}}(s)}{F_{\text{sys}}(s)} = \frac{s + 1}{s^2 + 3s + 1}$$

The forcing function, $f(t)$, acting on this system is

$$f(t) = \delta(t)$$
$$F(s) = \mathcal{L}(f(t)) = \mathcal{L}(\delta(t)) = 1$$
$$R(s) = T(s)F(s) = (T(s))(1)$$
$$= T(s)$$
$$= \boxed{\frac{s + 1}{s^2 + 3s + 1}}$$

8. Using the final value theorem, obtain the steady-state step response by substituting zero for s in the transfer function.

If $b_p \neq 0$,

$$\boxed{R(s) = T(0) = b_p/a_n}$$

If $b_p = 0$, the numerator is zero.

$$\boxed{R(s) = 0}$$

9. (a) Substitute $j\omega$ for s in the transfer function to obtain the steady-state response for a sinusoidal input.

$$R(s) = T(j\omega) = \frac{1 + j\omega}{1 + 2j\omega - 2\omega^2}$$

$$= \frac{1 + j\omega}{(1 - 2\omega^2) + 2j\omega}$$

The amplitude is the absolute value of $R(s)$.

$$|R(s)| = \sqrt{T^2(j\omega)} = \frac{\sqrt{1 + \omega^2}}{\sqrt{(1 - 2\omega^2)^2 + 4\omega^2}}$$

$$= \boxed{\sqrt{\frac{1 + \omega^2}{1 + 4\omega^4}}}$$

(b) The phase angle of the steady-state response is Arg $(R(s))$. Find this by substituting $T(j\omega)$ into the form $(a + jb)/c$, whose argument (Arg) is

$$\arctan \frac{\dfrac{b}{c}}{\dfrac{a}{c}} = \arctan \frac{b}{a}$$

First eliminate j from the denominator by multiplying by its complex conjugate $(1 - 2\omega^2) - 2j\omega$.

$$R(s) = T(j\omega) = \frac{(1 + j\omega)(1 - 2\omega^2 - 2j\omega)}{1 + 4\omega^4}$$

$$= \frac{1 - 2\omega^2 + 2\omega^2 + j(-\omega - 2\omega^3)}{1 + 4\omega^4}$$

$$= \frac{1 + j(-\omega - 2\omega^3)}{1 + 4\omega^4}$$

$$\text{Arg}(R(s)) = \text{Arg}(T(j\omega))$$

$$= \boxed{\arctan \frac{-(\omega + 2\omega^3)}{1}}$$

10. (a)

$$R(s) = T(s)F(s)$$
$$f(t) = \mu_0 \quad \text{[unit step]}$$
$$F(s) = \mathcal{L}(f(t)) = 1/s$$
$$R(s) = T(s)\left(\frac{1}{s}\right) = \frac{1}{s(s+a)(s+b)}$$

From the transform table,

$$\begin{aligned} r(t) &= \mathcal{L}^{-1}(R(s)) \\ &= \frac{1}{ab} + \frac{be^{-at} - ae^{-bt}}{ab(a-b)} \\ &= r_1 + r_2(t) \end{aligned}$$

The steady-state response is $\lim_{t\to\infty} r(t)$, and because $\lim_{t\to\infty} r_2(t) = 0$, r_1 is the steady-state response.

$$r_1 = \boxed{1/ab}$$

(b) The steady-state response in the frequency domain for a step input of magnitude h is $R(s) = hT(0)$. Substituting $h = 1$ and $s = 0$ in $T(s)$,

$$R(s) = h(T(0)) = \frac{1}{(0+a)(0+b)} = \boxed{1/ab}$$

11.

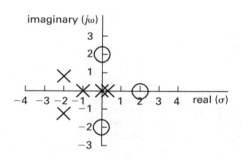

$$\frac{(s^2+4)(s-2)}{s^2(s^2+4s+5)(s+1)} = \frac{(s-2j)(s+2j)(s-2)}{s^2(s+1)(s+2-j)(s+2+j)}$$

For zeros, set the numerator equal to zero. For poles, set the denominator equal to zero.

12. The amplitude and phase of the steady-state response are given by $T(j\omega_0)$, where $T(s)$ is the transfer function. Since $j\omega_0$ is a zero of $T(s)$,

$$R(s) = T(j\omega_0) = 0$$

Therefore, the steady-state response is zero.

The system entirely blocks the angular frequency, ω_0.

13. (a)

$$\begin{aligned} T(s) &= \frac{3s+18}{s^2+12s+3200} \\ &= \frac{as+b}{s^2+(\text{BW})s+\omega_n^2} \end{aligned}$$
$$\text{BW} = \boxed{12 \text{ rad/s}}$$

(b)

$$\begin{aligned} \text{peak frequency} &= \omega_n \\ &= \sqrt{3200} \\ &= \boxed{56.57 \text{ rad/s}} \end{aligned}$$

(c)

$$\begin{aligned} \text{half-power points} &= \omega_n \pm \frac{\text{BW}}{2} \\ &= 56.57\,\frac{\text{rad}}{\text{s}} \pm \frac{12\,\dfrac{\text{rad}}{\text{s}}}{2} \\ &= \boxed{\begin{array}{c} 62.57 \text{ rad/s and} \\ 50.57 \text{ rad/s} \end{array}} \end{aligned}$$

(d)

$$\begin{aligned} \text{quality factor} = Q &= \frac{\omega_n}{\text{BW}} \\ &= \frac{56.57\,\dfrac{\text{rad}}{\text{s}}}{12\,\dfrac{\text{rad}}{\text{s}}} \\ &= \boxed{4.71} \end{aligned}$$

14. The system has a pole in the right half of the s-plane ($s = 1$).

$$s^2 + s - 2 = (s-1)(s+2)$$

$$\boxed{\text{The system is unstable.}}$$

Topic VIII: Protection and Safety

42 Protection and Safety

PRACTICE PROBLEMS

1. Universal relay characteristics result in the following general diagrams in the $R\text{-}X$ plane. Which represent(s) a directional relay?

I.
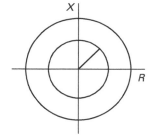

II.
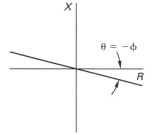

III.
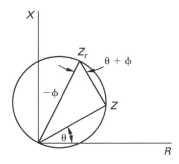

(A) I only

(B) II only

(C) I, II, and III

(D) II and III only

2. On a schematic of an electrical distribution system, a circle with the number 27 controls a device labeled 52. What does the circle represent?

(A) a time-delay relay operating an undervoltage device

(B) a time-delay relay operating a circuit breaker

(C) an interlocking relay between the two devices

(D) an undervoltage device for an AC circuit breaker

3. An ungrounded system is built for a military marine system with a normal voltage level, A. Following fault analysis design, the insulation of the distribution system is selected. What is most nearly the voltage level for which the insulation on the distribution system is designed?

(A) $(1/\sqrt{3})A$

(B) $\sqrt{2}\,A$

(C) $\sqrt{3}\,A$

(D) $2A$

4. A differential relay, which is designed to protect a zone of the distribution system, uses current transformers (CTs) connected to auxiliary CTs. The auxiliary CTs must match both the phase shift and the turns ratio of the differential relay CTs. If the values do not match, which of the following is the least likely outcome?

(A) The trip could occur at a lower level than expected.

(B) The trip could occur at a higher level than expected.

(C) The trip could occur under nonfault conditions.

(D) The trip will occur only when the CTs are wired in parallel.

5. What is the maximum allowable size of the overcurrent device normally allowed to protect an 12 AWG THHW conductor rated for 90°C used in an ambient temperature of 55°C? Only standard size overcurrent devices are available. Use the *National Electrical Code* (NEC).

(A) 10 A

(B) 15 A

(C) 20 A

(D) 30 A

6. Which property does NOT describe a ring-bus distribution system?

(A) reliability

(B) economy

(C) simple protection or relaying

(D) closed-loop arrangement of circuit breakers

7. Which statement about protective relay reliability is true?

(A) A dependable relay operates as required for fault conditions.

(B) A secure relay operates as required for fault conditions.

(C) As relay dependability increases, secure operation increases.

(D) All of the above are true.

8. A three-phase circuit is protected by an overcurrent relay with 600 primary turns and 5 secondary turns. Fault analysis indicates that the relay should trip to protect the system when the fault current on the primary side is 400 A. What is most nearly the relay tap setting that ensures protection?

(A) 2.0 A

(B) 2.5 A

(C) 3.0 A

(D) 3.5 A

9. A three-phase transformer is protected by a differential relay. The transformer is delta-connected on the primary side and wye-connected on the secondary side. The transformer's ratings are

$$20 \text{ MVA}$$
$$12\,500 \text{ V}/4160 \text{ V}$$
$$Z_{\text{pu}} = 6\%$$

The relay is set to trip on faults that draw 250% of the rated current. The primary CT current ratio is 400:5. The secondary CT current ratio is 1800:5. What is most nearly the relay current (tap) setting?

(A) 3.5 A

(B) 7.0 A

(C) 7.5 A

(D) 9.5 A

10. Which type of protection relay operates (i.e., protects) for faults at a given distance from the relay and is bidirectional?

(A) differential relay

(B) directional relay

(C) distance relay

(D) phase angle relay

11. *IEEE Guide for Performing Arc-Flash Hazard Calculations* (IEEE Std 1584) provides guidance for performing arc flash hazard analysis. Data are required regarding the short-circuit study, the conductor, the fault MVA or X/R ratio from the supplying utility, the instrument transformer, and the protectice device. The short-circuit study must include large motors, since such motors provide energy to short circuits. According to the standard, what is a large motor?

(A) 0.50 hp

(B) 1.0 hp

(C) 25 hp

(D) 50 hp

12. Instantaneous trips on circuit breakers are often used as protection against arc flash hazards. If such a trip is not available, the *National Electrical Code* (NEC) allows which method as a substitute?

(A) distance relay protection

(B) energy-reducing maintenance switching

(C) maintenance performed by qualified personnel

(D) maintenance performed by a qualified operator with an observer

13. An arc flash hazard analysis is NOT required if a circuit is rated for

(A) 240 V or less and is supplied by one transformer that is rated for less than 125 kVA

(B) 600 V or less and is supplied by one transformer that is rated for less than 125 kVA

(C) 600 V or less and is supplied by one transformer that is rated for less than 500 kVA

(D) 1000 V or less and is supplied by one transformer that is rated for less than 500 kVA

14. Which of the following should be used to determine the arc flash incident energy and boundaries for voltages above 15 kV?

(A) IEEE C2, *National Electrical Safety Code*

(B) a paper or article by R. L. Doughty and/or T. E. Neal

(C) IEEE Standard 1584

(D) a paper or article by Ralph Lee

15. Consider the generator capability shown.

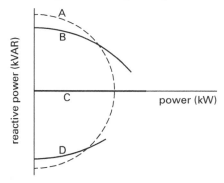

To set the protection setting for the generator correctly, the capability curves must be analyzed and limits set within their boundaries. Which curve or line indicates the rotor current limit?

(A) curve A

(B) curve B

(C) line C

(D) curve D

16. Consider a common performance/fault monitoring setup for a substation.

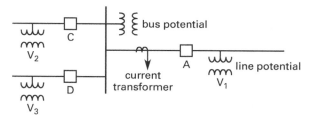

Since voltages are the same throughout the substation, a different phase is recorded on each line. V_1 may be designated phase A, while V_2 and V_3 may be labeled monitoring phases B and C, respectively. Current is monitored on phases A and C, at a minimum. From these inputs alone, all faults can be deduced from recordings of the signal traces. One such recording shows

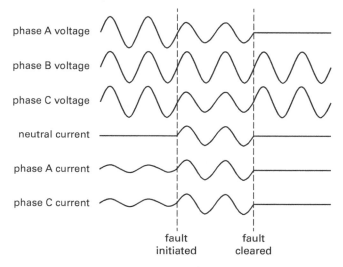

For the event recorded, which fault has occurred?

(A) phase A to phase C to ground

(B) phase A to phase B to ground

(C) phase B to phase C to ground

(D) phase to ground

17. A red lamp on a breaker generally indicates

(A) a breaker fault

(B) an open breaker

(C) a closed breaker

(D) a breaker that requires maintenance

18. Which of the following technologies defines the structure for output files of phasor measurement units (PMUs)?

   (A)   COMTRADE

   (B)   ETAP or SKM

   (C)   Modbus

   (D)   synchrophasor

19. Consider the schematic of a partial protective system shown.

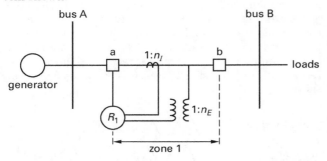

The loads draw a total of 400 A on an 11,000 V bus at a power factor of 0.8 lagging. The current transformer (CT) turns, $n_{\rm I}$, are 100. The voltage or potential transformer (PT) turns, $n_{\rm E}$, are 300.

The R-X diagram for protective relay no. 1 is shown. This maps the primary values as seen from the secondary to allow for analyzing the response of the protective relay in a visual manner.

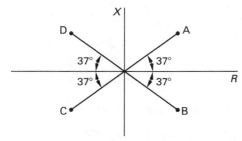

Where do the loads map and what is the approximate secondary impedance?

   (A)   point A, $7 + j5 \ \Omega$

   (B)   point B, $20 - j16 \ \Omega$

   (C)   point C, $7 + j5 \ \Omega$

   (D)   point D, $-60 - j50 \ \Omega$

SOLUTIONS

1. An impedance relay is shown in A, a directional relay in B, and a mho relay in C.

The answer is (B).

2. Per *IEEE Standard Electrical Power System Device Function Numbers, Acronyms, and Contact Designations* (IEEE Std C37.2) for power system device function, a time-delay relay is indicated by the number 2, an undervoltage relay by 27, an interlocking relay by 3, and an AC circuit breaker by 52. Therefore, a schematic with the numbers 27 and 52 on the device represents an undervoltage device for an AC circuit breaker.

The answer is (D).

3. When an ungrounded system suffers a fault (that is, a ground) on one phase, the other two phases become $\sqrt{3}$ larger than their normal value. (A designer should take the peak value as the "normal" value.) Consider the phasor diagram shown, which has a grounded phase A.

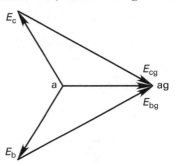

The answer is (C).

4. Depending upon the size of the auxiliary CTs, a trip may occur lower or higher than expected. Further, the trip could occur under normal (that is, nonfault) conditions. Answer options (A), (B), and (C) are possible outcomes.

The differential and auxiliary CTs will not serve their function if wired in parallel.

The answer is (D).

5. The overcurrent protection for conductors is specified by NEC Art. 240.4, which requires the overcurrent protection to be in accordance with the ampacity of the conductors as specified in NEC Art. 310.15, unless otherwise permitted by NEC Art. 240.4(A) through Art. 240.4(G).

NEC Art. 310.15 refers to NEC Table 310.16 to determine ampacity. According to NEC Table 310.16, a 12 AWG THHW rated for 90°C is rated for 30 A. The ambient temperature assumption for the table is 30°C. The correction factor for an ambient temperature of 55°C is 0.76. The maximum amperage allowed is

$$\begin{aligned} \text{ampacity} &= (\text{rating})(\text{correction factor}) \\ &= (30 \text{ A})(0.76) \\ &= 22.8 \text{ A} \end{aligned}$$

However, per NEC Art. 240.4(D)(5), the maximum overcurrent protection cannot exceed 20 A for a 12 AWG wire, unless the system qualifies for the exceptions. The system described does not qualify for the exceptions, so even though the conductor is rated to endure 22.8 A, NEC requires that overcurrent protection not exceed 20 A. If the ampacity had been calculated as less than 20 A, NEC would require the use of overcurrent protection that does not exceed that value.

The answer is (C).

6. A ring-bus distribution system arranges circuit breakers in a closed loop with the loads between circuit breakers. Since only one breaker is required to power a given load, the configuration is reliable. The ring-bus system is more economical than a multiple-bus configuration providing multiple breakers.

However, protection or relaying becomes more complex due to the multiple paths that power can take.

The answer is (C).

7. A relay is dependable if there is a measure of certainty that it will operate for designed faults. A relay is secure if there is a measure of certainty that it will operate correctly. As a relay becomes more dependable in its operation for a given fault condition, it becomes less secure, since its greater ability to operate under a fault increases the likelihood of false operation.

The answer is (A).

8. The turns ratio is

$$a = \frac{N_p}{N_s} = \frac{600}{5} = 120$$

With 400 A flowing in the primary, the secondary current is

$$I_s = \frac{I_p}{a} = \frac{400 \text{ A}}{120} = 3.33 \text{ A} \quad (3.0 \text{ A})$$

In order to provide protection, the tap must be set at 3.33 or lower.

The answer is (C).

9. The primary line current is

$$S = \sqrt{3} \, I_l V_l$$

$$I_{l,\text{primary}} = \frac{S}{\sqrt{3} \; V_l} = \frac{20 \times 10^6 \text{ VA}}{\sqrt{3} \, (12\,500 \times 10^3 \text{ V})}$$

$$= 923.76 \text{ A}$$

The voltage turns ratio of the transformer can be used to calculate the secondary line current.

$$\frac{V_p}{V_s} = \frac{I_s}{I_p}$$

$$I_s = \frac{V_p}{V_s} I_p = \left(\frac{12\,500 \text{ V}}{4160 \text{ V}} \right)(923.76 \text{ A})$$

$$= 2775.72 \text{ A}$$

Find the primary CT current from the primary current and the current turns ratio of the primary CT.

$$\begin{aligned} I_{\text{CT1}} &= \frac{I_{l,\text{primary}}}{a_p} = \frac{I_{l,\text{primary}}}{\dfrac{400}{5}} \\ &= \frac{923.76 \text{ A}}{80} \\ &= 11.55 \text{ A} \end{aligned}$$

Find the secondary CT current from the secondary current and the current turns ratio of the secondary CT.

$$\begin{aligned} I_{\text{CT2}} &= \frac{I_{l,\text{secondary}}}{a_s} = \frac{I_{l,\text{secondary}}}{\dfrac{1800}{5}} \\ &= \frac{2775.72 \text{ A}}{360} \\ &= 7.71 \text{ A} \end{aligned}$$

The difference current that feeds the differential relay at rated power is

$$\begin{aligned} I_{\text{diff}} &= |I_{\text{CT2}} - I_{\text{CT1}}| \\ &= |7.71 \text{ A} - 11.55 \text{ A}| \\ &= 3.84 \text{ A} \end{aligned}$$

The relay setting corresponding to 250% of this differential value is

$$\begin{aligned} I_{\text{setting}} &= 2.5 I_{\text{diff}} \\ &= (2.5)(3.84 \text{ A}) \\ &= 9.6 \text{ A} \quad (9.5 \text{ A}) \end{aligned}$$

The answer is (D).

Protection and Safety

10. A distance relay, also called an impedance or ratio relay, is bidirectional and operates on faults within a given distance from the relay.

A differential relay operates on a difference in currents from two separate sensing points and only responds to faults between those points. A directional relay operates on faults to either the left or the right of its location in a distribution system. A phase angle relay compares the angle between the current and the voltage to determine the direction of power flow. When the power flow is abnormal, the relay sends a protective signal.

The answer is (C).

11. A large motor is defined as 37 kW, which is 50 hp or larger per *IEEE Guide for Performing Arc-Flash Hazard Calculations* (IEEE Std 1584).

The answer is (D).

12. From NEC Art. 240.87, when an instantaneous trip circuit breaker is not available, equivalent means are approved: (a) zone-selective interlocking, (b) differential relaying, and (c) energy-reducing maintenance switching with a local status indicator that is used when a worker is within the arc flash boundary defined by NFPA 70E, *Standard for Electrical Safety in the Workplace.*

The answer is (B).

13. From NFPA 70E, *Standard for Electrical Safety in the Workplace*, Art. 130, Exception no. 2, an arc flash hazard analysis is not required when a circuit is rated for 240 V or less and is supplied by a single transformer that is rated less than 125 kVA.

The answer is (A).

14. All the answer options can be used to calculate arc flash incident energy and appropriate boundaries. Their assumptions and limitations are summarized in Annex D of NFPA 70E, *Standard for Electrical Safety in the Workplace.*

IEEE C2, *National Electrical Safety Code* is the source for information on high-voltage lines from 15 kV to 500 kV.

The answer is (A).

15. Curve B is the armature current limit. Line C is the true power line, where the power factor is equal to one, pf = 1. Curve D is the end ring limit, also known as the underexcitation limit.

The generator must operate within the limits of curves A, B, and D.

Curve A is the rotor current limit, also known as the overexcitation limit, and represents the maximum limit of the rotor current before heating effects become detrimental.

The answer is (A).

16. The recording, known as an oscillogram, shows an impact to phase A and phase C voltage. The fault includes those two phases. In addition, neutral current changed from 0 A before the fault to a given value. The fault occurred from phase A to phase C to ground.

The answer is (A).

17. Red is the industry-standard indication that a breaker is in service, or closed. Red, meaning "stop," indicates that the breaker is not safe for human interaction or maintenance. Because the breaker is closed, voltage or current is present.

The answer is (C).

18. COMTRADE is an IEEE standard for the data files of protection systems. ETAP and SKM are power system analysis tools. Modbus is a communication protocol. Synchrophasor is an IEEE standard for the output files of phasor measurement units (PMUs), whose purpose is to ensure interoperability among PMU manufacturers' units.

The answer is (D).

19. The *apparent impedance* is found from Ohm's law. This is the magnitude of impedance related to the apparent power.

$$V = IR = I|\mathbf{Z}|$$

$$\begin{aligned} Z_{\text{apparent}} &= \frac{V}{I} \\ &= \frac{11{,}000 \text{ V}}{400 \text{ A}} \\ &= 27.5 \ \Omega \end{aligned}$$

The primary impedance is then calculated as

$$\begin{aligned} \mathbf{Z}_p &= \mathbf{R}_p \mathbf{Z}_{0.8} \\ &= (27.5 \ \Omega)(0.8 + j0.6) \\ &= 22 + j16.5 \ \Omega \end{aligned}$$

The secondary impedance is then calculated using the turns ratios.

$$\mathbf{Z}_s = \frac{\mathbf{V}_s}{\mathbf{I}_s} = \frac{\mathbf{V}_p\left(\dfrac{1}{n_E}\right)}{\mathbf{I}_p\left(\dfrac{1}{n_1}\right)} = \mathbf{Z}_p\left(\frac{n_1}{n_E}\right)$$

$$= (22 + j16.5 \ \Omega)\left(\frac{100}{300}\right)$$

$$= 6.60 + j4.95 \ \Omega \quad (7 + j5 \ \Omega)$$

This lagging load, which shows a $+R$ and $+X$ value, maps to point A.

During a fault condition in the zone, power will flow from the loads (that is, from bus B toward the generator or bus A). This point maps to point C. Leading loads of the same magnitude map to point B, with their fault component mapping to point D.

The answer is (A).

Topic IX: Machinery and Devices

43

Rotating DC Machinery

PRACTICE PROBLEMS

1. A four-pole generator is turned at 3600 rpm. Each of its poles has a flux of 3.0×10^{-4} Wb. The armature has a total of 200 conductors and is simplex lap wound. What average voltage is generated at no load?

2. A DC generator has an armature resistance of $0.31 \ \Omega$. The shunt field resistance is $134 \ \Omega$. No-load voltage is 121 V at 1775 rpm. Full-load current is 30 A at 110 V. Assume a constant flux. What are the (a) no-load shunt field current, (b) full-load speed, and (c) copper power losses at full load?

3. (a) If 90 A flow through a $0.05 \ \Omega$ armature resistance, what is the back emf of a 10 hp shunt motor operating on 110 V terminals? (b) If the field resistance is $60 \ \Omega$, what is the line current?

4. The nameplate of a 240 V DC motor states that the full-load line current is 67 A. During a no-load test at rated voltage, 3.35 A armature current and 3.16 A field current are drawn. The no-load armature resistance is $0.207 \ \Omega$. The no-load brush drop is 2 V. What are the (a) horsepower and (b) efficiency at full load?

5. A DC generator has a 250 V output at a 50 A load current and a 260 V output at no load. Estimate the armature resistance.

6. A separately excited DC generator is rated at 10 kW, 240 V, and 2800 rpm for an input power of 15 hp. The no-load voltage is 260 V. Determine the (a) armature resistance and (b) mechanical power loss at 2800 rpm.

7. A DC shunt motor is running at rated values when the line voltage suddenly drops by 10%. Determine the resulting changes in input power and motor speed if (a) the load torque is constant and (b) the load torque is proportional to the speed.

8. A DC series motor is running at rated speed, voltage, and current when the load torque increases by 25%. Determine the effects of this change, assuming that the line voltage remains the same.

9. A DC shunt motor receives 200 V and 25 A at 1000 rpm and 5 hp. It has an armature resistance of $0.5 \ \Omega$ and a shunt field resistance of $100 \ \Omega$. Determine the (a) rotational losses and (b) no-load speed at 200 V.

10. A DC shunt motor is rated at 7.5 hp, 60 A, 1000 rpm, and 120 V. Assume the load torque is constant, regardless of speed. What field resistance values are required to change the speed to (a) 1150 rpm and (b) 750 rpm?

11. In the figure shown, which line represents the characteristics of a series motor?

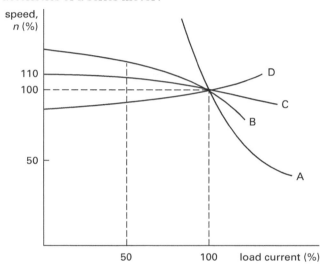

(A) A

(B) B

(C) C

(D) D

12. A conductor, a magnetic field, and relative motion are required to generate a voltage. In the illustration shown, a conductor is moved through a magnetic field.

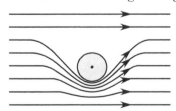

What direction of motion is required for the conductor to generate the voltage that results in the current moving out of the paper as shown?

(A) down

(B) left

(C) right

(D) up

13. A conductor, a magnetic field, and a current on the conductor are required to generate a force. In the illustration shown, a conductor has a current applied into the paper, resulting in two magnetic fields.

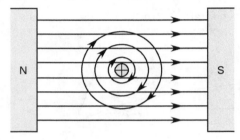

What direction of force on the conductor results from the direction of the current and the direction of each of the magnetic fields?

(A) down

(B) left

(C) right

(D) up

14. A DC machine is shown with the direction of motion of the machine and the direction of the current. The velocity of the current is out of the paper, with the magnetic field to the right.

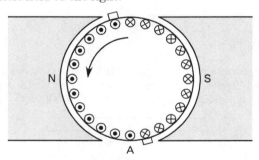

What type of machine is shown?

(A) generator

(B) motor

(C) either a generator or a motor

(D) not enough information

15. The equivalent circuit for a shunt-wired DC motor is shown. The applied voltage at the terminals is 120 V, the field resistance is 10 Ω, the armature resistance is 0.1 Ω, and the terminal current is 15 A.

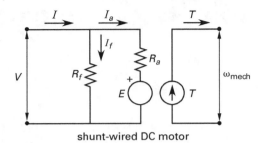

shunt-wired DC motor

Most nearly, what is the counter emf (CEMF) at this steady-state point?

(A) 0 V

(B) 80 V

(C) 120 V

(D) 240 V

SOLUTIONS

1. For conductors, z, parallel armature paths, a, poles, p, a magnetic flux, Φ, and a rotational speed, n, Eq. 43.13 is used.

$$V = \left(\frac{z}{a}\right)p\Phi\left(\frac{n}{60}\right)$$

For simplex lap winding, $a = p$.

$$V = \left(\frac{200}{4}\right)(4)(3.0 \times 10^{-4}\ \text{Wb})\left(\frac{3600\ \dfrac{\text{rev}}{\text{min}}}{60\ \dfrac{\text{sec}}{\text{min}}}\right)$$

$$= \boxed{3.6\ \text{V}}$$

2. (a) The parameters pertain to full load, fl; no load, nl; field, f; and armature, a. Using Ohm's law (Eq. 26.11),

$$I_{f,\text{nl}} = \frac{V_{\text{nl}}}{R_f} = \frac{121\ \text{V}}{134\ \Omega}$$

$$= \boxed{0.903\ \text{A}}$$

(b) Voltage is proportional to rotational speed.

$$n_{\text{fl}} = \left(\frac{V_{\text{fl}} + R_a\left(I_{\text{fl}} + \dfrac{V_{\text{fl}}}{R_f}\right)}{V_{\text{nl}} + R_a\left(\dfrac{V_{\text{nl}}}{R_f}\right)}\right)n_{\text{nl}}$$

$$= \left(\frac{110\ \text{V} + (0.31\ \Omega)\left(30\ \text{A} + \dfrac{110\ \text{V}}{134\ \Omega}\right)}{121\ \text{V} + (0.31\ \Omega)\left(\dfrac{121\ \text{V}}{134\ \Omega}\right)}\right)\left(1775\ \frac{\text{rev}}{\text{min}}\right)$$

$$= \boxed{1750\ \text{rev/min}\quad(1750\ \text{rpm})}$$

(c)

$$P = \frac{V_{\text{fl}}^2}{R_f} + \left(I_{\text{fl}} + \frac{V_{\text{fl}}}{R_f}\right)^2 R_a$$

$$= \frac{(110\ \text{V})^2}{134\ \Omega} + \left(30\ \text{A} + \frac{110\ \text{V}}{134\ \Omega}\right)^2(0.31\ \Omega)$$

$$= \boxed{385\ \text{W}}$$

3. (a) From Eq. 43.31,

$$E = V_{\text{fl}} - I_{a,\text{load}}R_a$$

$$= 110\ \text{V} - (90\ \text{A})(0.05\ \Omega)$$

$$= \boxed{105.5\ \text{V}}$$

(b)

$$I_{\text{fl}} = \frac{V_{\text{fl}}}{R_f} + I_{a,\text{load}} = \frac{110\ \text{V}}{60\ \Omega} + 90\ \text{A}$$

$$= \boxed{91.8\ \text{A}}$$

4.

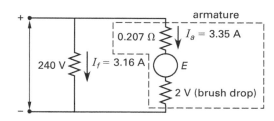

Using Eq. 26.12, the total no-load power input to the armature is

$$P_a = I_a V = (3.35\ \text{A})(240\ \text{V})$$

$$= 804\ \text{W}$$

The I^2R power loss in the armature resistance is

$$P_{\text{resistance}} = I_a^2 R = (3.35\ \text{A})^2(0.207\ \Omega)$$

$$= 2.323\ \text{W}$$

The armature brush power loss is

$$P_{\text{brush}} = I_a V_{\text{brush}}$$

$$= (3.35\ \text{A})(2\ \text{V})$$

$$= 6.7\ \text{W}$$

The no-load rotational loss is

$$P_{\text{rotational}} = P_a - P_{\text{resistance}} - P_{\text{brush}}$$

$$= 804\ \text{W} - 2.323\ \text{W} - 6.7\ \text{W}$$

$$= 794.9\ \text{W}\quad(795.0\ \text{W})$$

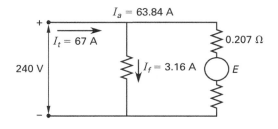

At full load, the total input power is

$$P_{in} = I_{fl} V_{nameplate}$$
$$= (67\ A)(240\ V)$$
$$= 16,080\ W$$

The field current is the same as for the no-load condition.

$$I_f = 3.16\ A$$

The field power loss is

$$P_f = I_f V = (3.16\ A)(240\ V)$$
$$= 758.4\ W$$
$$I_a = I_{fl} - I_f = 67\ A - 3.16\ A$$
$$= 63.84\ A$$
$$P_{resistance} = I_a^2 R = (63.84\ A)^2(0.207\ \Omega)$$
$$= 843.6\ W$$

Using the no-load data, the equivalent brush resistance is

$$R_{brush} = \frac{V_{brush}}{I} = \frac{2\ V}{3.35\ A} = 0.597\ \Omega$$

The armature brush power loss is

$$P_{brush} = I_a^2 R_{brush} = (63.84\ A)^2(0.597\ \Omega)$$
$$= 2433.1\ W$$

The rotational loss is the same as for the no-load condition.

$$P_{rotational} = 795.0\ W$$

The net power to the load is

$$P_{net} = P_{in} - P_f - P_{resistance} - P_{brush} - P_{rotational}$$
$$= 16,080\ W - 758.4\ W - 843.6\ W$$
$$\quad -2433.1\ W - 795.0\ W$$
$$= 11,249.9\ W$$

(a) The motor output horsepower is

$$P = \frac{(11,249.9\ W)\left(1.341\ \dfrac{hp}{kW}\right)}{1000\ \dfrac{W}{kW}}$$
$$= \boxed{15.09\ hp}$$

(b) The efficiency is

$$\eta = \frac{P_{net}}{P_{in}} = \frac{11,249.9\ W}{16,080\ W}$$
$$= \boxed{0.70\quad (70.0\%)}$$

5. The generated voltage, E_g, supplies the load, as shown.

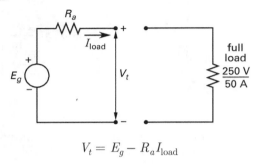

$$V_t = E_g - R_a I_{load}$$

At no load,

$$V_t = E_g = 260\ V$$

At full load,

$$V_t = E_g - (50\ A) R_a$$
$$250\ V = E_g - (50\ A) R_a$$
$$R_a = \frac{V_{nl} - V_{fl}}{I_{fl}}$$
$$= \frac{260\ V - 250\ V}{50\ A}$$
$$= \boxed{0.2\ \Omega}$$

6. (a)

$$I_{fl} = \frac{10 \times 10^3\ W}{240\ V} = 41.67\ A$$
$$R_a = \frac{V_{nl} - V_{fl}}{I_{fl}}$$
$$= \frac{260\ V - 240\ V}{41.67\ A}$$
$$= \boxed{0.48\ \Omega}$$

(b)

$$P_{in} = P_{mech.loss} + I_a^2 R_a + p_{out}$$
$$= (15 \text{ hp})\left(745.7 \ \frac{W}{hp}\right)$$
$$= 11.186 \times 10^3 \text{ W} \quad (11.186 \text{ kW})$$
$$I_a^2 R_a = (41.67 \text{ A})^2 (0.48 \ \Omega)$$
$$= 0.833 \times 10^3 \text{ W} \quad (0.833 \text{ kW})$$
$$P_{out} = 10 \times 10^3 \text{ W} \quad (10 \text{ kW})$$
$$P_{mech.loss} = 11.186 \times 10^3 \text{ W} - 10.833 \times 10^3 \text{ W}$$
$$= \boxed{353 \text{ W}}$$

7. For field flux, Φ is proportional to line voltage.

$$\Phi = aV$$
$$\Omega_r = \text{rotor speed} \ [\text{rad/s}]$$
$$E_g = k\Phi\Omega_r = kaV\Omega_r$$
$$T = k\Phi I_a = kaV I_a \quad [\text{torque in N·m}]$$
$$V = E_g + I_a R_a$$
$$P_{in} \approx V I_a$$
$$P_{out} \approx T\Omega_r$$

The parameters pertain to the field, f; armature, a; generator, g; and rotor, r.

(a) When V is decreased by 10%, to keep T constant, the armature current, I_a, is increased by approximately 10%, and the generated emf, E_g, is decreased by 10% with the line voltage, V.

$$P_{in} = \boxed{V I_a \approx \text{ constant}}$$
$$P_{out} = P_{in} - I_a^2 R_a - \text{mechanical loss}$$

Because I_a has increased by 10%, I_a^2 is increased by 20%. $I_a^2 R_a$ is on the order of 5% of the load power, so it only causes a 1% decrease in P_{out}.

E_g decreases 10% with V, but

$$\boxed{\Omega_r \text{ decreases by only 1\%.}}$$

(b) For $T \propto \Omega_r$, in a first approximation,

$$E_g \approx V$$
$$E_g \approx kaV\Omega_r$$

$ka\Omega_r \approx 1$, and there is no change in speed.

In a second approximation,

$$I_a = \frac{V - E_g}{R_a}$$
$$T = kaV I_a$$

The air-gap power is

$$P_{air\,gap} = E_g I_a = T\Omega_r$$
$$T = m\Omega_r$$
$$T\Omega_r = m\Omega_r^2$$
$$E_g = kaV\Omega_r$$
$$E_g I_a = \frac{kaV^2\Omega_r(1 - ka\Omega_r)}{R_a} = m\Omega_r^2$$
$$\frac{kaV^2}{R_a} = \frac{m\Omega_r}{1 - ka\Omega_r} = \frac{m}{\frac{1}{\Omega_r} - ka}$$

When V decreases by 10%, $1/\Omega_r - ka$ must increase by 21%, so Ω_r must decrease some amount. If under normal conditions $I_a R_a = 0.05\,V$,

$$I_a R_a = V - E_{g,nl} = V(1 - ka\Omega_{r,nl})$$
$$1 - ka\Omega_{r,nl} = 0.05$$
$$ka = \frac{0.95}{\Omega_{r,nl}}$$

When the line voltage, V, decreases by 10%,

$$\frac{m\Omega_r}{1 - ka\Omega_r} \rightarrow (0.81)\left(\frac{m\Omega_{r,nl}}{1 - ka\Omega_{r,nl}}\right)$$
$$\frac{\Omega_r}{1 - (0.95)\left(\frac{\Omega_r}{\Omega_{r,nl}}\right)} = \left(\frac{0.81}{0.05}\right)\Omega_{r,nl} = 16.2\Omega_{r,nl}$$
$$\frac{\Omega_r}{\Omega_{r,nl}} = (16.2)\left(1 - (0.95)\left(\frac{\Omega_r}{\Omega_{r,nl}}\right)\right)$$
$$= \frac{16.2}{1 + (0.95)(16.2)}$$
$$= 0.99$$

$$\boxed{\Omega_r \text{ decreases by about 1\%.}}$$

$$I_a R_a = V - E_g = V - ka\Omega_r$$
$$T = m\Omega_r = kaV I_a$$

Eliminating Ω_r gives

$$I_a R_a = \frac{V}{1 + \dfrac{(kaV)^2}{mRa}}$$

The original value of $I_a R_a$ is

$$I_{a,\mathrm{nl}} R_a \approx 0.05 V_{\mathrm{nl}}$$

V_0 is the initial voltage. Then,

$$1 + \frac{(ka V_{\mathrm{nl}})^2}{m R_a} \approx 20$$

When the voltage decreases by 10%,

$$1 + \frac{(ka(0.9) V_{\mathrm{nl}})^2}{m R_a} = 16.39$$

The resulting armature IR drop is

$$I_a R_a = \frac{0.9 V_{\mathrm{nl}}}{16.39} = 0.0549 V_{\mathrm{nl}}$$

Omitting V_{nl}, the current changes by

$$\frac{I_a R_a - I_{a,\mathrm{nl}} R_a}{I_{a,\mathrm{nl}} R_a} = \frac{0.0549 - 0.05}{0.05}$$
$$= 0.098 \text{ pu}$$

The input power also changes.

$$P_{\mathrm{in}} = (0.9 V_{\mathrm{nl}})(1.098 I_{a,\mathrm{nl}}) = 0.988 P_{\mathrm{nl}}$$

$$\boxed{\text{The power decreases by about 1\%.}}$$

8.

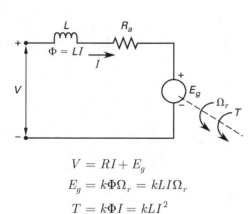

$$V = RI + E_g$$
$$E_g = k\Phi\Omega_r = kLI\Omega_r$$
$$T = k\Phi I = kLI^2$$

If the torque, T, increases by 25%, so does I^2, and $\sqrt{1.25} = 1.11$. So,

$$\boxed{I \text{ is up 11\%.}}$$

Using subscript zero for the initial values,

$$E_{g0} = V - I_0 R_a$$

When the current, I, increases by 11%, and when using subscript 1 for this condition,

$$\boxed{\text{The generated emf, } E_g, \text{ decreases by about 1\%.}}$$

$$E_{g1} = 0.99 E_{g0}$$
$$E_{g0} = kLI_0\Omega_{r0}$$
$$E_{g1} = 0.99kLI_0\Omega_{r0}$$
$$= kLI_1\Omega_{r1}$$
$$\Omega_{r1} = \left(\frac{0.99 I_0}{I_1}\right)\Omega_{r0}$$
$$= \left(\frac{0.99}{1.11}\right)\Omega_{r0}$$
$$= 0.89\Omega_{r0}$$

$$\boxed{\text{The speed also decreases by about 11\%.}}$$

The output power is

$$P_2 = (1.25 T_0)(0.89\Omega_{r0}) = 1.11 P_0$$

$$\boxed{\text{The air-gap power increases by 11\%.}}$$

9. (a)

$$I_t = 25 \text{ A} = I_a + I_f$$
$$I_f = \frac{V}{R} = \frac{200 \text{ V}}{100 \text{ }\Omega} = 2 \text{ A}$$
$$I_a = I_t - I_f = 25 \text{ A} - 2 \text{ A} = 23 \text{ A}$$
$$E_g = V - I_a R_a = 200 \text{ V} - (23 \text{ A}) R_a$$
$$= 188.5 \text{ V}$$
$$P_{\mathrm{air\,gap}} = E_g I_a = (188.5 \text{ V})(23 \text{ A})$$
$$= 4335.5 \text{ W}$$
$$P_{\mathrm{load}} = (5 \text{ hp})\left(745.7 \frac{\text{W}}{\text{hp}}\right) = 3728.5 \text{ W}$$
$$P_{\mathrm{rotational\,losses}} = P_{\mathrm{air\,gap}} - P_{\mathrm{load}} = 4335.5 \text{ W} - 3728.5 \text{ W}$$
$$= \boxed{607 \text{ W}}$$

Machinery and Devices

(b) As shown in Eq. 43.30 for generated voltage, the speed, n, is proportional to the counter voltage.

$$E_g = k\Phi n$$

From the information in Part (a), $k\Phi$ is determined to be

$$k\Phi = \dfrac{188.5 \text{ V}}{1000 \; \dfrac{\text{rev}}{\text{min}}}$$

If the rotational losses are fixed at 607 W,

$$E_g I_a = 607 \text{ W}$$
$$I_a = \dfrac{607 \text{ W}}{E_g}$$

The armature current at the specified 200 V is

$$I_a = \dfrac{200 \text{ V} - E_g}{0.5 \; \Omega}$$

Substituting for the armature current gives

$$\dfrac{607 \text{ W}}{E_g} = \dfrac{200 \text{ V} - E_g}{0.5 \; \Omega}$$

Rearranging gives the quadratic equation.

$$E_g^2 - 200 E_g + 303.5 = 0$$

Solving the quadratic equation gives

$$E_g = \dfrac{-(-200 \text{ V}) \pm \sqrt{(-200 \text{ V})^2 - (4)(1)(303.5 \text{ V}^2)}}{(2)(1)}$$
$$= 198.5 \text{ V}$$

Comparing voltage to rpm ratios gives

$$\dfrac{n}{198.5 \text{ V}} = \dfrac{1000 \; \dfrac{\text{rev}}{\text{min}}}{188.5 \text{ V}}$$
$$n = \boxed{1053 \text{ rev/min} \quad (1053 \text{ rpm})}$$

10.

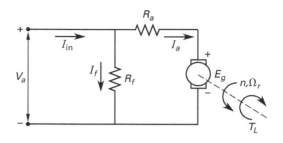

$$P_{\text{in}} = V_a I_{\text{in}} = (60 \text{ A})(120 \text{ V}) = 7200 \text{ W}$$
$$P_{\text{out}} = (7.5 \text{ hp})\left(745.7 \; \dfrac{\text{W}}{\text{hp}}\right) = 5593 \text{ W}$$

Because load torque is constant,

$$P_{\text{out}} = k_T n$$
$$k_T = \dfrac{5593 \text{ W}}{1000 \; \dfrac{\text{rev}}{\text{min}}}$$
$$= 5.593 n \quad [n \text{ in rev/min}]$$

The losses are the difference between P_{in} and P_{out}.

$$P_{\text{loss}} = P_{\text{in}} - P_{\text{out}} = 7200 \text{ W} - 5593 \text{ W}$$
$$= 1607 \text{ W}$$

Because no information is given, these losses are assigned to the rotational losses, the armature losses $I_a^2 R_a$, and the field losses $I_f^2 R_f$, making all three equal in magnitude.

$$I_a^2 R_a = I_f^2 R_f = P_{\text{rotational losses}} = 535 \text{ W}$$
$$I_f^2 R_f = \dfrac{V_a^2}{R_f} = 535 \text{ W}$$
$$R_f = \dfrac{(120 \text{ V})^2}{535 \text{ W}} = 26.9 \; \Omega$$
$$I_f = \dfrac{120 \text{ V}}{26.9 \; \Omega} = 4.46 \text{ A}$$

Therefore,

$$I_a = I_{\text{in}} - I_f = 60 \text{ A} - 4.46 \text{ A}$$
$$= 55.54 \text{ A}$$
$$R_a = \dfrac{P_{\text{rotational losses}}}{I_a^2} = \dfrac{535 \text{ W}}{(55.54 \text{ A})^2}$$
$$= 0.173 \; \Omega$$

Because of rounding, the powers are recalculated and the remainder are assigned to rotational losses.

$$P_{\text{rotational losses}} = P_{\text{loss}} - \frac{V_a^2}{R_f} - R_a I_a^2$$

$$= 1607 \text{ W} - \frac{(120 \text{ V})^2}{26.9 \text{ }\Omega}$$

$$- (0.173 \text{ }\Omega)(55.54 \text{ A})^2$$

$$= 538 \text{ W}$$

The air-gap power must then be

$$P_{\text{air gap}} = P_{\text{out}} + P_{\text{rotational losses}}$$

$$= 5.593n + 538 \text{ W}$$

The air-gap power is also $E_g I_a$, and E_g is proportional to I_f and n.

$$E_g = k' I_f n = k'' \left(\frac{n}{R_f} \right)$$

At $n = 1000$ rpm and $R_f = 26.9 \text{ }\Omega$,

$$E_g = 120 \text{ V} - (55.54 \text{ A})(0.173 \text{ }\Omega)$$

$$= 110.4 \text{ V}$$

$$k'' = \frac{(110.4 \text{ V})(26.9 \text{ }\Omega)}{1000 \ \frac{\text{rev}}{\text{min}}}$$

$$= 2.97 \ \frac{\text{V}\cdot\Omega}{\frac{\text{rev}}{\text{min}}}$$

$$E_g = 2.97 \left(\frac{n}{R_f} \right)$$

$$I_a = \frac{120 \text{ V} - E_g}{R_a}$$

$$= \frac{120 \text{ V} - (2.97)\left(\dfrac{n}{R_f} \right)}{0.173 \text{ }\Omega}$$

$$= 694 \text{ A} - (17.2)\left(\frac{n}{R_f} \right)$$

$$P_{\text{air gap}} = E_g I_a$$

$$= (2061)\left(\frac{n}{R_f} \right) - (51.08)\left(\frac{n}{R_f} \right)^2$$

Equating the electrical side to the mechanical side,

$$(51.08)\left(\frac{n}{R_f} \right)^2 + \left(5.593 - \frac{2061}{R_f} \right)n + 538 \text{ W} = 0$$

$$0 = R_f^2 - \left(\frac{2061n}{5.593n + 538 \text{ W}} \right)(R_f) + \frac{51.08n^2}{5.593n + 538 \text{ W}}$$

(a) For $n = 1150$ rpm,

$$R_f^2 - (340.0 \text{ }\Omega)R_f + 9692 \text{ }\Omega^2 = 0 \text{ }\Omega^2$$

$$R_f = \boxed{31.4 \text{ }\Omega}$$

(b) For $n = 750$ rpm,

$$R_f^2 - (326.6 \text{ }\Omega)R_f + 6071 \text{ }\Omega^2 = 0 \text{ }\Omega^2$$

$$R_f = \boxed{19.8 \text{ }\Omega}$$

11. A series motor generates large torque at low speeds and overspeeds at low loads. The graph line that shows those properties is A.

The answer is (A).

12. The $\mathbf{v} \times \mathbf{B}$ portion of the force equation (Eq. 43.9) or generated voltage equation (Eq. 43.4) is used to to determine the direction. In this case, the velocity is the velocity of the conductor, which must be $\boxed{\text{down}}$ to provide the current out of the paper. This is shown by the head of the arrow. The right-hand rule is used to confirm.

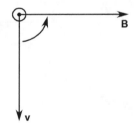

The answer is (A).

13. The $\mathbf{v} \times \mathbf{B}$ portion of the force equation (Eq. 43.9) or generated voltage equation (Eq. 43.4) is used to determine the direction. In this case, the velocity is the velocity of the current flow. This is the direction of positive or conventional current flow, not the electron flow. Given that the velocity of the current is into the paper, and using the right-hand rule, the force is exerted downward.

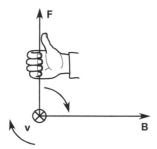

The net result is shown.

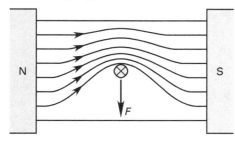

The answer is (A).

14. The $\mathbf{v} \times \mathbf{B}$ portion of the force equation (Eq. 43.9) or generated voltage equation (Eq. 43.4) is used to determine the direction.

The velocity of the current flow is used. This is the direction of positive or conventional current flow, not the electron flow. Using the right-hand rule, the force is upward, which means the direction is expected to be upward, resulting in clockwise rotation. However, the net force on the conductor is downward, resulting in counterclockwise rotation, indicating that the machine is operating as a generator. The force in question is the counterforce or countertorque.

The result can be confirmed using the velocity of the conductor itself. The velocity of the conductor is downward, resulting in a counterclockwise rotation with the magnetic field to the right, as shown. Using the right-hand rule, the current should flow out of the paper. The net force on the conductor results in conductor movement, such that it generates a voltage that results in a current out of the paper.

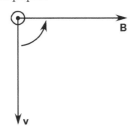

The machine is a generator.

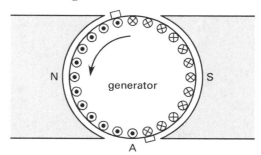

The answer is (A).

15. The electrical portion of the circuit is shown.

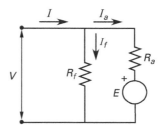

Using Ohm's law, calculate the field current applied to the field.

$$V = I_f R_f$$
$$I_f = \frac{V}{R_f}$$
$$= \frac{120 \text{ V}}{10 \text{ }\Omega}$$
$$= 12 \text{ A}$$

Using Kirchhoff's current law (KCL) at the field/armature node, calculate the armature current.

$$I - I_f - I_a = 0$$
$$I_a = I - I_f$$
$$= 15 \text{ A} - 12 \text{A}$$
$$= 3 \text{A}$$

Substituting the armature current into the equation for Kirchhoff's voltage law (KVL) gives

$$V - I_a R_a - E_{\text{CEMF}} = 0$$
$$E_{\text{CEMF}} = V - I_a R_a$$
$$= 120 \text{ V} - (3 \text{ A})(0.1 \text{ }\Omega)$$
$$= 119.7 \text{ V} \quad (120 \text{ V})$$

The answer is (B).

44 Rotating AC Machinery

PRACTICE PROBLEMS

1. An old single-phase generator has four poles. The armature is simplex lap wound and has a total of 240 armature conductors. The effective armature length and radius are 16.7 cm and 5 cm, respectively. The armature rotates at 1800 rpm. (a) If the uniform magnetic flux density per pole is 1.0 T, what is the maximum voltage induced? (b) What is the horsepower of the driving motor if the rated output is 1 kW and the conversion efficiency is 90%?

2. (a) What is the rotational speed of a 24-pole alternator that produces a 60 Hz sinusoid? (b) If the effective emf is 2200 V and each phase coil has 20 turns in series, what is the maximum flux?

3. Calculate the full-load phase current drawn by a 440 V (rms), 20 hp (per phase) induction motor having a full-load efficiency of 86% and a full-load power factor of 76%.

4. A 200 hp, three-phase, four-pole, 60 Hz, 440 V (rms) squirrel-cage induction motor operates at full load with an efficiency of 85%, a power factor of 91%, and 3% slip. Find the (a) speed in rpm, (b) torque developed, and (c) line current.

5. A factory's induction motor load draws 550 kW at 82% power factor. What size synchronous motor is required to carry 250 hp and raise the power factor to 95%? The line voltage is 220 V (rms).

6. The nameplate of an induction motor lists 960 rpm as the full-load speed. For what frequency was the motor designed?

7. A three-phase, 440 V (line-to-line) synchronous motor has a synchronous reactance (X_s) of 20 Ω and an armature resistance of 0.5 Ω per phase. The synchronous speed is 1800 rpm. The no-load field current is adjusted for minimum armature current at a field current (I_f) of 5 A DC, at which time the input power is 900 W. This machine is to provide 10 kVAR of power factor correction when running lightly loaded. Determine the rotor current required.

8. A 440 V, three-phase synchronous motor is driving a 45 hp pump. Assume the motor is 93% efficient. The rotor current is adjusted until the line current is at a minimum, occurring when the rotor current is 8.2 A

DC. Determine the rotor current necessary to provide a leading power factor of 0.8.

9. A 50 hp, 50 kVA, 1746 rpm squirrel-cage induction motor has phase windings rated at 440 V. With the windings connected in a delta configuration, a blocked rotor test is carried out under rated current conditions. The result in per-phase volt-amperes is found to be $P + jQ = 1097 + j\,2791$ VA. This machine is fed from a three-phase, 50 kVA, 440 V transformer that has 10% reactance per phase (on its own base). Determine the dip in the line voltage when the motor is started. The transformer can be modeled as a wye connection of 254 V sources in series with the 10% reactance on each phase.

10. A 50 hp, 50 kVA, 1746 rpm squirrel-cage induction motor has phase windings rated at 440 V and is normally connected in delta. Its starting current is 367 A and 38.3 kW per phase when started directly across the line, and under those conditions it develops a starting torque of 190 ft-lbf. A blocked rotor test at the rated current results in a voltage of 78.6 V and a power of 1.1 kW per phase. A no-load test at the rated voltage results in a current of 25.9 A and 1.1 kW per phase. If this machine is started in the wye configuration, determine its starting current.

11. What is the synchronous speed of a two-pole 400 Hz generator?

(A) 120 rpm

(B) 200 rpm

(C) 7200 rpm

(D) 24,000 rpm

12. An AC induction motor has the torque versus speed characteristic curve shown.

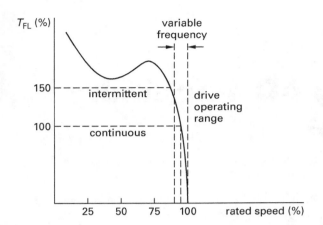

An expansion of the continuous and intermittent operating range is shown, with 1800 rpm as the 100% rated speed.

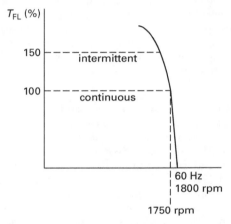

What is the percent slip and over what frequency range must the variable frequency drive (VFD) operate to control the speed?

(A) 2.8%; 1.67 Hz

(B) 2.8%; 1.71 Hz

(C) 3.0%; 1.67 Hz

(D) 3.0%; 1.71 Hz

13. In rotating machinery, which occurrence causes zero sequence reactances?

(A) A rotor moves synchronously with the magnetomotive force (mmf) produced by the steady-state armature current.

(B) Due to switching events or terminal short circuits, armature mmf changes very rapidly.

(C) Due to unbalanced loading or faults, armature mmf rotates in the direction opposite the rotor at synchronous speed.

(D) Due to a ground fault, all three phases of the armature have currents that are identical in magnitude and phase.

14. Consider the following sources of or causes for reactances that occur in rotating machinery.

I. a rotor that is moving in synchronism with the magnetomotive force (mmf) produced by the steady-state armature current

II. armature mmf that is changing with respect to time due to electromechanical transients

III. armature mmf that is changing very rapidly with respect to time due to switching events or terminal short circuits

IV. reactance that occurs when the armature mmf rotates in the opposite direction of the rotor at synchronous speed due to unbalanced loading or faults

V. reactance that occurs when all three phases of the armature have identical currents in magnitude and phase due to a ground fault

Which description applies to direct synchronous reactances and quadratone synchronous reactances?

(A) I

(B) II

(C) III

(D) IV and V

15. Consider the following sources of or causes for reactances that occur in rotating machinery.

I. a rotor that is moving in synchronism with the magnetomotive force (mmf) produced by the steady-state armature current

II. armature mmf that is changing with respect to time due to electromechanical transients

III. armature mmf that is changing very rapidly with respect to time due to switching events or terminal short circuits

IV. reactance that occurs when the armature mmf rotates in the opposite direction of the rotor at synchronous speed due to unbalanced loading or faults

V. reactance that occurs when all three phases of the armature have identical currents in magnitude and phase due to a ground fault

Which description applies to direct subtransient reactances and quadratone subtransient reactances?

(A) I

(B) II

(C) III

(D) IV and V

16. Consider the following sources of or causes for reactances that occur in rotating machinery.

I. a rotor that is moving in synchronism with the magnetomotive force (mmf) produced by the steady-state armature current

II. armature mmf that is changing with respect to time due to electromechanical transients

III. armature mmf that is changing very rapidly with respect to time due to switching events or terminal short circuits

IV. reactance that occurs when the armature mmf rotates in the opposite direction of the rotor at synchronous speed due to unbalanced loading or faults

V. reactance that occurs when all three phases of the armature have identical currents in magnitude and phase due to a ground fault

Which description applies to direct transient reactances and quadratone transient reactances?

(A) I

(B) II

(C) III

(D) IV and V

17. Consider the following sources of or causes for reactances that occur in rotating machinery.

I. a rotor that is moving in synchronism with the magnetomotive force (mmf) produced by the steady-state armature current

II. armature mmf that is changing with respect to time due to electromechanical transients

III. armature mmf that is changing very rapidly with respect to time due to switching events or terminal short circuits

IV. reactance that occurs when the armature mmf rotates in the opposite direction of the rotor at synchronous speed due to unbalanced loading or faults

V. reactance that occurs when all three phases of the armature have identical currents in magnitude and phase due to a ground fault

Which description applies to negative sequence reactances, X_2?

(A) I

(B) II and III

(C) IV

(D) V

18. A basic, nonelectronic synchronism check device circuit is shown.

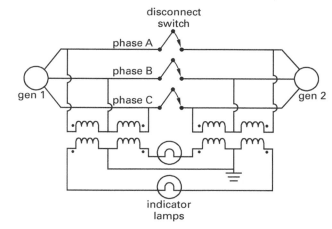

In order to parallel two generators properly, the voltages and frequencies must be matched and the phase sequences must be identical. When these conditions are met, the indicator lamps of the circuit are

(A) bright at different times

(B) bright at the same time

(C) dark at different times

(D) dark at the same time

19. A generic pulse width modulation (PWM) circuit is shown.

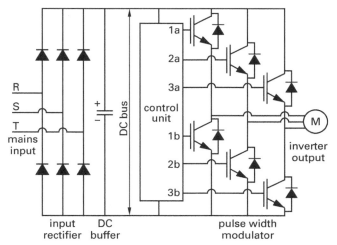

The signal output of the PWM is shown. The signal output is used to control the speed and power of the motor.

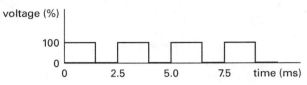

What are the design frequency applied to the motor and the duty cycle in use?

(A) 60 Hz; 40%

(B) 60 Hz; 60%

(C) 400 Hz; 40%

(D) 400 Hz; 60%

20. A 230 V motor designed to operate at 60 Hz is attached to a variable frequency drive (VFD) that maintains a constant V/Hz ratio from 0 Hz to 60 Hz. The motor, though manufactured in the United States, is to be used on a European transmission system operating at 50 Hz. For proper operation of the motor on the European system, the VFD maintains constant

(A) power at an approximate ratio of 3.8 V/Hz

(B) power at an approximate ratio of 4.6 V/Hz

(C) torque at an approximate ratio of 3.8 V/Hz

(D) torque at an approximate ratio of 4.6 V/Hz

SOLUTIONS

1. (a) The maximum voltage is

$$V_m = \frac{\pi}{2} V_{\text{ave}}$$

In terms of the rotational speed, n, the maximum voltage is given in Table 44.1, using the total number of poles, p; the number, N, of series armature paths; area, A; and magnetic flux density, B.

$$V_m = \frac{\pi p n N A B}{60}$$
$$= \pi p \left(\frac{n}{60} \right) \left(\frac{z}{a} \right) A B$$

For a simplex lap-wound armature, $a = p$.

$$V_m = 4\pi \left(\frac{1800 \ \frac{\text{rev}}{\text{min}}}{60 \ \frac{\text{s}}{\text{min}}} \right) \left(\frac{240}{4} \right)$$
$$\times (0.10 \ \text{m})(0.167 \ \text{m}) \left(1.0 \ \frac{\text{Wb}}{\text{m}^2} \right)$$
$$= \boxed{377.8 \ \text{V}}$$

(b)

$$P_{\text{in}} = \frac{P_{\text{out}}}{\text{efficiency}} = \frac{(1 \ \text{kW}) \left(\frac{1 \ \text{hp}}{0.7457 \ \text{kW}} \right)}{0.9}$$
$$= \boxed{1.49 \ \text{hp}}$$

2. (a)

$$n = 120 \frac{f}{p}$$
$$n = \left(2 \ \frac{\text{pole·rev}}{\text{cycle}} \right) \left(60 \ \frac{\text{s}}{\text{min}} \right) \left(\frac{f}{p} \right)$$
$$= \left(2 \ \frac{\text{pole·rev}}{\text{cycle}} \right) \left(60 \ \frac{\text{s}}{\text{min}} \right) \left(\frac{60 \ \text{Hz}}{24 \ \text{poles}} \right)$$
$$= \boxed{300 \ \text{rpm}}$$

(b)

$$BA = \frac{\sqrt{2} \ V_{\text{eff}}}{N\omega} = \frac{\sqrt{2} \ (2200 \ \text{V})}{(20)(2\pi)(60 \ \text{Hz})}$$
$$= \boxed{0.413 \ \text{Wb}}$$

3.

$$I_p = \frac{P_p}{\eta V_p \cos \phi}$$

$$= \frac{(20 \text{ hp})\left(745.7 \ \frac{\text{W}}{\text{hp}}\right)}{(0.86)(440 \text{ V})(0.76)}$$

$$= \boxed{51.86 \text{ A}}$$

4. (a)

$$n_r = \left(\frac{2f}{p}\right)(1 - s)$$

$$= \left(\frac{(2)\left(60 \ \frac{\text{s}}{\text{min}}\right)(60 \text{ Hz})}{4}\right)(1 - 0.03)$$

$$= \boxed{1746 \text{ rpm}}$$

(b)

$$T = \frac{P}{\Omega} = \frac{(200 \text{ hp})\left(550 \ \frac{\frac{\text{ft-lbf}}{\text{s}}}{\text{hp}}\right)}{2\pi\left(1746 \ \frac{\text{rev}}{\text{min}}\right)\left(\frac{1 \text{ min}}{60 \text{ s}}\right)}$$

$$= \boxed{602 \text{ ft-lbf}}$$

(c)

$$I_l = \left(\frac{1}{\sqrt{3}}\right)\left(\frac{P}{\eta V_l \cos \phi}\right)$$

$$= \left(\frac{1}{\sqrt{3}}\right)\left(\frac{(200 \text{ hp})\left(745.7 \ \frac{\text{W}}{\text{hp}}\right)}{(0.85)(440 \text{ V})(0.91)}\right)$$

$$= \boxed{253 \text{ A}}$$

5.

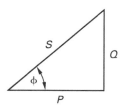

The original power angle is

$$\phi_1 = \arccos 0.82 = 34.92°$$
$$P_1 = 550 \text{ kW}$$
$$Q_i = P_1 \tan \phi_1$$
$$= (550 \text{ kW})\tan 34.92°$$
$$= 384.0 \text{ kVAR}$$

The new conditions are

$$P_2 = (250 \text{ hp})\left(0.7457 \ \frac{\text{kW}}{\text{hp}}\right)$$
$$= 186.4 \text{ kW}$$
$$\phi_f = \arccos 0.95 = 18.19°$$

Because both motors perform real work,

$$P_f = P_1 + P_2 = 550 \text{ kW} + 186.4 \text{ kW}$$
$$= 736.4 \text{ kW}$$

The new reactive power is

$$Q_f = P_f \tan \phi_f = (736.4 \text{ kW})\tan 18.19°$$
$$= 242.0 \text{ kVAR}$$

The change in reactive power is

$$\Delta Q = Q_i - Q_f$$
$$= 384.0 \text{ kVAR} - 242.0 \text{ kVAR} = 142 \text{ kVAR}$$

Synchronous motors used for power factor correction are rated by apparent power.

$$S = \sqrt{(\Delta P)^2 + (\Delta Q)^2}$$
$$= \sqrt{(186.4 \text{ kW})^2 + (142 \text{ kVAR})^2}$$
$$= \boxed{234.3 \text{ kVA}}$$

6.

$$f = \frac{pn_s}{120}$$
$$= \frac{pn}{120(1-s)}$$

The slip and number of poles are unknown. Assume $s = 0$ and $p = 4$.

$$f = \frac{(4)\left(960 \dfrac{\text{rev}}{\text{min}}\right)}{120 \dfrac{\text{s}}{\text{min}}} = 32 \text{ Hz}$$

This is not close to anything in commercial use. Try $p = 6$.

$$f = \frac{(6)\left(960 \dfrac{\text{rev}}{\text{min}}\right)}{120 \dfrac{\text{s}}{\text{min}}} = 48 \text{ Hz}$$

With a 4% slip, $f = 50$ Hz.

$$\boxed{50 \text{ Hz (European)}}$$

7. The required volt-amps per phase are determined by

$$\frac{Q}{3 \text{ phases}} = V_a I_a = \frac{V_a(E_g - V_a)}{X_s}$$

$$E_g = \left(\frac{Q}{3 \text{ phases}}\right)\left(\frac{X_s}{V_a}\right) + V_a$$

$$= \left(\frac{10^4 \text{ VA}}{3 \text{ phases}}\right)\left(\frac{20 \text{ }\Omega}{254 \text{ V}}\right) + \frac{440 \text{ V}}{\sqrt{3}}$$

$$= 516 \text{ V}$$

$$I_f = (516 \text{ V})\left(\frac{5 \text{ A}}{254 \text{ V}}\right)$$

$$I_r = E_g\left(\frac{I_f}{V_a}\right)$$

$$= \boxed{10.16 \text{ A}}$$

8.

$$P_{\text{in}} = \frac{P_{\text{out}}}{\text{efficiency}} = \frac{(45 \text{ hp})\left(745.7 \dfrac{\text{W}}{\text{hp}}\right)}{0.93}$$

$$= 36.1 \times 10^3 \text{ W}$$

At the minimum line current, the current is in phase with the voltage.

$$I = \frac{\dfrac{36.1 \times 10^3 \text{ W}}{3 \text{ phases}}}{\dfrac{440 \text{ V}}{\sqrt{3}}} = 47.37 \text{ A}$$

$$Z_{\text{base}} = \frac{254 \text{ V}}{47.37 \text{ A}} = 5.36 \text{ }\Omega$$

A synchronous reactance of 100% is common and is assumed.

$$X_s = 5.36 \text{ }\Omega$$
$$I_a X_s = (5.36 \text{ }\Omega)(47.37 \text{ A}) = 254 \text{ V}$$

The emf is

$$E_g = V_a - jX_s I_a = (254 \text{ V})(1 - j)$$
$$|E_g| = 359 \text{ V}$$

The rotor current is given as $I_r = 8.2$ A.

To provide a leading power factor of 0.8, it is necessary to retain the same power.

$$|I_a| \cos\theta = 47.37 \text{ A}$$

$$|I_a| = \frac{47.37 \text{ A}}{0.8} = 59.2 \text{ A}$$

$$I_a = (59.2 \text{ A})(\cos\theta + j\sin\theta) = 47.37 \text{ A} + j35.52 \text{ A}$$
$$= 59.2 \text{ A} \angle 36.87°$$

$$E_g = 254 \text{ V} - (j5.36 \text{ }\Omega)I_a$$
$$= 512 \text{ V} \angle -29.74°$$

Using a ratio gives

$$I_r = \left(\frac{8.2 \text{ A}}{359 \text{ V}}\right)(512 \text{ V})$$

$$= \boxed{11.7 \text{ A}}$$

9. The bases for the motor and transformer are the same.

$$I_{\text{rated}} = \frac{\dfrac{50 \times 10^3 \text{ VA}}{3 \text{ phases}}}{254 \text{ V}} = 65.62 \text{ A}$$

$$Z_{\text{base}} = \frac{254 \text{ V}}{65.62 \text{ A}} = 3.87 \text{ }\Omega$$

From the blocked rotor test,

$$P + jQ = I^2(R_e + jX_e)$$
$$1097 \text{ W} + j2791 \text{ VAR} = (65.62 \text{ A})^2(R_e + jX_e)$$
$$R_e + jX_e = 0.2548 + j0.6482 \ \Omega$$

On a per-unit basis,

$$R_e + jX_e = 0.0658 + j0.1675 \text{ pu}$$

The per-unit circuit for starting is

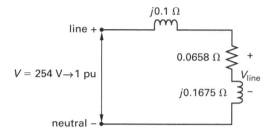

By voltage division,

$$|V_{\text{line}}| = \left| \frac{0.0658 + j0.1675 \ \Omega}{0.0658 + j0.2675 \ \Omega} \right| = 0.653 \text{ pu}$$
$$\text{dip} = \boxed{1 - 0.653 \text{ pu } (35\%)}$$

10. The impedance for wye starting is three times that for delta starting, so the current will be only one-third the delta current.

$$I_{\text{starting}} = \frac{367 \text{ A}}{3} = \boxed{122.3 \text{ A}}$$

11. The synchronous speed is given by the following equation.

$$n_{\text{sync}} = \frac{120f}{p} = \frac{(120)(400 \text{ Hz})}{2 \text{ poles}}$$
$$= \boxed{24{,}000 \text{ rpm}}$$

The answer is (D).

12. Calculate the slip, s, and the frequency range, Δf.

$$s = \frac{n_s - n}{n_s}$$
$$= \frac{1800 \ \dfrac{\text{rev}}{\text{min}} - 1750 \ \dfrac{\text{rev}}{\text{min}}}{1800 \ \dfrac{\text{rev}}{\text{min}}}$$
$$= 0.0278 \quad (2.8\%)$$

$$\Delta f = s f_{\text{FL}} = (0.0278)(60 \text{ Hz})$$
$$= \boxed{1.67 \text{ Hz}}$$

Theoretically, the 1.67 Hz difference is all that is required to generate 100% torque. During startup the VFD can be used to limit the frequency of the applied signal to ensure that a surge of current does not occur or is minimal. For example, at approximately 2.5 Hz the motor is producing nearly 150% of rated torque, which should be adequate to start a properly matched load without the surge of five to seven times the full load current normally expected.

The answer is (A).

13. Zero sequence reactances are components that are in phase and of equal magnitude in all three phases.

The answer is (D).

14. The direct and quadrature synchronous reactances are positive sequence reactances generated during steady-state synchronous conditions.

The answer is (A).

15. Subtransient reactances are positive sequence reactances that occur over a very short period of time due to rapid switching events or terminal short circuits.

The answer is (C).

16. Transient reactances are positive sequence reactances that occur when the armature mmf is changing relatively slowly due to electromechanical transients, such as rapid load shifts. The transient quadrature reactance is very small and usually assumed to be equal to the quadrature reactance, X_q.

The answer is (B).

17. Negative sequence reactances, X_2, occur when the armature mmf is moving in a direction opposite the rotor at synchronous speed. This occurs when loads are unbalanced or faults have occurred on the system.

The answer is (C).

18. From the polarity marks shown, when the voltages are of the same magnitude, there will be 0 V across the indicator lamps, and the lamps will be dark.

If the phase sequences are different (e.g., a-b-c on generator 1 and c-b-a on generator 2), the peak voltage across each indicator lamp will occur at a different time. When the phase sequences are the same (e.g., a-b-c on generator 1 and a-b-c on generator 2), the indicator lamps are both bright or both dark at the same time, since the peak voltages occur on the same phases at the same time.

The answer is (D).

19. The design frequency can be calculated from the period.

$$T = \frac{1}{f}$$

$$f = \frac{1}{T} = \frac{1}{2.5 \times 10^{-3} \text{ s}}$$
$$= 400 \text{ Hz}$$

The duty cycle is the percentage of total time the circuit provides a signal (is *on*).

$$T_{\text{duty cycle}} = \frac{T_{\text{on}}}{T_{\text{cycle}}}$$

$$= \frac{1.5 \times 10^{-3} \text{ s}}{2.5 \times 10^{-3} \text{ s}}$$

$$= \boxed{0.60 \quad (60\%)}$$

The answer is (D).

20. Maintaining the V/Hz ratio results in constant torque from 0 Hz to 60 Hz. Above this range, the voltage is at a maximum and the field must be weakened, so the V/Hz ratio drops. This is called the flux weakening region or the constant horsepower region.

In the United States, the V/Hz ratio at rated conditions is

$$R_{\text{V/Hz}} = \frac{230 \text{ V}}{60 \text{ Hz}}$$
$$= 3.8 \text{ V/Hz}$$

To use the motor in Europe, software programming must change to provide the V/Hz ratio at rated conditions.

$$R_{\text{V/Hz}} = \frac{230 \text{ V}}{50 \text{ Hz}}$$

$$= \boxed{4.6 \text{ V/Hz}}$$

The answer is (D).

Topic X: Electronics

Chapter

45 Electronics Fundamentals

PRACTICE PROBLEMS

1. Intrinsic germanium has a concentration of electrons of 2.5×10^{13} cm^{-3}. Most nearly, what is the concentration of holes?

(A) 1.2×10^{13} cm^{-3}

(B) 1.5×10^{13} cm^{-3}

(C) 2.0×10^{13} cm^{-3}

(D) 2.5×10^{13} cm^{-3}

2. What type of semiconductor material results if phosphorus is the only impurity?

(A) acceptor

(B) donor

(C) n-type

(D) p-type

3. A silicon semiconductor is doped with 10^{15} atoms/cm^{-3} of phosphorus. What is the concentration of electrons?

(A) 1.6×10^{-5} cm^{-3}

(B) 2.6×10^{5} cm^{-3}

(C) 0.5×10^{15} cm^{-3}

(D) 1.0×10^{15} cm^{-3}

4. A silicon semiconductor is doped with a concentration of 10^{15} atoms/cm^{-3} of phosphorus. What is the concentration of holes?

(A) 1.6×10^{-5} cm^{-3}

(B) 2.6×10^{5} cm^{-3}

(C) 0.5×10^{15} cm^{-3}

(D) 1.0×10^{15} cm^{-3}

5. A germanium semiconductor is doped with an impurity having three valence electrons. What are the majority carriers?

(A) conduction band electrons

(B) conduction band holes

(C) valence band electrons

(D) valence band holes

6. A discrete silicon semiconductor diode is forward biased at 0.75 V. Assume a typical reverse saturation current. What is the approximate current?

(A) 2.0 mA

(B) 3.0 nA

(C) 1.5 A

(D) 3.0 MA

7. Gallium phosphide (GaP), with an energy gap of 2.26 eV, is to be used in an LED. What is the expected wavelength of the emitted photons?

(A) 8.79×10^{-26} m

(B) 5.49×10^{-7} m

(C) 3.42×10^{12} m

(D) 5.46×10^{14} m

8. At 20°C, a germanium diode passes 70 μA when the reverse bias is -1.5 V. (a) What is the saturation current? (b) What is the current that flows when a forward bias of $+0.2$ V is applied at 20°C, and (c) at 40°C?

9. A voltage of $20 + 5\sqrt{2}\sin 60t$ V is applied to the circuit shown. The diode characteristics are a static forward resistance, r_f, of 120 Ω and a dynamic resistance, r_p, of 100 Ω. What is the voltage across the inductance?

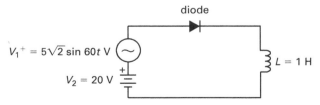

10. At 25°C, a germanium diode shows a saturation current of 100 μA. What current is expected at 100°C, when the diode becomes "useless"?

11. At 25°C, a germanium diode shows a saturation current of 50 μA. What current is expected at 0°C?

12. At 25°C, a germanium diode shows a saturation current of 80 μA. At 25°C, what current is predicted for a voltage of -0.5 V?

13. The output of a single-phase transformer may be used in a switch-mode power supply in a three-phase, four-wire system to provide regulated output to modern low-power office equipment as well as lighting. Such a power supply generates which type of harmonics on the system?

    (A)   negative-sequence harmonics

    (B)   positive-sequence harmonics

    (C)   zero-sequence harmonics

    (D)   all of the above

14. A variable frequency drive (VFD) requires an electronic device to shape an incoming waveform so that only a portion of the power is passed to a downstream motor. Furthermore, the amount of the waveform passed must be controllable in order to vary the motor speed. Which electronic component is widely used in VFDs?

    (A)   insulated gate bipolar junction transistor (IGBT)

    (B)   Schottky diode

    (C)   silicon-controlled rectifier (SCR)

    (D)   either an IGBT or SCR

15. A common emitter (CE) circuit is shown.

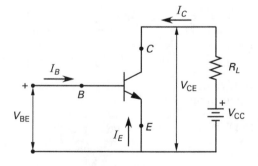

The silicon bipolar junction transistor (BJT) used in the circuit has a collector supply voltage of 12 V. Which load line shown corresponds to the circuit?

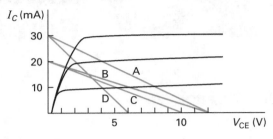

    (A)   load line A

    (B)   load line B

    (C)   load line C

    (D)   load line D

SOLUTIONS

1. In intrinsic semiconductors, $n = p$.

$$n_i^2 = np = n^2$$

$$n_i = \boxed{2.5 \times 10^{13} \text{ cm}^{-3}}$$

The answer is (D).

2. Phosphorus is from group VA on the periodic table, with five valence electrons. Therefore, it contributes a free electron. The resulting semiconductor would be n-type.

Acceptor and *donor* describe types of impurities.

The answer is (C).

3. Space-charge neutrality gives

$$N_A + n = N_D + p$$

Since $N_A = 0$ and $n \gg p$,

$$n \approx N_D = \boxed{1.0 \times 10^{15} \text{ cm}^{-3}}$$

The answer is (D).

4. From Prob. 3, $N_D \approx n$. The law of mass action becomes

$$n_i^2 = np = N_D p$$

$$p = \frac{n_i^2}{N_D} = \frac{(1.6 \times 10^{10} \text{ cm}^{-3})^2}{10^{15} \text{ cm}^{-3}}$$

$$= \boxed{2.56 \times 10^5 \text{ cm}^{-3} \quad (2.6 \times 10^5 \text{ cm}^{-3})}$$

The answer is (B).

5. Germanium has four valence electrons. The impurity has three valence electrons and adds a hole. The majority carriers will be holes.

The answer is (D).

6. A reverse saturation current of 10^{-9} A is typical for silicon. For a silicon diode, the constant $\eta = 2$. The thermal voltage, V_T, at room temperature is

$$V_T = \frac{\kappa T}{q}$$

$$= \frac{\left(1.381 \times 10^{-23} \frac{\text{J}}{\text{K}}\right)(300\text{K})}{1.6 \times 10^{-19} \text{ C}}$$

$$= 0.026 \text{ V}$$

The diode current is

$$I = I_s \left(e^{V/\eta V_T} - 1\right) = (10^{-9} \text{ A})\left(e^{0.75 \text{ V}/(2)(0.026 \text{ V})} - 1\right)$$

$$= \boxed{1.84 \times 10^{-3} \text{ A} \quad (2.0 \text{ mA})}$$

The answer is (A).

7. The wavelength is

$$\lambda = \frac{hc}{E_G}$$

$$= \frac{(6.626 \times 10^{-34} \text{ J·s})\left(3.00 \times 10^8 \frac{\text{m}}{\text{s}}\right)}{(2.26 \text{ eV})\left(1.602 \times 10^{-19} \frac{\text{J}}{\text{eV}}\right)}$$

$$= \boxed{5.49 \times 10^{-7} \text{ m}}$$

h is Planck's constant, and c is the speed of light. The wavelength is in the visible spectrum.

The answer is (B).

8. (a) For a germanium diode, the constant $\eta \approx 1$. The absolute temperature is $T_K = 20°C + 273° = 293K$. The saturation current is

$$I_s = \frac{I}{e^{qV/\eta\kappa T} - 1}$$

$$= \frac{70 \times 10^{-6} \text{ A}}{e^{\left[\frac{(1.6 \times 10^{-19} \text{ C})(-1.5 \text{ V})}{(1)\left(1.38 \times 10^{-23} \frac{\text{J}}{\text{K}}\right)(293\text{K})}\right]} - 1}$$

$$\approx \boxed{\frac{70 \times 10^{-6} \text{ A}}{-1} \quad (-70 \text{ μA})}$$

The reverse saturation current is constant for reverse voltages up to the breakdown voltage.

(b) The current will be

$$I \approx I_s(e^{40 \text{ V}/\eta} - 1)$$

$$= (70 \times 10^{-6} \text{ A})\left(e^{(40 \text{ V}^{-1})(0.2 \text{ V})/1} - 1\right)$$

$$= \boxed{0.209 \text{ A}}$$

(c) The saturation current doubles for every $10°C$.

$$\frac{I_{s,40°C}}{I_{s,20°C}} = (2)^{(40°C-20°C)/10°C}$$

$$(2)^2 = 4$$

$$I_{s,40°C} = 4I_{s,20°C} = (4)(70 \times 10^{-6} \text{ A})$$

$$= 280 \times 10^{-6} \text{ A}$$

The absolute temperature is

$$T = 40°C + 273°$$
$$= 313\text{K}$$

The current will be

$$I = I_s(e^{qV/\eta\kappa T} - 1)$$
$$= (280 \times 10^{-6} \text{ A})$$

$$\times \left(e^{\left[\frac{(1.6 \times 10^{-19} \text{ C})(0.2 \text{ V})}{(1)\left(1.38 \times 10^{-23} \frac{\text{J}}{\text{K}}\right)(313\text{K})} \right]} - 1 \right)$$

$$= \boxed{0.4617 \text{ A}}$$

9. Because $20 \text{ V} > 5\sqrt{2} \text{ V}$, the applied voltage is never negative. Therefore, the diode does not rectify the applied voltage and serves no purpose. Model the circuit as an ideal diode with a $100 \text{ }\Omega$ resistance. (The forward resistance could also be used.)

$$X_L = \omega L = \left(60 \frac{\text{rad}}{\text{s}}\right)(1 \text{ H}) = 60 \text{ }\Omega$$

$$Z = \sqrt{R^2 + X_L^2} = \sqrt{(100 \text{ }\Omega)^2 + (60 \text{ }\Omega)^2}$$
$$= 116.6 \text{ }\Omega$$

$$\phi = \arctan \frac{X_L}{R} = \arctan \frac{60 \text{ }\Omega}{100 \text{ }\Omega} = 30.96°$$

$$\mathbf{I} = \frac{\mathbf{V}}{\mathbf{Z}} = \frac{20 \text{ V} + 5\sqrt{2} \text{ V} \angle 0°}{116.6 \text{ }\Omega \angle 30.96°}$$

The DC voltage across the inductor is zero. Therefore, the 20 V bias is not used.

$$\mathbf{V}_L = \mathbf{IZ}_L = \left(\frac{5\sqrt{2} \text{ V} \angle 0°}{116.6 \text{ }\Omega \angle 30.96°} \right)(60 \text{ }\Omega \angle 90°)$$

$$= \boxed{3.639 \text{ V} \angle 59.04°}$$

10. The saturation current doubles for every $10°C$.

$$\frac{I_{s2}}{I_{s1}} = (2)^{\Delta T/10°C} = (2)^{(100°C-25°C)/10°C}$$

$$= (2)^{7.5} = 181$$

The saturation current at $100°C$ is

$$I_{s2} = (181)(100 \times 10^{-6} \text{ A})$$

$$= \boxed{18{,}100 \times 10^{-6} \text{ A} \quad (18.1 \text{ mA})}$$

11. At $0°C$, the saturation current is

$$I_{s2} = (2)^{(0°C-25°C)/10°C} I_{s1}$$
$$= (0.177)(50 \times 10^{-6} \text{ A})$$
$$= \boxed{8.85 \times 10^{-6} \text{ A} \quad (8.85 \text{ }\mu\text{A})}$$

12. The predicted current is

$$I_{-0.5} \approx I_s(e^{40V} - 1)$$
$$= (80 \times 10^{-6} \text{ A})(e^{(40 \text{ V}^{-1})(-0.5 \text{ V})} - 1)$$
$$= \boxed{-80 \times 10^{-6} \text{ A} \quad (-80 \text{ }\mu\text{A})}$$

13. A switch-mode power supply uses transformer AC output, rectifies it to DC, and uses it to charge a capacitor. The capacitor output current is then drawn as required to maintain a DC output voltage. The capacitor is recharged when its output falls below a regulated limit. When charged, the capacitor draws no current. Therefore the capacitor uses only a portion of the incoming AC signal, resulting in pulses on the source line. The load on the capacitor is nonlinear.

These single-phase, nonlinear loads generate currents through the neutral of the three-phase, four-wire system supplying the power. These currents are third-order, *zero-sequence harmonics* in odd multiples, that is, the 3rd, 9th, 15th, 21st, and so on.

These zero-sequence currents can be nearly as large as the fundamental current (the 60 Hz current) being supplied to a given load, so they must be accounted for. Because of this, the *National Electrical Code* (NEC) specifies limits on circuits with nonlinear loads.

The answer is (C).

14. SCRs are used in power applications to allow small voltages and currents to control much larger electrical quantities. IGBTs are now commonly used as well because of their ability to handle large currents and the ability to precisely control their output.

Schottky diodes are used in applications where high switching speed and low capacitance are needed, such as RF circuits, but they are not used in VFD circuits.

The answer is (D).

15. In the cutoff condition, the BJT is modeled as open and $I_C = 0$ A.

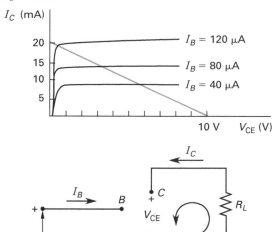

Use this information and KVL to calculate V_{CE}.

$$V_{\text{CC}} - I_C R_L - V_{\text{CE}} = 0$$
$$12 \text{ V} - (0 \text{ A})(400 \text{ }\Omega) - V_{\text{CE}} = 0$$
$$V_{\text{CE}} = 12 \text{ V}$$

Plot the point (I_C, V_{CE}), or $(0, 12)$, on the x-axis. Line A or line B is an answer.

Short the terminals, which represent the saturation condition, and use Ohm's law to determine the value of I_C when V_{CE} is zero.

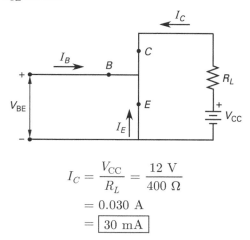

$$I_C = \frac{V_{\text{CC}}}{R_L} = \frac{12 \text{ V}}{400 \text{ }\Omega}$$
$$= 0.030 \text{ A}$$
$$= \boxed{30 \text{ mA}}$$

Plot the point (I_C, V_{CE}), or $(30, 0)$, on the y-axis. Line A or line D is an answer. Because line A is common to both points, it is the final answer.

The answer is (A).

46 Junction Transistors

PRACTICE PROBLEMS

1. Consider the common base (CB) circuit shown.

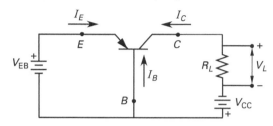

If the collector supply voltage has a magnitude of 1 V and the load is 50 Ω, draw the load line.

2. A silicon *npn* transistor is used in the circuit shown. The transistor parameters are $I_{CBO} = 10^{-6}$ A and $\alpha_0 = 0.99$. The base-emitter voltage is 0.6 V. What is the quiescent collector current, I_C?

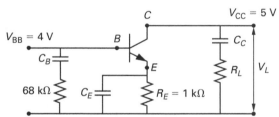

3. A silicon *npn* transistor is shown. Use $\alpha = 0.95$, $I_{CO} = 10^{-6}$ A, and $V_{BE} = 0.6$ V. What current, I_C, will result from this arrangement?

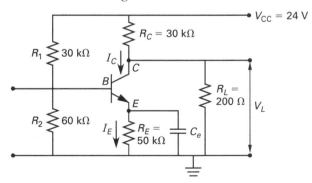

4. The equivalent circuit *h*-parameters for a bipolar junction transistor operating near the center of its active region in a common emitter configuration are $h_{ie} = 500$ Ω and $h_{fe} = 100$. Determine the value of h_{ib}.

5. The equivalent circuit *h*-parameters for a bipolar junction transistor operating near the center of its active region in a common emitter configuration are $h_{ie} = 500$ Ω and $h_{fe} = 50$. Determine the value of h_{fc}.

6. The equivalent circuit *h*-parameters for a bipolar junction transistor operating near the center of its active region in a common emitter configuration are $h_{ie} = 600$ Ω and $h_{fe} = 100$. Determine the value of h_{ic}.

7. The equivalent circuit *h*-parameters for a bipolar junction transistor operating near the center of its active region in a common emitter configuration are $h_{ie} = 500$ Ω and $h_{fe} = 200$. Determine the value of h_{fb}.

8. The equivalent circuit common base *h*-parameters for a bipolar junction transistor are $h_{ib} = 21.5$ Ω, $h_{fb} = -0.95$, $h_{rb} = 0.0031$, and $h_{ob} = 4.7 \times 10^{-7}$ 1/Ω. Determine the value of h_{ie}.

9. The *h*-parameters in the transistor depicted are $h_{ie} = 2200$ Ω, $h_{re} = 3.6 \times 10^{-4}$, $h_{fe} = 55$, and $h_{oe} = 12.5$ μS. (a) What load resistance is needed for a signal output voltage of 0.1 V with an input signal current of 0.5 μA? (b) What input voltage is required?

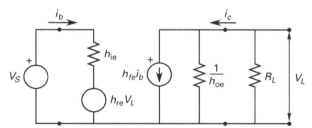

10. Consider the characteristics of a bipolar junction transistor (BJT) configured as a common emitter.

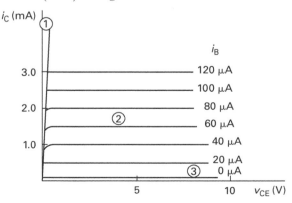

The BJT is to be used in a TTL (transistor-transistor logic) circuit. What operating region(s) will be utilized?

(A) 1

(B) 2

(C) 3

(D) 1 and 3

SOLUTIONS

1. Redraw the circuit as shown.

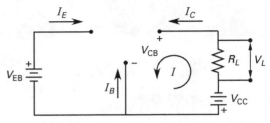

Write Kirchhoff's voltage law (KVL) around the indicated loop.

$$-V_{CC} - I_C R_L - V_{CB} = 0 \text{ V}$$

Because $I_C = 0$ A and $V_{CC} = -1$ V,

$$V_{CC} = V_{CB} = -1 \text{ V}$$

Redraw the current as shown.

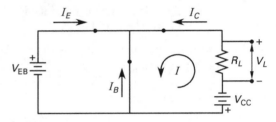

Using Ohm's law and the given information,

$$I_C = \frac{V_{CC}}{R_L} = \frac{-1 \text{ V}}{50 \text{ }\Omega} = -20 \times 10^{-3} \text{ A} \quad (-20 \text{ mA})$$

To draw the load line, plot V_{CC} and I_C, and connect them with a straight line.

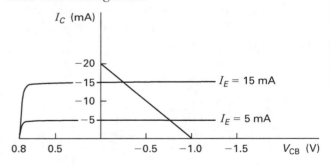

2.

$$I_C = \alpha I_E - I_{\text{CBO}}$$

$$= \alpha\left(\frac{V_{\text{BB}} - V_{\text{BE}}}{R_E}\right) - I_{\text{CBO}}$$

$$= (0.99)\left(\frac{4\text{ V} - 0.6\text{ V}}{1000\ \Omega}\right) - 10^{-6}\text{ A}$$

$$= \boxed{3.365 \times 10^{-3}\text{ A} \quad (3.365\text{ mA})}$$

Alternate Solution

Writing KVL around the base loop,

$$V_{\text{BB}} - V_{\text{BE}} - (I_C + I_B)R_E = 0$$

$$\frac{I_C}{I_B} = \frac{\alpha}{1 - \alpha} = \frac{0.99}{1 - 0.99} = 99$$

$$4\text{ V} - V_{\text{BE}} - \left(I_C + \frac{I_C}{99}\right)(1000\ \Omega) = 0$$

$$4\text{ V} - 0.6\text{ V} - (1010\ \Omega)I_C = 0$$

$$I_C = \boxed{3.366 \times 10^{-3}\text{ A} \ (3.366\text{ mA})}$$

3. R_1 and R_2 form a voltage divider.

$$V_B = \left(\frac{R_2}{R_1 + R_2}\right)V_{\text{CC}}$$

$$= \left(\frac{60\text{ k}\Omega}{30\text{ k}\Omega + 60\text{ k}\Omega}\right)(24\text{ V})$$

$$= 16\text{ V}$$

For the purpose of the signal, R_1 and R_2 both connect to ground and are in parallel.

$$R_{\text{in}} = \frac{R_1 R_2}{R_1 + R_2} = \frac{(30\text{ k}\Omega)(60\text{ k}\Omega)}{30\text{ k}\Omega + 60\text{ k}\Omega}$$

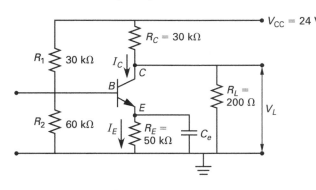

$$\beta = \frac{I_C}{I_B} = \frac{\alpha}{1 - \alpha} = \frac{0.95}{1 - 0.95}$$

$$= 19$$

So, $I_B = I_C/19$.

Writing KVL around the base circuit,

$$V_B - V_{\text{BE}} - (I_C + I_B)R_E - I_B R_{\text{in}} = 0\text{ V}$$

$$16\text{ V} - 0.6\text{ V} - \left(I_C + \frac{I_C}{19}\right)(50\text{ k}\Omega)\left(1000\ \frac{\Omega}{\text{k}\Omega}\right)$$

$$- \left(\frac{I_C}{19}\right)(20\text{ k}\Omega)\left(1000\ \frac{\Omega}{\text{k}\Omega}\right) = 0\text{ V}$$

$$I_C = 2.869 \times 10^{-4}\text{ A} \ (0.2869\text{ mA})$$

I_{CO} is given, so it should be used even though its effect is small.

$$I'_C = I_C + I_{\text{CO}} = 2.869 \times 10^{-4}\text{ A} + 10^{-6}\text{ A}$$

$$= \boxed{2.879 \times 10^{-4}\text{ A} \quad (0.2879\text{ mA})}$$

The exact solutions, based on node-voltage analysis, are

$$V_B = 15.70\text{ V}$$

$$I_B = 0.0151\text{ mA}$$

$$I_C = 0.2869\text{ mA}$$

4.

$$h_{\text{ib}} = \frac{h_{\text{ie}}}{1 + h_{\text{fe}}} = \frac{500\ \Omega}{1 + 100}$$

$$= \boxed{4.95\ \Omega}$$

5.

$$h_{\text{fc}} = -1 - h_{\text{fe}} = -1 - 50$$

$$= \boxed{-51}$$

6.

$$h_{\text{ic}} = h_{\text{ie}}$$

$$= \boxed{600\ \Omega}$$

7.

$$h_{\text{fb}} = \frac{-h_{\text{fe}}}{1 + h_{\text{fe}}} = \frac{-200}{1 + 200}$$

$$= \boxed{-1}$$

8.

$$h_{\mathrm{ie}} = \frac{h_{\mathrm{ib}}}{1 + h_{\mathrm{fb}}} = \frac{21.5\ \Omega}{1 + (-0.95)}$$

$$= \boxed{430\ \Omega}$$

9.

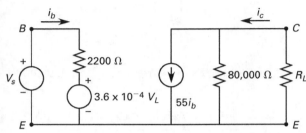

(a)

$$-i_c R_L = V_L$$

$$i_b = 0.5 \times 10^{-6}\ \mathrm{A}$$

$$(55i_b)\left(\frac{(80{,}000\ \Omega)R_L}{R_L + 80{,}000\ \Omega}\right)$$

$$= V_L = 0.1\ \mathrm{V}$$

$$(80{,}000\ \Omega)R_L = (R_L + 80{,}000\ \Omega)$$

$$\times \left(\frac{0.1\ \mathrm{V}}{(55)(0.5 \times 10^{-6}\ \mathrm{A})}\right)$$

$$= (3636.36\ \Omega)R_L + 290.91 \times 10^6\ \Omega$$

$$R_L = \boxed{3809\ \Omega}$$

(b)

$$V_s = i_b(2200\ \Omega) + (3.6 \times 10^{-4})\,V_L$$

$$= (0.5 \times 10^{-6}\ \mathrm{A})(2200\ \Omega)$$

$$+ (3.6 \times 10^{-4})(0.1\ \mathrm{V})$$

$$= \boxed{1.136 \times 10^{-3}\ \mathrm{V} \quad (1.136\ \mathrm{mV})}$$

10. Logic circuits require the transistor to operate as a switch, thereby providing two outputs (logic 0 and logic 1). Logic circuit operation requires the transistor to operate between the saturation region (1) and the cutoff region (3).

The answer is (D).

47 Field Effect Transistors

PRACTICE PROBLEMS

1. An AC-coupled JFET amplifier circuit is shown.

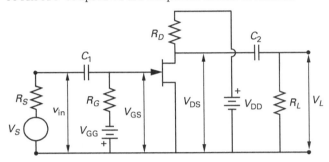

Assume the drain bias voltage is $V_{DD} = 20$ V. If I_D at $V_{DS} = 0$ V is 10 mA, draw the load line.

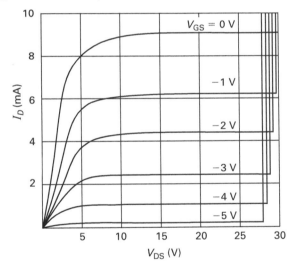

2. A FET with the characteristics shown is to be used in the circuit in Prob. 1. $R_D = 1500$ Ω, $V_{DD} = 20$ V, and $V_{GSQ} = -2$ V. Draw the DC load line and locate the Q-point.

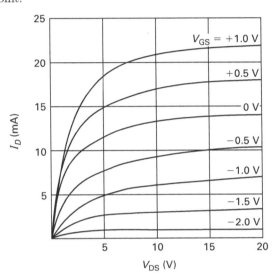

3. The manufacturer of a MOSFET has provided the following information: $V_{DS,max} = 20$ V, $V_{GS,min} = -8$ V, $I_{D,max} = 25$ mA, and $P_{max} = 330$ mW. The MOSFET is to be used to replace the JFET in Prob. 1 with $R_D = 800$ Ω and $R_L = 800$ Ω. Draw the DC load line and a Q-point to ensure linear amplification. Use the characteristic curves from Prob. 2.

4. A voltage source produces a triangular wave shape as shown.

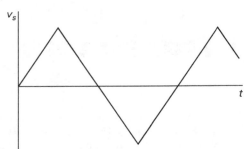

An output wave shape of the indicated form is desired.

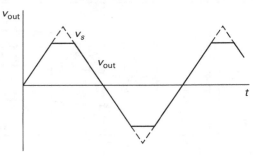

Consider the four circuits shown. Which of the indicated circuits can provide the desired output?

(A)

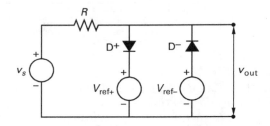

(B)

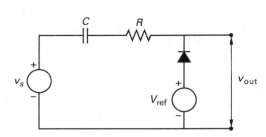

(C)

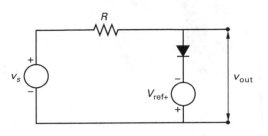

(D)

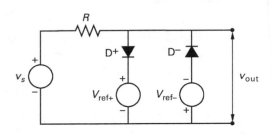

5. Consider the following figure of a MOSFET circuit.

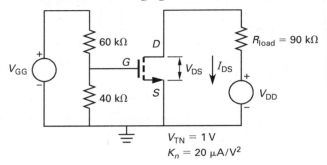

What type of device is the MOSFET?

    (A) n-channel enhancement

    (B) n-channel depletion

    (C) p-channel enhancement

    (D) p-channel depletion

SOLUTIONS

1.

$$P_1 : (V_{DS} = 20 \text{ V}, I_D = 0 \text{ A})$$
$$P_2 : (V_{DS} = 0 \text{ V}, I_D = 10 \text{ mA})$$

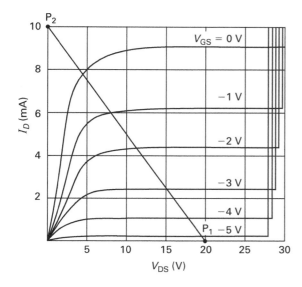

2.

$$P_1 : (V_{DS} = 20 \text{ V}, 0)$$
$$P_2 : \left(0, I_D = \frac{20 \text{ V}}{1500 \text{ }\Omega} = 0.01333 \text{ A} \quad (13.33 \text{ mA})\right)$$

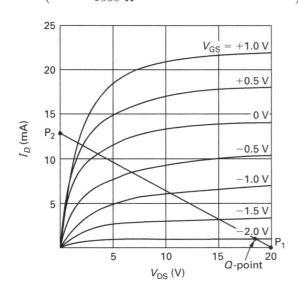

3. The highly negative V_{GS} curves are not shown and are, presumably, crowded together near the $I_D = 0$ A axis. Because of this, if the Q-point were in the highly negative region, positive swings in V_{GS} would accompany large increases in I_D, but negative swings in V_{GS} would decrease I_D very little. Therefore, the quiescent operating point needs to be near the $V_{GS} = 0$ V line. In this position, changes in I_D and V_{GS} are approximately proportional. Power dissipation is another consideration. It should be evaluated at the extreme points of the signal swing, not only at the Q-point.

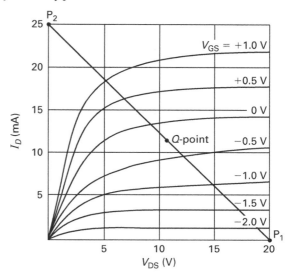

4. A clipping (limiting) circuit that clamps the voltage on both the positive and negative portion of the cycles is needed. Therefore, only choices (A) and (D) are possibilities.

Consider both the figures for choices (A) and (D). During the positive half cycle, diode D^+ (associated with V_{ref^+}) conducts once the source voltage is greater than the reference voltage. This means that in a portion of the positive half cycle, the diode V_{ref^+} conducts, limiting the voltage drop to the reference value. Diode D^- (associated with V_{ref^-}) is reverse biased and does not conduct.

Consider the negative half cycle. In choice (A), diode D^- conducts almost immediately, thereby clamping the output voltage to near zero. In choice (D), diode D^- conducts once the source voltage is less than the reference voltage. Or, one can say that diode D^- conducts when it overcomes V_{ref^-}. (A symmetrical wave is generated when $|V_{ref^+}| = |V_{ref^-}|$.)

The answer is (D).

5. The dashed line on the MOSFET indicates an *enhancement* device. Some symbology uses a single connected line for an enhancement device and a thicker line for a depletion device. The thicker line is meant to indicate that a channel is already present. The alternate symbology is shown.

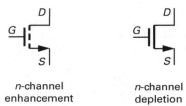

| *n*-channel enhancement | *n*-channel depletion |

The arrow indicates the direction of conventional current flow. Therefore, the current flows from the drain to the source. That is, positive charges flow from the drain to the source, meaning that negative charges flow from the source (of the majority charges) to the drain. The channel is composed of these negative charges, so this is an *n*-channel enhancement device.

The answer is (A).

48 Electrical and Electronic Devices

PRACTICE PROBLEMS

1. An operational amplifier has a gain of 10^5. The frequencies at the half-power points are $f_L = 500$ Hz and $f_H = 1000$ Hz. What is the figure of merit?

(A) 1×10^6 rad/s

(B) 5×10^7 rad/s

(C) 3×10^8 rad/s

(D) 6×10^8 rad/s

2. During testing of an operational amplifier, a signal of equal magnitude is applied to the two terminals. That is, $v^+ = -v^-$. What gain is being measured?

(A) A_{cm}

(B) A_{dm}

(C) A_V

(D) G_P

3. An operational amplifier data sheet indicates a voltage gain of 10^5 and a power supply voltage of 12 V. What is the maximum voltage difference between the input terminals that will ensure linear operation?

(A) 60 μV

(B) 90 μV

(C) 120 μV

(D) 150 μV

4. Consider the circuit shown. The feedback resistance is 10 kΩ, and the input resistance is 100 Ω. If the op amp gain is 10^5, what is the gain of the circuit?

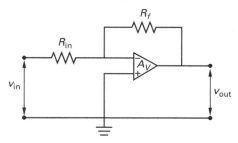

(A) -100

(B) -99.9

(C) $+100$

(D) 10^5

5. An operational amplifier has an input resistance of 10^5 Ω and a bandwidth of 1000 Hz, and is at a room temperature of 300K. If 10 dB above the noise is required for proper operation, what is the required signal power?

(A) 0.10 μV

(B) 2.0 μV

(C) 4.0 μV

(D) 16 μV

6. Consider the circuit shown. (a) If $R_f = 1$ MΩ and $R_{in} = 50$ Ω, what is the gain? (b) If $v_{in} = 1$ mV, what is v_{out}?

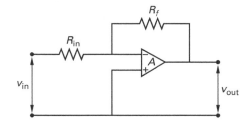

7. For the circuit shown, $v_1 = 10 \sin 200t$ and $v_2 = 15 \sin 200t$. What is v_{out}? The op amp is ideal and has infinite gain.

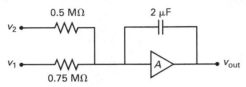

8. In the circuit shown, if v_{in} is $\sin 30t$, what is v_{out}? The op amp is ideal and has infinite gain.

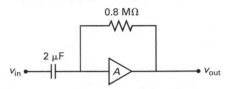

9. In the circuit shown, if $v_1 = v_2 = v_3 = 12$ V, what is v_{out}? The op amp is ideal and has infinite gain.

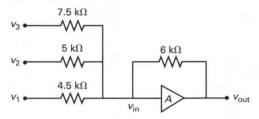

10. Given the following circuit, (a) what value of R_f will give an output of -8 V? (b) If the maximum of each input is 10 V and the output must not exceed -15 V, what value must be used for R_f? The op amp is ideal and has infinite gain.

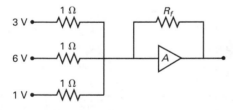

11. An integrator is used in a device for measuring displacement from a fixed location. The scale of the device is 1 V to 1 in. The switch is closed to set the output for initial displacement and then opened at $t = 0$ s to initiate measurement. If the displacement is initially $d_0 = 12$ in, what is the output voltage for an input of $8 \cos 10t$ V? The op amp is ideal and has infinite gain.

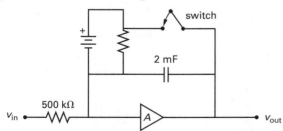

12. What differential equation is solved by the following circuit? The op amp is ideal and has infinite gain.

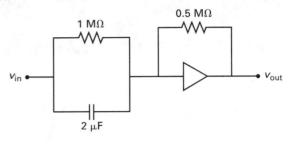

13. For the circuit shown, what is the differential equation that writes v_{out} in terms of v_{in}? The op amp is ideal and has infinite gain.

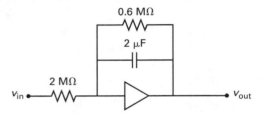

14. For the circuit shown, determine the output voltage as a function of the input voltage.

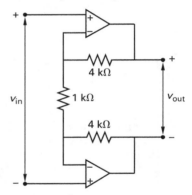

15. The op amp circuit shown is designed to provide a characteristic similar to a relay. When the output is negative and the input current increases, the output will go to the positive limit of $+12$ V when the input voltage exceeds $+1$ V. It will remain there until the input voltage becomes less than -1 V, at which time the output will go to the negative limit of -12 V. Determine an appropriate value for R.

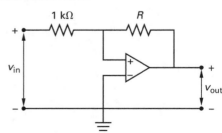

16. A germanium diode with an offset voltage of 0.2 V is used in a precision diode circuit. The power supply voltages are ±15 V. The amplifier voltage gain is 10^7. What is the threshold voltage of the precision diode?

(A) 20 nV

(B) 100 nV

(C) 1.5 μV

(D) 9.0 μV

The following circuit is applicable to Prob. 17 through Prob. 20. The output is to be maintained at 24 V, while the input varies from 120 V to 126 V with a load current ranging from 0 A to 5 A. Assume ideal conditions, that is, the keep-alive current is zero and the zener resistance, R_Z, is negligible.

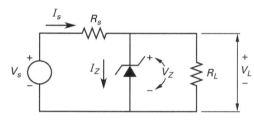

17. What is the required supply resistance to ensure proper operation of the zener regulator circuit?

(A) 19 Ω

(B) 20 Ω

(C) 24 Ω

(D) 25 Ω

18. What is the maximum zener current?

(A) 4.1 A

(B) 4.3 A

(C) 5.1 A

(D) 5.3 A

19. What is the minimum diode power rating that can be used?

(A) 100 W

(B) 150 W

(C) 300 W

(D) 600 W

20. What is the minimum power rating of the supply resistor?

(A) 400 W

(B) 450 W

(C) 500 W

(D) 550 W

The following circuit is applicable to Prob. 21 through Prob. 24. The circuit is to be operated at low frequencies. The transistor has the following h-parameter values: $h_{ie} = 4 \times 10^3 \, \Omega$, $h_{fe} = 200$, $h_{oe} = 100 \times 10^{-6} \, S$, and $h_{re} = 4 \times 10^3$.

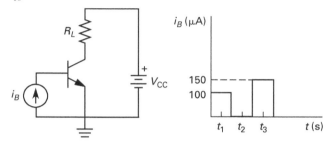

21. If the load resistance is 400 Ω, what collector supply voltage is required for the transistor to be on at time t_1?

(A) 4 V

(B) 8 V

(C) 12 V

(D) 16 V

22. If the collector supply voltage is 12 V and the load resistor is 400 Ω, what is the condition of the transistor at time t_2?

(A) active

(B) off

(C) on

(D) saturated

23. The selected collector supply voltage is 12 V. The transistor must be on for pulses of 150 μA or greater only. What is the required value of the load resistance?

(A) 200 Ω

(B) 300 Ω

(C) 400 Ω

(D) 600 Ω

24. For the circuit parameters in Prob. 23, what is the actual saturation current for a silicon transistor?

(A) 20.0 mA

(B) 25.5 mA

(C) 29.5 mA

(D) 30.0 mA

25. Consider the inverter, or NOT gate, circuit shown.

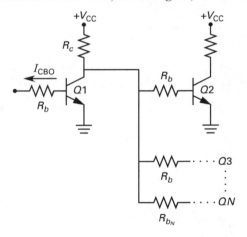

Parameters taken from a manufacturer's data sheet indicate that $V_{CC} = 5.0$ V, $I_{B,\text{sat}} > 0.5$ mA, $V_{BE,\text{sat}} < 1.5$ V, $R_c = 500$ Ω, and $R_b = 1000$ Ω. By how much does a reverse saturation current of 1 mA lower the fan-out of $Q1$?

(A) 2

(B) 4

(C) 10

(D) 12

26. For the transistor circuit shown, determine (a) the necessary base current for this gate to sink five identical gates connected to its output, and (b) the minimum size of R_b to allow this gate to source five identical gates.

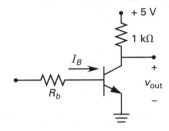

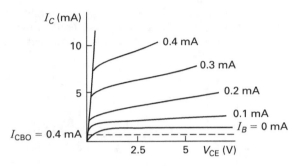

27. A particular logic gate has the following parameters: $t_d = 5$ ns, $t_r = 50$ ns, $t_s = 30$ ns, and $t_f = 30$ ns. Determine the maximum frequency at which this gate can operate.

28. Use an operational amplifier to convert a rectangular wave voltage of ± 5 V and a period of 10 ms to a triangular wave with a peak output of ± 5 V with the same period.

29. For the circuit shown, determine the transfer characteristic of the output voltage versus the input voltage. The diodes are ideal.

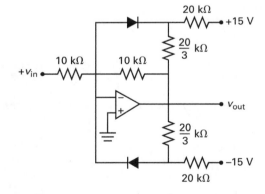

30. Determine the power requirement for the DTL circuit shown for a fan-out of five.

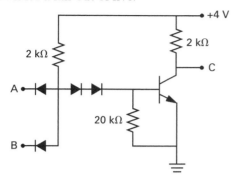

31. A portion of the program coding for a PLC application is shown. The input line is hot.

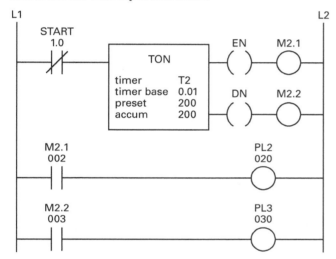

Given the status of the timer, which relays are energized?

(A) M2.1, PL2

(B) M2.1, PL3

(C) M2.2, PL3

(D) M2.1, M2.2, PL2, PL3

32. A portion of the program coding for a PLC application is shown. The input line is hot.

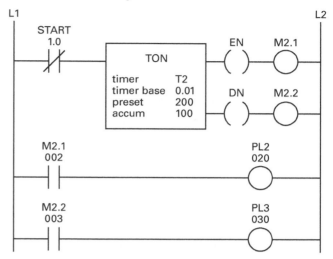

Given the status of the timer, the remaining time before the contact at address 002 closes is most nearly

(A) 0.01 sec

(B) 1.00 sec

(C) 2.00 sec

(D) 3.00 sec

33. A circuit diagram of a digital switch is shown. The control voltage is 0 V.

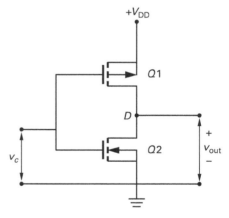

Which CMOS device, if any, is conducting (ON) and what is the output voltage?

(A) Q1, 0 V

(B) Q1, +V$_{DD}$

(C) Q2, 0 V

(D) Q2, +V$_{DD}$

34. The linear frequency sweep of a phase-lock loop (PLL) detector used as input for determining the frequency error is given as

$$\omega = \frac{d\phi}{dt}$$
$$= 60t \text{ rad/sec}^2$$

The number of cycles per second swept by the detector in using a 34 msec sweep time is most nearly

(A) 0.280 cycles/sec

(B) 0.325 cycles/sec

(C) 10.0 cycles/sec

(D) 325 cycles/sec

35. A portion of a ground detection circuit is shown.

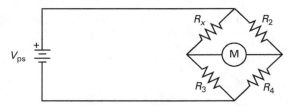

Resistor R_2 has a value of 1 kΩ. Resistors R_3 and R_4 have values of 10 kΩ and 2 kΩ, respectively. The power-supply voltage of the circuit is 24 V. The meter is designed to deflect fully to the right on the scale face, indicating "satisfactory" (i.e., no ground) when 0 A passes through the meter.

What is most nearly the resistance value, R_x, necessary to result in a no-ground condition?

(A) 200 Ω

(B) 1000 Ω

(C) 5000 Ω

(D) 10 000 Ω

SOLUTIONS

1. The figure of merit is

$$F_m = A_{\text{ref}}(\text{BW}) = A_{\text{ref}}(2\pi f_H - 2\pi f_L)$$
$$= (10^5)2\pi(1000 \text{ Hz} - 500 \text{ Hz})$$
$$= \boxed{3.14 \times 10^8 \text{ rad/s} \quad (3 \times 10^8 \text{ rad/s})}$$

The answer is (C).

2. The differential-mode voltage is

$$v_{\text{dm}} = v^+ - v^-$$

Therefore, with $v^+ = -v^-$,

$$v_{\text{dm}} = v^+ + v^+ = 2v^+$$

The common-mode voltage is

$$v_{\text{cm}} = \tfrac{1}{2}(v^+ + v^-) = \tfrac{1}{2}\big(v^+ + (-v^-)\big) = 0 \text{ V}$$

Therefore, only the difference is being amplified, and the gain is A_{dm}.

The answer is (B).

3. The maximum voltage difference is

$$|v^+ - v^-| < \frac{V_{\text{DC}} - 3 \text{ V}}{A_V}$$
$$|\Delta V| < \frac{12 \text{ V} - 3 \text{ V}}{10^5} = \frac{9 \text{ V}}{10^5}$$
$$= \boxed{90 \times 10^{-6} \text{ V} \quad (90 \text{ } \mu\text{V})}$$

The answer is (B).

4. Assume the indicated current directions at the negative terminal.

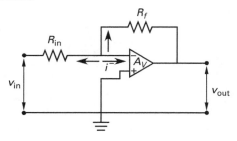

Write Kirchhoff's current law (KCL) at the inverting terminal node.

$$\frac{V^- - v_{\text{in}}}{R_{\text{in}}} + \frac{V^- - v_{\text{out}}}{R_f} + i^- = 0$$

Applying the ideal op amp assumptions means $i^- = 0$ A and $v^- = 0$ V. Substituting and solving for the gain results in

$$\frac{0 - v_{\text{in}}}{R_{\text{in}}} + \frac{0 - v_{\text{out}}}{R_f} + 0 = 0$$

$$\frac{v_{\text{out}}}{v_{\text{in}}} = \frac{-R_f}{R_{\text{in}}} = A_{V,\text{ideal}}$$

$$= \frac{-10 \times 10^3 \ \Omega}{100 \ \Omega} = -100$$

The op amp is real and has a finite gain. The constraint equation (that is, the fundamental op amp equation) rearranged is

$$V^- - V^+ = \Delta = \frac{-v_{\text{out}}}{A_V}$$

Since V^+ is at 0 V,

$$V^- = \frac{-v_{\text{out}}}{A_V}$$

Substitute into the original equation.

$$\frac{\dfrac{-v_{\text{out}}}{A_V} - v_{\text{in}}}{R_{\text{in}}} + \frac{\dfrac{-v_{\text{out}}}{A_V} - v_{\text{out}}}{R_f} + 0 = 0$$

The input current is still small enough to be ignored. Rearranging,

$$\frac{v_{\text{out}}}{v_{\text{in}}} = -A_V\left(\frac{R_f}{R_f + R_{\text{in}} + A_V R_{\text{in}}}\right)$$

When the gain is large, this expression reduces to the expression for gain in the ideal case.

Substitute the given values.

$$\frac{v_{\text{out}}}{v_{\text{in}}} = (-10^5)\left(\frac{10 \times 10^3 \ \Omega}{\begin{array}{c}10 \times 10^3 \ \Omega + 100 \ \Omega \\ + (10^5)(100 \ \Omega)\end{array}}\right)$$

$$= \boxed{-99.9}$$

The answer is (B).

5. The signal power is found from the signal to noise ratio (SNR).

$$\text{SNR} = 10\log\frac{P_s}{P_n}$$

$$\text{antilog}\frac{\text{SNR}}{10} = \frac{P_s}{P_n}$$

$$P_s = P_n \ \text{antilog}\frac{\text{SNR}}{10}$$

The noise power is unknown. The noise power is given in terms of the Boltzmann constant, k; temperature, T, in kelvins; and the bandwidth, BW, in hertz.

$$P_n = \frac{V_{\text{rms}}^2}{4R} = (\text{BW})kT$$

$$= (1000 \ \text{Hz})\left(1.3805 \times 10^{-23} \ \frac{\text{J}}{\text{K}}\right)(300\text{K})$$

$$= 4.14 \times 10^{-18} \ \text{W}$$

Substituting,

$$P_s = (4.14 \times 10^{-18} \ \text{W})\,\text{antilog}\frac{10}{10}$$

$$= 41.4 \times 10^{-18} \ \text{W}$$

The voltage is found from

$$P_s = \frac{V_{\text{rms}}^2}{4R}$$

$$V_{\text{rms}} = \sqrt{4RP_s} = \sqrt{(4)(10^5 \ \Omega)(41.4 \times 10^{-18} \ \text{W})}$$

$$= \boxed{4.07 \times 10^{-6} \ \text{V} \quad (4.0 \ \mu\text{V})}$$

The answer is (C).

6. (a) The gain is

$$A = \frac{-R_f}{R_{in}} = \frac{-10^6 \ \Omega}{50 \ \Omega} = \boxed{-20,000}$$

(b) $v_{out} = Av_{in} = (-20,000)(0.001 \ V) = \boxed{-20 \ V}$

7. This is a summing circuit.

$$v_{out} = -\left(\frac{Z_f}{Z_1}\right)v_1 - \left(\frac{Z_f}{Z_2}\right)v_2$$

$$= -\left(\frac{1}{sC}\right)\left(\frac{v_1}{R_1} + \frac{v_2}{R_2}\right)$$

$$= \frac{-1}{(2 \times 10^{-6} \ F)(0.75 \times 10^6 \ \Omega)}\int 10\sin 200t \ dt$$

$$\qquad - \frac{1}{(2 \times 10^{-6} \ F)(0.5 \times 10^6 \ \Omega)}\int 15\sin 200t \ dt$$

$$= \left(\frac{2}{3}\right)\left(\frac{1}{200}\right)(10)\cos 200t$$

$$\qquad + \left(\frac{1}{1}\right)\left(\frac{15}{200}\right)\cos 200t + C$$

$$= \boxed{\left(\frac{1}{30}\right)\cos 200t + \left(\frac{3}{40}\right)\cos 200t + C \ V}$$

8. The circuit is a differentiator.

$$v_{out} = -RC\frac{dv_{in}}{dt}$$

$$= -(0.8 \times 10^6 \ \Omega)(2 \times 10^{-6} \ F)\left(\frac{d(\sin 30t)}{dt}\right)$$

$$= \boxed{-48\cos 30t \ V}$$

9. This is a summing circuit.

$$v_{out} = \left(\frac{-6 \ k\Omega}{4.5 \ k\Omega}\right)v_1 - \left(\frac{6 \ k\Omega}{5 \ k\Omega}\right)v_2$$

$$\qquad - \left(\frac{6 \ k\Omega}{7.5 \ k\Omega}\right)v_3$$

$$= (-12 \ V)\left(\frac{6 \ k\Omega}{4.5 \ k\Omega} + \frac{6 \ k\Omega}{5 \ k\Omega} \atop + \frac{6 \ k\Omega}{7.5 \ k\Omega}\right)$$

$$= \boxed{-40 \ V}$$

10. (a) This is a summing circuit.

$$v_{out} = -\left(\left(\frac{R_f}{1 \ \Omega}\right)(1 \ V) + \left(\frac{R_f}{1 \ \Omega}\right)(6 \ V) \atop + \left(\frac{R_f}{1 \ \Omega}\right)(3 \ V)\right) = -8 \ V$$

$$R_f = \boxed{0.8 \ \Omega}$$

(b)

$$v_{out} = -\left(\left(\frac{R_f}{1 \ \Omega}\right)(10 \ V) + \left(\frac{R_f}{1 \ \Omega}\right)(10 \ V) \atop + \left(\frac{R_f}{1 \ \Omega}\right)(10 \ V)\right) = -15 \ V$$

$$R_f = \boxed{0.5 \ \Omega}$$

11.

$$v_{out} = -\frac{1}{RC}\int v_{in} dt$$

$$= -\left(\frac{1}{(500,000 \ \Omega) \atop \times (2 \times 10^{-3} \ F)}\right)\int 8\cos 10t \ dt$$

$$= (-8 \times 10^{-4})\sin 10t + V_0$$

At $t = 0$ s, $d_0 = 12$ in.

$$V_0 = (12 \ in)\left(\frac{1 \ V}{1 \ in}\right) = 12 \ V$$

$$v_{out} = \boxed{(-8 \times 10^{-4})\sin 10t + 12 \ V}$$

12.

$$\frac{v_{out}}{v_{in}} = -\frac{Z_f}{Z_{in}} = \frac{-500{,}000 \ \Omega}{\dfrac{(10^6 \ \Omega)\left(\dfrac{1}{s(2 \times 10^{-6})} \ \Omega\right)}{10^6 \ \Omega + \dfrac{1}{s(2 \times 10^{-6})} \ \Omega}}$$

$$= -\frac{\dfrac{1}{2}}{\dfrac{1}{2s+1}} = -\left(s + \frac{1}{2}\right)$$

$$v_{out} = -\left(s + \frac{1}{2}\right)v_{in}$$

$$= \boxed{-\frac{dv_{in}}{dt} - \frac{v_{in}}{2}}$$

13.

$$\frac{v_{out}}{v_{in}} = -\frac{Z_f}{Z_{in}}$$

$$= -\frac{\dfrac{(600{,}000 \ \Omega)\left(\dfrac{1}{s(2 \times 10^{-6}) \ \text{F}}\right)}{600{,}000 \ \Omega + \left(\dfrac{1}{s(2 \times 10^{-6}) \ \text{F}}\right)}}{2{,}000{,}000 \ \Omega}$$

$$= -\frac{0.3}{1.2s + 1}$$

$$v_{out}(1.2s + 1) = -0.3v_{in}$$

$$(1.2)\left(\frac{dv_{out}}{dt}\right) + v_{out} = -0.3v_{in}$$

$$\boxed{(1.2)\left(\frac{dv_{out}}{dt}\right) + v_{out} + 0.3v_{in} = 0}$$

14.

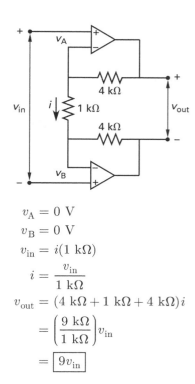

$$v_A = 0 \ \text{V}$$

$$v_B = 0 \ \text{V}$$

$$v_{in} = i(1 \ \text{k}\Omega)$$

$$i = \frac{v_{in}}{1 \ \text{k}\Omega}$$

$$v_{out} = (4 \ \text{k}\Omega + 1 \ \text{k}\Omega + 4 \ \text{k}\Omega)i$$

$$= \left(\frac{9 \ \text{k}\Omega}{1 \ \text{k}\Omega}\right)v_{in}$$

$$= \boxed{9v_{in}}$$

15. Except while slewing from one saturation level to the other, the op amp is not operating in its linear region.

The current flowing in the resistors is

$$i_{in} = \frac{v_{in} - v_{out}}{1 \ \text{k}\Omega + R}$$

The voltage of the positive terminal is

$$v_{in} - i_{in}(1 \ \text{k}\Omega) = v_{in} - \frac{(1 \ \text{k}\Omega)v_{in} - (1 \ \text{k}\Omega)v_{out}}{1 \ \text{k}\Omega + R}$$

$$v^+ = \frac{Rv_{in} + (1 \ \text{k}\Omega)v_{out}}{R + 1 \ \text{k}\Omega}$$

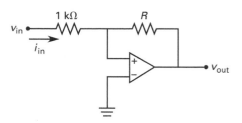

For v^+ positive, v_{out} is $+12$ V. v_{out} will reverse when v^+ becomes negative.

$$v_{in} = (-1 - \delta) \ \text{V}$$

$$-R(1 \ \text{V} + \delta) + (1 \ \text{k}\Omega)(12 \ \text{V}) < 0$$

$$-R + 12 \ \text{k}\Omega - \delta R < 0$$

Let δ approach zero. Then,

$$R = \boxed{12\text{ k}\Omega}$$

16. The threshold voltage is reduced by an amount equal to the second term in Eq. 48.17, which is the transfer equation for a precision diode. The forward threshold voltage drop is V_F, and A_V is the voltage gain.

$$V_{\text{out}} = V_{\text{in}} - \frac{V_F}{A_V}$$

Substituting the given information gives

$$\frac{V_F}{A_V} = \frac{0.2\text{ V}}{10^7} = \boxed{20 \times 10^{-9}\text{ V} \quad (20\text{ nV})}$$

The answer is (A).

17. For a zener regulator circuit, the supply resistance is given by Eq. 48.20. V_{ZM} is the voltage zener at maximum rated current. $I_{L,\max}$ is the maximum load current.

$$
\begin{aligned}
R_s &= \frac{V_{\text{in,min}} - V_{\text{ZM}}}{I_{L,\max}} \\
&= \frac{120\text{ V} - 24\text{ V}}{5\text{ A}} \\
&= \boxed{19.2\ \Omega \quad (19\ \Omega)}
\end{aligned}
$$

The answer is (A).

18. The maximum zener current is determined from Eq. 48.21.

$$
\begin{aligned}
I_{Z,\max} &= \left(\frac{V_{\text{in,max}} - V_{\text{ZM}}}{V_{\text{in,min}} - V_{\text{ZM}}}\right)I_{L,\max} \\
&= \left(\frac{126\text{ V} - 24\text{ V}}{120\text{ V} - 24\text{ V}}\right)(5\text{ A}) \\
&= \boxed{5.3\text{ A}}
\end{aligned}
$$

The answer is (D).

19. The diode power is given by Eq. 48.22.

$$
\begin{aligned}
P_D &= I_{Z,\max}V_{\text{ZM}} = (5.3\text{ A})(24\text{ V}) \\
&= 127.2\text{ W}
\end{aligned}
$$

> The minimum diode power rating that can be used is 150 W.

The answer is (B).

20. The minimum power rating of the supply resistor is given by Eq. 48.23.

$$
\begin{aligned}
P_{R_s} &= I_{Z,\max}^2 R_s \\
&= (5.3\text{ A})^2(19.2\ \Omega) \\
&= \boxed{539.3\text{ W} \quad (550\text{ W})}
\end{aligned}
$$

The answer is (D).

21. At low frequencies h_{fe} is approximately equal to h_{FE}, so the small-signal parameter given can be used. The base current, I_B, necessary for saturation is given by Eq. 48.37. V_{CC} is the collector supply voltage. R_L is the load resistance.

$$I_B \geq \frac{V_{\text{CC}}}{h_{\text{FE}}R_L}$$

For a load resistance of 400 Ω, the collector supply voltage is

$$
\begin{aligned}
V_{\text{CC}} &\leq I_{B_{t1}}h_{\text{FE}}R_L \\
&\leq (100 \times 10^{-6}\text{ A})(200)(400\ \Omega) \\
&\leq \boxed{8\text{ V}}
\end{aligned}
$$

The answer is (B).

22. If the base current is zero, the transistor is off.

The answer is (B).

23. The minimum base current, which will be 150 μA to ensure that the transistor is not on at t_1, is given by Eq. 48.37.

$$I_B \geq \frac{V_{\text{CC}}}{h_{\text{FE}}R_L} \approx \frac{V_{\text{CC}}}{h_{\text{fe}}R_L}$$

Rearranging gives

$$
\begin{aligned}
R_L &\geq \frac{V_{\text{CC}}}{h_{\text{fe}}I_B} \\
&\geq \frac{(12\text{ V})\left(1 \times 10^6\ \dfrac{\mu\text{A}}{\text{A}}\right)}{(200)(150\ \mu\text{A})} \\
&\geq \boxed{400\ \Omega}
\end{aligned}
$$

The answer is (C).

24. The actual saturation current is given by Eq. 48.36. $V_{CE,sat}$ is the collector-emitter voltage at saturation.

$$I_{C,sat} = \frac{V_{CC} - V_{CE,sat}}{R_L}$$
$$= \frac{12 \text{ V} - 0.2 \text{ V}}{400 \text{ }\Omega}$$
$$= \boxed{29.5 \times 10^{-3} \text{ A} \quad (29.5 \text{ mA})}$$

If the collector-emitter voltage drop is ignored, the saturation current will be 30 mA. Except for exacting calculations, the ON transistor can be modeled as a short circuit.

The answer is (C).

25. A transistor inverter is a common-emitter configured transistor as shown.

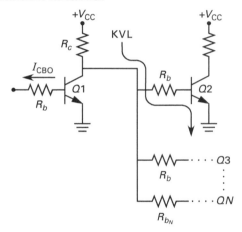

Initially ignoring the reverse saturation current, I_{CBO}, means $Q2$ can be modeled as an open circuit. Then, writing Kirchhoff's voltage law (KVL) from the reference ground at $+V_{CC}$ (not shown) back to the reference ground (shown at $Q2$), which comprises a loop, gives

$$V_{CC} - NI_{B,min}R_c - I_{B,min}R_b - V_{BE2} = 0$$

The minimum base current is the current necessary to cause $Q2$ and other load transistors to operate in the saturation region. This is given as $I_{B,sat} > 0.5$ mA. The base-emitter voltage is the voltage of the transistor while saturated, given as $V_{BE,sat} < 1.5$ V. Substituting and solving for the fan-out gives

$$N = \frac{V_{CC} - V_{BE,sat} - I_{B,min}R_b}{I_{B,min}R_c}$$
$$= \frac{5 \text{ V} - 1.5 \text{ V} - (0.5 \times 10^{-3} \text{ A})(1000 \text{ }\Omega)}{(0.5 \times 10^{-3} \text{ A})(500 \text{ }\Omega)}$$
$$= 12$$

To account for I_{CBO}, add the additional voltage drop $I_{CBO}R_c$ to the original KVL equation, or use Eq. 48.45 directly.

$$V_{CC} - NI_{B,min}R_c - I_{CBO}R_c - I_{B,min}R_b - V_{BE2} = 0$$
$$N + \frac{R_b}{R_c} = \frac{V_{CC} - V_{BE2} - I_{CBO}R_c}{I_{B,min}R_c}$$

Substituting and solving gives

$$N = 10$$

The reverse saturation current lowers the fan-out by

$$12 - 10 = \boxed{2}$$

The answer is (A).

26. The load line with no loads has $I_C \approx 4.9$ mA, which requires I_B to be slightly more than 0.3 mA.

(a) To sink five additional loads, because $I_{CBO} = 0.4$ mA, I_C must increase by $(5)(0.4 \text{ mA}) = 2$ mA to approximately 6.9 mA. This will require $I_B \approx 0.4$ mA.

$$I_{B,low \text{ output}} = \boxed{0.4 \text{ mA}}$$

(b) When this transistor is off, the five loads draw 0.4 mA each, while the bases have 0.7 V. The 0.4 mA sourcing transistor's reverse saturation current must also be accounted for, as shown.

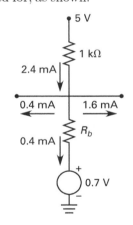

$$5 \text{ V} = (2.4 \text{ mA})(1 \text{ kohms})$$
$$+ (0.4 \text{ mA})R_b + 0.7 \text{ V}$$
$$R_b = \frac{5 \text{ V} - 0.7 \text{ V} - 2.4 \text{ V}}{(0.4 \text{ mA})\left(0.001 \dfrac{\text{A}}{\text{mA}}\right)}$$
$$= \boxed{4750 \text{ }\Omega \quad (4.75 \text{ k}\Omega)}$$

27.

$$\tau_{\min} = \tau_d + \tau_r + \tau_s + \tau_f = 115 \text{ ns}$$
$$= 5 \text{ ns} + 50 \text{ ns} + 30 \text{ ns} + 30 \text{ ns}$$
$$f_{\max} = \frac{1}{\tau_{\min}} = \frac{10^9 \dfrac{\text{ns}}{\text{s}}}{115 \text{ ns}}$$
$$= \boxed{8.7 \times 10^6 \text{ Hz} \quad (8.7 \text{ MHz})}$$

28. The desired conversion is

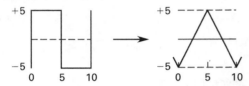

The required circuit is an integrator, shown in the following illustration.

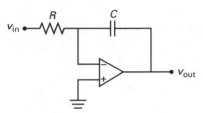

Use Kirchhoff's current law (KCL) at the inverting terminal to get

$$\frac{0 \text{ V} - v_{\text{in}}}{R} + C \frac{d(0 \text{ V} - v_{\text{out}})}{dt} = 0$$

Rearranging and integrating gives

$$v_{\text{out}} = -\frac{1}{RC} \int v_{\text{in}} dt + \kappa$$

At half the period, 5 ms, the necessary initial conditions are the level of the voltage and the time frame.

$$v_{\text{out}} = -\frac{1}{RC} \int_{0\,\text{s}}^{0.005\,\text{s}} 5\,dt + 5 \text{ V} = -5 \text{ V}$$
$$= -\frac{5 \text{ V}}{RC} t \Big|_{0\,\text{s}}^{0.005\,\text{s}} + 5 \text{ V} = -5 \text{ V}$$
$$= -\frac{5 \text{ V}}{RC}(0.005 \text{ s}) = -10 \text{ V}$$

Solve for the combination RC.

$$-\frac{5 \text{ V}}{RC}(0.005 \text{ s}) = -10 \text{ V}$$
$$RC = -\frac{(5 \text{ V})(0.005 \text{ s})}{-10 \text{ V}}$$
$$= \boxed{2.5 \times 10^{-3} \text{ s}}$$

Therefore, choosing an RC combination value equal to the calculated value provides the transient necessary to effect the desired conversion.

29. With both diodes open,

$$\frac{v_{\text{out}}}{v_{\text{in}}} = -1$$

With the upper diode conducting,

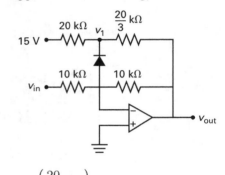

$$\frac{\left(\dfrac{20}{3} \text{ k}\Omega\right)(10 \text{ k}\Omega)}{\left(\dfrac{20}{3} \text{ k}\Omega\right) + 10 \text{ k}\Omega} = 4 \text{ k}\Omega$$
$$\frac{15 \text{ V}}{20 \text{ k}\Omega} + \frac{v_{\text{in}}}{10 \text{ k}\Omega} + \frac{v_{\text{out}}}{4 \text{ k}\Omega} = 0$$
$$v_{\text{out}} = -0.4 v_{\text{in}} - 3 \text{ V}$$

For the lower diode conducting,

$$v_{\text{out}} = -0.4 v_{\text{in}} + 3 \text{ V}$$

For both diodes open,

$$v_1 = (15 \text{ V})\left(\frac{20 \text{ k}\Omega}{\frac{20}{3} \text{ k}\Omega + 20 \text{ k}\Omega}\right)$$

$$+ v_{\text{out}}\left(\frac{20 \text{ k}\Omega}{\frac{20}{3} \text{ k}\Omega + 20 \text{ k}\Omega}\right)$$

$$= \frac{(15 \text{ V})\left(\frac{20}{3} \text{ k}\Omega\right) + v_{\text{out}}(20 \text{ k}\Omega)}{\frac{20}{3} \text{ k}\Omega + 20 \text{ k}\Omega}$$

$$= \frac{15 \text{ V} + 3v_{\text{out}}}{4}$$

$$= \frac{15 \text{ V} - 3v_{\text{in}}}{4}$$

Because the negative terminal is at virtual ground, for the upper diode to be off, $v_1 > 0$ V, so

$$15 \text{ V} - 3v_{\text{in}} > 0 \text{ V}$$
$$v_{\text{in}} < 5 \text{ V}$$

For the lower diode,

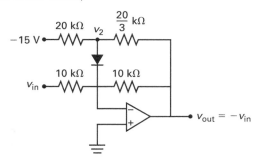

For both diodes open,

$$v_2 = (-15 \text{ V})\left(\frac{\frac{20}{3} \text{ k}\Omega}{\frac{20}{3} \text{ k}\Omega + 20 \text{ k}\Omega}\right)$$

$$- v_{\text{in}}\left(\frac{\frac{20}{3} \text{ k}\Omega}{\frac{20}{3} \text{ k}\Omega + 20 \text{ k}\Omega}\right)$$

$$= \frac{(-15 \text{ V})\left(\frac{20}{3} \text{ k}\Omega\right) - v_{\text{in}}(20 \text{ k}\Omega)}{\frac{20}{3} \text{ k}\Omega + 20 \text{ k}\Omega}$$

$$= \frac{-15 \text{ V} - 3v_{\text{in}}}{4} < 0 \text{ V}$$

$$v_{\text{in}} > -5 \text{ V}$$

For $-5 \text{ V} < v_{\text{out}} < 5 \text{ V}$, $v_{\text{out}} = -v_{\text{in}}$
$$v_{\text{in}} < -5 \text{ V}, \quad v_{\text{out}} = -0.4v_{\text{in}} + 3 \text{ V}$$
$$v_{\text{in}} > 5 \text{ V}, \quad v_{\text{out}} = -0.4v_{\text{in}} - 3 \text{ V}$$

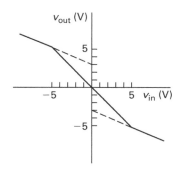

30. With at least one input low, the collector is high and no current flows in the collector circuit.

One or more low inputs of 0.2 V, when combined with a diode drop of 0.7 V, results in a current through the left resistor of

$$I = \frac{\Delta V}{R}$$

$$I = \frac{4 \text{ V} - 0.7 \text{ V} - 0.2 \text{ V}}{(2 \text{ k}\Omega)\left(1000 \ \frac{\Omega}{\text{k}\Omega}\right)}$$

$$= 1.55 \times 10^{-3} \text{ A}$$

The power for high output is

$$P = IV$$
$$= (1.55 \times 10^{-3} \text{ A})(4 \text{ V})$$
$$= 6.2 \times 10^{-3} \text{ W}$$

With both inputs high,

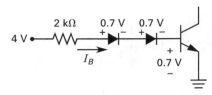

$$I = \frac{\Delta V}{R}$$

$$I_B = \frac{4\text{ V} - (3)(0.7\text{ V})}{(2\text{ k}\Omega)\left(1000\ \dfrac{\Omega}{\text{k}\Omega}\right)}$$

$$= 0.95 \times 10^{-3}\text{ A}$$

This saturates the transistor, so

$$I = \frac{\Delta V}{R}$$

$$I_C = \frac{4\text{ V} - 0.2\text{ V}}{(2\text{ k}\Omega)\left(1000\ \dfrac{\Omega}{\text{k}\Omega}\right)}$$

$$= 1.9 \times 10^{-3}\text{ A}$$

The power supply provides

$$\begin{aligned}
P_{\text{supply}} &= IV = (I_C + I_B)V \\
&= (1.9 \times 10^{-3}\text{ A} + 0.95 \times 10^{-3}\text{ A})(4\text{ V}) \\
&= 11.4 \times 10^{-3}\text{ W}
\end{aligned}$$

However, the transistor must absorb power from the loads, which must be calculated from a current of five times the input current for the high output. This excess collector current is

$$I_c = 5I = (5)(1.55 \times 10^{-3}\text{ A}) = 7.75 \times 10^{-3}\text{ A}$$

Assuming the transistor remains saturated, the excess transistor power can be as high as

$$\begin{aligned}
P_{\text{excess}} &= IV \\
&= (7.75 \times 10^{-3}\text{ A})(0.2\text{ V}) \\
&= 1.55 \times 10^{-3}\text{ W}
\end{aligned}$$

The maximum total power is

$$\begin{aligned}
P_{\text{total}} &= P_{\text{supply}} + P_{\text{excess}} \\
&= 1.55 \times 10^{-3}\text{ W} + 11.4 \times 10^{-3}\text{ W} \\
&= \boxed{12.95 \times 10^{-3}\text{ W} \quad (12.95\text{ mW})}
\end{aligned}$$

31. The on delay timer (TON) shows an accumulator count of 200 with a preset of 200. This indicates the count is complete; therefore, the enable (EN) line is energized (ON or High) and the done (DN) line is energized (ON or High). The EN line energizes relay M2.1, while the DN line energizes relay M2.2. The M relays close their respective contacts at address locations 002 and 003, energized PL2 and PL3, respectively.

All relays are energized.

The answer is (D).

32. The timer base is 0.01 sec, meaning the timer counts up 1 for every hundredth of a second. The accumulator shows that 100 counts, or 1 sec, has elapsed, with 100 counts, or 1 sec, remaining.

Once the preset count is reached, the enable (EN) line energizes M2.1, which closes contact M2.1 at address 002.

Therefore, 1.00 sec remains before the indicated contact closes.

The answer is (B).

33. Transistor Q1 is a PMOS enhancement device, meaning the conducting channel majority carriers are holes. Therefore, a negative gate to source voltage is required to turn Q1 ON. Since the control voltage is 0 V and the source voltage is given as $+V_{DD}$, the gate-source voltage is negative, positive (hole) carriers are drawn into the channel, and Q1 is ON. Consider the arrow as pointing in the direction of conventional current flow, that is, from positive to negative. Thus, the channel is positive with respect to the substrate at the arrow head.

By contrast, transistor Q2 is a NMOS enhancement device (remember that the arrow points to the direction of conventional current flow) and the gate-source voltage is 0 V. The source is connected to the arrow. Thus in this case, the source is the ground connection (0 V) and the control voltage (as given) is 0 V. A NMOS device requires a positive gate-source voltage to draw electrons into the channel to make it conduct. Since such a voltage is not present, Q2 is OFF.

The net result is that the $+V_{DD}$ is felt at point D and V_{out} (from point D to the ground) takes on the value of $+V_{DD}$.

The answer is (B).

34. A cycle is measured in the time it takes to complete it. The phase of a signal at any given time within a cycle can be measured in portions of a cycle (with 1.0 being one complete cycle), radians, or degrees, all of which are used in PLL design. This is further complicated by the common practice of not showing units for Laplace transforms, derivatives, and integrals. Additionally, a radian is not a unit, per se; neither is a cycle a unit.

To obtain the answer, carrying the units—and non-units—is recommended.

$$\omega = \frac{d\phi}{dt}$$

$$= 60t \text{ rad/sec}^2$$

$$f = \left(60t \; \frac{\text{radians}}{\text{sec}^2}\right)\left(\frac{1 \text{ cycle}}{2\pi \text{ radians}}\right)$$

$$= \left(9.549t \; \frac{\text{cycles}}{\text{sec}^2}\right)\left(\frac{\text{Hz}}{\text{cycle/sec}}\right)$$

$$= 9.549t \; \frac{\text{Hz}}{\text{sec}}$$

$$= \frac{\left(9.549 \; \frac{\text{Hz}}{\text{sec}}\right)(34 \text{ msec})}{1000 \; \frac{\text{msec}}{\text{sec}}}$$

$$= 0.325 \text{ Hz}$$

Be sure to use radians inside the argument for trigonometric functions, or use this type of unit conversion and analysis to ensure the correct answer is obtained.

The answer is (B).

35. The ground detection circuit, with the given values inserted, is shown.

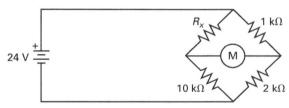

The meter experiences 0 A of current when the bridge is balanced. The relationship among the resistors in this Wheatstone bridge can be used to find the unknown resistance.

$$\frac{R_x}{R_3} = \frac{R_2}{R_4}$$

$$R_x = \frac{R_3 R_2}{R_4} = \frac{(10 \text{ k}\Omega)(1 \text{ k}\Omega)}{2 \text{ k}\Omega}$$

$$= 5 \text{ k}\Omega \quad (5000 \; \Omega)$$

The answer is (C).

49 Digital Logic

PRACTICE PROBLEMS

1. What is the state of a TTL cell at 4.0 V?

(A) 0

(B) indeterminate

(C) 1

(D) saturated

2. A computer accesses a register in 4-bit packets, sometimes called *nibbles*. What is the total number of values that can be represented in a nibble?

(A) 4

(B) 8

(C) 15

(D) 16

3. A computer data sheet indicates a 64K byte RAM. How many bytes of memory are available?

(A) 8 bytes

(B) 1024 bytes

(C) 64,000 bytes

(D) 65,536 bytes

4. How many bits of information are contained within a 64K byte RAM?

(A) 64 bits

(B) 8192 bits

(C) 512,000 bits

(D) 524,288 bits

5. If $x_1 = 0$, $x_2 = 1$, and $x_3 = 0$, what is the value of the following expression?

$$f(x_1, x_2, x_3) = \overline{x}_1 x_2 x_3 + x_1 \overline{x}_2 x_3 \cdot x_1 x_2 \overline{x}_3$$

6. A four-variable function has a minterm in row five. That minterm is given by which of the following expressions?

(A) $\overline{A}\,B\overline{C}$

(B) $A\overline{B}\,C$

(C) $\overline{A}\,B\overline{C}D$

(D) $A\overline{B}\,C\overline{D}$

The following truth table is applicable to Prob. 7 and Prob. 8.

A	B	C	$F(A,B,C)$
0	0	0	0
0	0	1	0
0	1	0	0
0	1	1	1
1	0	0	0
1	0	1	0
1	1	0	0
1	1	1	1

7. What is the canonical sum-of-products (SOP) expression for the function in the truth table shown?

(A) $\overline{A}\,\overline{B}\,\overline{C} + \overline{A}\,\overline{B}\,C + \overline{A}\,B\overline{C}$

(B) $A\overline{B}\,\overline{C} + A\overline{B}\,\overline{C} + AB\overline{C}$

(C) (A) and (B) combined

(D) $\overline{A}\,BC + ABC$

8. What maxterms are necessary to create the canonical product-of-sums (POS) expression for the function in the truth table shown?

(A) $M_0 M_1 M_2$

(B) $M_4 M_5 M_6$

(C) (A) and (B) combined

(D) $M_3 M_7$

The following truth table is applicable to Prob. 9 and Prob. 10.

A	B	$F(A,B)$
-5 V	-5 V	-5 V
-5 V	0	0
0	-5 V	0
0	0	-5 V

9. If positive logic is applied to the truth table, what logic operation is obtained?

    (A)   coincidence

    (B)   NOR

    (C)   XOR

    (D)   XNOR

10. If negative logic is applied to the truth table, what logic operation is obtained?

    (A)   coincidence

    (B)   NOR

    (C)   OR

    (D)   XOR

SOLUTIONS

1. The state of a cell is its binary value. For TTL, any voltage between 2.0 V and 5.0 V is state 1.

The answer is (C).

2. The total number of values in any n-tuple logic expression is given by Eq. 49.1.

$$N = 2^n = 2^4 = \boxed{16}$$

The answer is (D).

3. The K indicates $2^{10} = 1024$. So,

$$64\text{K} = (64)(2)^{10} = \boxed{65{,}536 \text{ bytes}}$$

The answer is (D).

4. One byte is 8 bits. Therefore,

$$(65{,}536 \text{ bytes})\left(8 \ \frac{\text{bits}}{\text{byte}}\right) = \boxed{524{,}288 \text{ bits}}$$

The answer is (D).

5. The function is

$$f(x_1, x_2, x_3) = \overline{x}_1 x_2 x_3 + x_1 \overline{x}_2 x_3 \cdot x_1 x_2 \overline{x}_3$$

Substituting the given values and using the order of precedence rules gives

$$\begin{aligned}
f(0,1,0) &= (\overline{0} \cdot 1 \cdot 0) + (0 \cdot \overline{1} \cdot 0) \cdot (0 \cdot 1 \cdot \overline{0}) \\
&= (1 \cdot 1 \cdot 0) + (0 \cdot 0 \cdot 0) \cdot (0 \cdot 1 \cdot 1) \\
&= 0 + 0 \cdot 0 \\
&= \boxed{0}
\end{aligned}$$

6. A function with four variables (that is, with $n = 4$) with a value of one in row five has a minterm in row five. Changing row five from decimal to binary gives

$$5_{10} = 0101_2$$

The minterm is deduced from its binary representation as

$$\boxed{\overline{A}\,B\overline{C}D = m_5}$$

The answer is (C).

7. The minterms are in rows three and seven—that is, where the function equals one. The canonical sum-of-products form is

$$F(A, B, C) = m_3 + m_7 = \boxed{\overline{A}BC + ABC}$$

The answer is (D).

8. The maxterms are in the rows where the function has a value of zero.

$$F(A, B, C) = \boxed{\prod M(0,1,2,4,5,6)}$$

The answer is (C).

9. If positive logic is applied, the high voltage has a value of one. Rewrite the truth table with -5 V as logic 0 and 0 V as logic 1.

A	B	$F(A,B)$
0	0	0
0	1	1
1	0	1
1	1	0

This is the exclusive OR (XOR) operation.

The answer is (C).

10. If negative logic is applied, the low voltage has a value of one. Rewrite the truth table with -5 V as logic 1 and 0 V as logic 0.

A	B	$F(A,B)$
1	1	1
1	0	0
0	1	0
0	0	1

Reverse the order to place the table in standard format.

A	B	$F(A,B)$
0	0	1
0	1	0
1	0	0
1	1	1

This is the XNOR, or coincidence, logic.

The answer is (A).

Topic XI: Special Applications

Chapter

Special
Applications

50 Lightning Protection and Grounding

PRACTICE PROBLEMS

1. What range of voltage does lightning span?

(A) 1 MV to 2 MV

(B) 2 MV to 5 MV

(C) 5 MV to 10 MV

(D) 10 MV to 1000 MV

2. What range of current does lightning span?

(A) 100 A to 1000 A

(B) 1000 A to 10,000 A

(C) 1000 A to 100,000 A

(D) 1000 A to 200,000 A

3. What is the name for an insulation level that is specified in terms of the crest value of the standard lightning impulse?

(A) BIL

(B) lightning impulse insulation level

(C) TIL

(D) withstand voltage

4. The minimum voltage that a power system can be designed for and be protected against direct lightning strikes by its own insulation is approximately

(A) 15 kV

(B) 60 kV

(C) 230 kV

(D) 920 kV

5. An arrester provides a protection level of 38 kV to equipment with a 95 kV withstand voltage. If the minimum protective margin allowed by standards is 20%, is the equipment adequately protected?

6. What is the name for an insulation level that is specified in terms of the crest value of the withstand voltage?

(A) basic lightning impulse insulation level (BIL)

(B) NEC level

(C) NESC level

(D) transient insulation level (TIL)

7. An arrester with a protection level of 100 kV is used in a system with a withstand voltage of 120 kV. Most nearly, what is the protective margin?

(A) −20%

(B) 17%

(C) 20%

(D) 83%

8. What is the theoretical maximum protection quality index?

(A) 0

(B) 1

(C) 100

(D) 1000

9. A small shed covers a pump and is less than 1 ft high. A metal mast with a strike termination device mounted 200 ft high is to be installed to protect the pump from sideflash. Most nearly, what is the minimum distance between the mast and the pump shed?

(A) 25 ft

(B) 35 ft

(C) 100 ft

(D) 200 ft

10. In short-circuit analysis, which statement is NOT true of the fusing current and ampacity of a conductor?

(A) Ampacity is lower than the fusing current.

(B) The fusing current depends on the area of the conductor.

(C) The fusing current results in a change of state of the conductor.

(D) The fusing current's generated heat is transmitted immediately to its surroundings.

SOLUTIONS

1. Lightning voltage varies widely but is in the range of 10 MV to 1000 MV.

The answer is (D).

2. The current of lightning ranges from 1000 A to 200,000 A.

The answer is (D).

3. The voltage level that insulation can withstand, specified in terms of the crest value of a standard lightning impulse, is called the basic lightning impulse level (BIL).

The answer is (A).

4. Systems of 230 kV or less cannot be insulated against the overvoltage caused by direct lightning strikes and must use a combination of shielding and grounding.

The answer is (C).

5. The protective margin is

$$
\begin{aligned}
\text{PM} &= \frac{V_w - V_p}{V_p} \\
&= \frac{95 \text{ kV} - 38 \text{ kV}}{38 \text{ kV}} \\
&= 1.5 \quad (150\%)
\end{aligned}
$$

This is well over 20%, so the equipment is adequately protected.

6. The voltage level that insulation can withstand, specified in terms of the withstand voltage, is called the transient insulation level (TIL).

The answer is (D).

7. Use the equation for the protective margin.

$$
\begin{aligned}
\text{PM} &= \frac{V_w - V_p}{V_p} \\
&= \frac{120 \text{ kV} - 100 \text{ kV}}{100 \text{ kV}} \\
&= \boxed{0.20 \quad (20\%)}
\end{aligned}
$$

The answer is (C).

8. The equation for the protection quality index (PQI) is

$$\mathrm{PQI} = \frac{V_r}{V_p}$$

An arrester operates at a given *voltage protection level*, V_p, and then reseals at a lower *reseal level*, V_r. The theoretical maximum protection quality index is one.

The answer is (B).

9. A strike termination device (or lightning rod) mounted on a metal mast for the purpose of lightning protection is considered to be an air terminal. A sideflash is an electrical spark that occurs between a conductive metal body and a component of the lightning protective system. The minimum distance a sideflash could occur is governed by NFPA 780. For the indicated condition,

$$D = \frac{h}{6} = \frac{200 \text{ ft}}{6}$$
$$= \boxed{33.33 \text{ ft} \quad (33 \text{ ft})}$$

Since the shed is low-profile, it can be located anywhere within this range. (In actuality, the entire height of the shed needs to be within this range.)

The answer is (A).

10. The fusing current is the current at which the conductor melts. This depends upon the area, A, the current level, I, the amount of time that current is applied, t, and the type of material used, k. Use the following formula to calculate the time it takes a conductor to melt.

$$t = \frac{A^2 k}{I^2}$$

The fusing current may also be understood as the current level at which the conductor undergoes fundamental changes (i.e., changes phases), which makes the conductor unable to handle the same current level repeatedly.

Ampacity is always lower than the fusing current. The formula for the fusing current shows that it depends upon the area of the conductor. When the fusing current occurs, the conductor material melts, producing a temporary change of phase from solid to liquid. The formulas and calculations for fusing current assume that the heat generated by the current is stored in the conductor and that no heat reduction occurs through heat transfer to the surroundings.

The answer is (D).

51 Illumination

PRACTICE PROBLEMS

1. What is the wavelength of a 60 Hz signal (using vacuum as the reference)?

(A)   0.016 m

(B)   60 m

(C) 5.0×10^6 m

(D) 3.0×10^8 m

2. What is the energy contained in a single quantum of visible light at a wavelength of 5.55×10^{-7} m?

(A) 3.7×10^{-40} J

(B) 3.6×10^{-19} J

(C) 5.6×10^{-7} J

(D)   170 J

3. Which type of light may be used to destroy bacteria?

(A)   infrared

(B)   laser

(C)   visible

(D)   ultraviolet

4. Which of the following statements about visible light is NOT true?

(A)   The index of refraction is 1.5 in air.

(B)   Visible light consists of many colors.

(C)   The average frequency of visible light is approximately 5.4×10^{14} Hz.

(D)   The wavelengths are approximately 380–780 nm.

5. Kirchhoff's law of thermal radiation specifies that the emissitivity, ϵ, equals the spectral absorption, α, in

(A)   a blackbody

(B)   a greybody

(C)   a selective radiator

(D)   none of the above

6. What is the maximum wavelength expected for a light source at 3000K?

(A) 9.7×10^{-7} m

(B) 3.3×10^{-4} m

(C) 2.9×10^{-3} m

(D) 3.0×10^{-3} m

7. What wavelength of light is emitted if an electron experiences a 2.2 V transition in a semiconductor?

(A)   220 nm

(B)   300 nm

(C)   555 nm

(D)   560 nm

8. The maximum lumens available per watt at a wavelength of 555 nm is ___ lm, though an ideal white source provides a maximum of ___ lm per watt.

(A)   555, 220

(B)   683, 220

(C)   683, 683

(D)   683, 432

9. The illuminance required at desktop level for a given task is 500 lx. Manufacturer data for the light source indicates an illuminance of 3500 lx at 1 m. The situation is as shown.

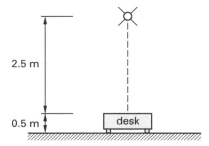

What is the illuminance level at the desk top?

(A) 80 lx

(B) 560 lx

(C) 1400 lx

(D) 14 000 lx

10. Consider the lighting configuration shown. The light sources all have the same intensity, $I = 500$ cd. What is the contribution of light source 2 at point A?

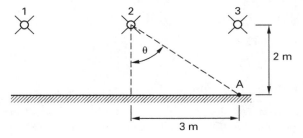

(A) 21 lx

(B) 42 lx

(C) 70 lx

(D) 130 lx

11. The required light level at a 36-inch high work plane is 50 fc. The luminaires hang 6 inches from the ceiling and have a coefficient of utilization (CU) of 0.8. The entire room is considered a work area. The room is 20 ft × 20 ft. What is most nearly the total lumens required from the luminaires to achieve the desired light level? Do not consider light loss factors.

(A) 50 lm

(B) 150 lm

(C) 6200 lm

(D) 25,000 lm

12. A public space requires a floor-level illuminance of at least 30 lx. An omnidirectional light source provides 10,000 lm. Evaluating the light as a point source, what is most nearly the maximum distance from the floor that the light can be installed?

(A) 5 m

(B) 25 m

(C) 80 m

(D) 300 m

13. A light source with a luminous intensity of 200 lm, approximated as a point source, is positioned 1 m above the floor (measured perpendicularly). What is the approximate illuminance 1.5 m from this perpendicular point on the floor?

(A) 35 lx

(B) 90 lx

(C) 110 lx

(D) 120 lx

14. A single-mode optical fiber has an index of refraction of 1.5. What is most nearly the critical incident angle?

(A) −41.8°

(B) −31.5°

(C) 41.8°

(D) 90.0°

15. A factory working space must maintain a minimum illuminance of 30 lx over the life of its lighting sources. The area of the space is 100 m², the coefficient of utilization is 0.8, the light loss factor is 0.8, and the lights' data sheets indicate they each provide 95 lm. How many lights are required in the space to maintain the required illuminance?

(A) 3

(B) 30

(C) 45

(D) 50

SOLUTIONS

1. The wavelength (with an index of refraction equal to one—that of a vacuum) is derived from the relation of the speed of light, c, to its wavelength, λ, and its frequency, ν.

$$c = \lambda\nu$$

$$\lambda = \frac{c}{\nu} = \frac{2.9979 \times 10^8 \, \frac{\text{m}}{\text{s}}}{60 \text{ Hz}}$$

$$= \boxed{4.9665 \times 10^6 \text{ m} \quad (5 \times 10^6 \text{ m})}$$

The wavelength is 5000 km, or approximately 3100 mi. Because the index of refraction in air is nearly equal to the refraction index in a vacuum, this is also the wavelength in air. Extremely low frequency (ELF) waves (such as a 60 Hz signal) travel in the cavity between the Earth's surface and the ionosphere.

The answer is (C).

2. The human retina is most sensitive to visible light at 555 nm (5.55×10^{-7} m). Using Planck's constant, h, the energy of a single quantum of light at this wavelength is

$$E = h\nu = \frac{hc}{\lambda}$$

$$= \frac{(6.6256 \times 10^{-34} \text{ J·s})\left(2.9979 \times 10^8 \, \frac{\text{m}}{\text{s}}\right)}{5.55 \times 10^{-7} \text{ m}}$$

$$= \boxed{3.56 \times 10^{-19} \text{ J} \quad (3.6 \times 10^{-19} \text{ J})}$$

The answer is (B).

3. Ultraviolet light causes numerous biological effects, one of which is the destruction of various forms of bacteria.

The answer is (D).

4. The wavelengths of visible light are approximately 380–780 nm. Visible light is polychromatic, that is, it consists of many colors. The human retina is most sensitive to light at 555 nm. The index of refraction in air is approximately one. (The actual difference from one is so slight that it is unimportant in most calculations.)

The answer is (A).

5. A blackbody radiator has the property that

$$\epsilon = \alpha = 1$$

The answer is (A).

6. According to the Wien displacement law, the maximum wavelength depends only on the temperature, and is given by

$$\lambda_{\max} = \frac{b}{T}$$

$$= \frac{2.8978 \times 10^{-3} \text{ m·K}}{3000 \text{K}}$$

$$= \boxed{9.7 \times 10^{-7} \text{ m}}$$

The constant b is determined by experiment.

The answer is (A).

7. The wavelength of emitted radiation in terms of the potential difference is

$$\lambda_{\text{nm}} = \frac{1239.76 \text{ nm·V}}{\Delta V}$$

$$= \frac{1239.76 \text{ nm·V}}{2.2 \text{ V}}$$

$$= \boxed{563 \text{ nm} \quad (560 \text{ nm})}$$

The answer is (D).

8. The theoretical maximum number of lumens available per watt of energy is 683 lm. Because of inefficiencies and unavoidable losses, an ideal white source can only provide 220 lm.

The answer is (B).

9. The inverse square law gives the illuminance from a point source at a straight-line distance, r, from the source as

$$E_1 r_1^2 = E_2 r_2^2$$

$$E_2 = \frac{E_1 r_1^2}{r_2^2}$$

$$= \frac{(3500 \text{ lx})(1 \text{ m})^2}{(2.5 \text{ m})^2}$$

$$= \boxed{560 \text{ lx}}$$

The answer is (B).

10. By the cosine-cubed law, the illuminance E, for a light source of intensity I, at point A, is

$$E = \frac{I \cos^3 \theta}{h^2} \qquad [I]$$

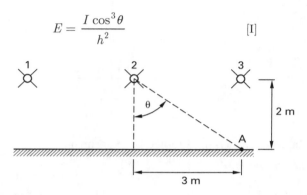

The variable h is the height. The only unknown is the angle θ.

$$\theta = \arctan \frac{\text{opposite}}{\text{adjacent}}$$
$$= \arctan \frac{3 \text{ m}}{2 \text{ m}}$$
$$= 56.3°$$

Substitute the value of θ into Eq. [I].

$$E = \frac{I \cos^3 \theta}{h^2}$$
$$= \frac{(500 \text{ cd}) \cos^3 56.3°}{(2 \text{ m})^2}$$
$$= \boxed{21.35 \ \frac{\text{cd}}{\text{m}^2} \quad (21 \text{ lx})}$$

The answer is (A).

11. The light level is given by the following equation. The luminance is labeled initial since light loss factors (LLF) were not considered.

$$E_{\text{initial}} = \frac{L_{\text{total}} \text{CU}}{A_{\text{workplane}}}$$

Rearrange the equation to solve for the total lamp lumens, L_{total}. A footcandle (fc) is one lumen per square foot.

$$L_{\text{total}} = \frac{E_{\text{initial}} A_{\text{workplane}}}{\text{CU}}$$
$$= \frac{(50 \text{ fc})(20 \text{ ft})(20 \text{ ft})}{0.8}$$
$$= 25{,}000 \text{ lm}$$

The answer is (D).

12. The distance can be calculated using the equation for illuminance, which assumes a spherical receptor. (Multiple sources would be required to maintain this illuminance over a flat floor space.)

$$E = \frac{\Phi_t}{4\pi r^2}$$
$$r^2 = \frac{\Phi_t}{4\pi E}$$
$$r = \sqrt{\frac{\Phi_t}{4\pi E}} = \sqrt{\frac{10{,}000 \text{ lm}}{4\pi(30 \text{ lx})}}$$
$$= \boxed{5.15 \text{ m} \quad (5 \text{ m})}$$

The answer is (A).

13. Draw a diagram of the light source in relation to the floor.

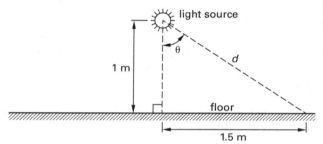

Use Lambert's law and its extension, the cosine-cubed law.

$$E = \frac{I}{d^2} \cos \theta = \frac{I \cos^3 \theta}{h^2}$$
$$= \frac{(200 \text{ lm}) \left(\cos^3 \left(\arctan \frac{1.5 \text{ m}}{1.0 \text{ m}} \right) \right)}{(1.0 \text{ m})^2}$$
$$= \boxed{34.14 \text{ lx} \quad (35 \text{ lx})}$$

The answer is (A).

14. The critical incident angle is the angle at which an optically transparent surface becomes totally reflecting.

$$\sin \theta_c = \frac{1}{n}$$
$$\theta_c = \arcsin \frac{1}{n}$$
$$= \arcsin \frac{1}{1.5}$$
$$= \boxed{41.8°}$$

The answer is (C).

15. From the maintained illuminance, solve for the number of required lights.

$$E_{\text{maintained}} = \frac{N_{\text{lights}} L_{\text{per light}}(\text{CU})(\text{LLF})}{A_w}$$

$$N_{\text{lights}} = \frac{E_{\text{maintained}} A_w}{L_{\text{per light}}(\text{CU})(\text{LLF})}$$

$$= \frac{(30 \text{ lx})(100 \text{ m}^2)}{(95 \text{ ln})(0.8)(0.8)}$$

$$= \boxed{49.34 \quad (50)}$$

The answer is (D).

Special
Applications

52 Power System Management

PRACTICE PROBLEMS

1. A SCADA system configuration uses input from an instrument CT with a turns ratio of 100:5. The transducer supplying the SCADA system has a full-scale AC current value of 2.5 A. What is most nearly the maximum current in the primary side of the CT?

(A) 20 A

(B) 25 A

(C) 50 A

(D) 70 A

2. When generators are in parallel, their loads are shared. What electrical parameter is used to control the reactive load sharing between generators?

(A) current

(B) frequency

(C) speed

(D) voltage

3. Stability studies are used to determine the impact of disturbances on an electrical system. Which of the following assumptions are common to nearly all stability studies?

(A) Unbalanced faults are calculated using symmetrical components.

(B) Generated voltage is not affected by the speed of the prime mover.

(C) Harmonics are ignored.

(D) All of the above are common.

4. A main function of the SCADA system is to ensure economical operation of the electrical distribution system. This function is accomplished by evaluating the load, load forecast, fuel cost, and usage for each particular type of machine. What is the term for this function?

(A) energy management

(B) Modbus® adjustment

(C) power flow study

(D) scheduling

5. A pair of single-phase transformers is placed in parallel. (The transformers have the same voltage ratings.) The applicable data for the two transformers follows.

$$T_1: \quad S_1 = 2000 \text{ kVA}$$
$$Z_{T_1} = 3\%$$
$$T_2: \quad S_2 = 4000 \text{ kVA}$$
$$Z_{T_2} = 4.5\%$$

What is most nearly the maximum load allowed without overloading either of the transformers?

(A) 2000 kVA

(B) 3500 kVA

(C) 4600 kVA

(D) 6000 kVA

6. The diagram shows a portion of a power plant electrical distribution system.

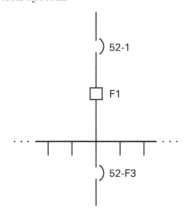

The time current characteristics (TCC) shown display a coordination problem.

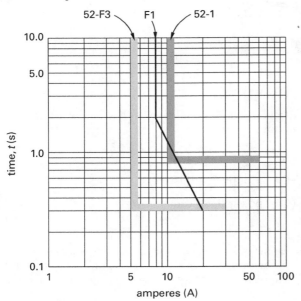

What is a valid solution to the coordination problem?

(A) Adjust the TCC curve on 52-F3 to the left.

(B) Adjust or replace F1 so that its curve falls between 52-1 and 52-F3.

(C) Adjust the TCC curve on 52-1 to the left of the F1 curve.

(D) Adjust the TCC curve on 52-F3 to the right.

7. A coordination problem is discovered during breaker replacement in a power plant. Changing the time current characteristic (TCC) curve can solve the problem. The four possible TCC curves shown allow for proper coordination.

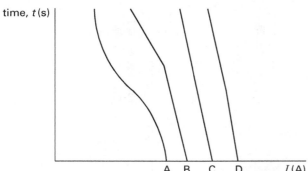

Which curve represents the greatest arc flash hazard?

(A) curve A

(B) curve B

(C) curve C

(D) curve D

8. Several software products can be used to perform power flow studies, also called load studies. Three general bus types are used in such studies.

I. generator bus

II. load bus

III. slack bus

To which bus type(s) is a generator directly attached?

(A) I only

(B) II only

(C) III only

(D) I and III only

9. The fuel use per electrical power output for two generators is shown.

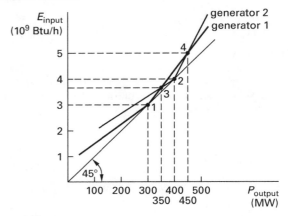

At point 1, generator 1 is tangent to the line from the origin.

At point 2, generator 2 is tangent to the line from the origin.

At points 3 and 4, the curves for the fuel consumption per output power of generators 1 and 2 cross.

At what point is generator 1 most efficient, and what is the fuel requirement for generator 2 at its most fuel-efficient point?

(A) point 1; 1×10^7 Btu/MW·h

(B) point 2; 3×10^9 Btu/MW·h

(C) point 3; 1×10^7 Btu/MW·h

(D) point 4; 5×10^9 Btu/MW·h

10. Power systems are monitored, not just during faults, but also during normal events. A circuit breaker opening event is recorded. The voltage trace across the main contacts is shown.

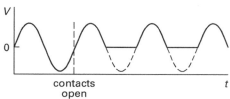

Which of the following conditions might be indicated by the trace shown?

   I.   arc flash

  II.   circuit breaker re-strike

 III.   main contacts out of alignment

  (A)   I only

  (B)   III only

  (C)   I or II only

  (D)   I, II, or III

SOLUTIONS

1. SCADA refers to the Supervisory Control and Data Acquisition subsystem of the EMS. The relationship for an ideal transformer's primary and secondary currents (which will provide the maximum current) is given by

$$\frac{I_{\text{primary}}}{I_{\text{secondary}}} = \frac{n_{\text{secondary}}}{n_{\text{primary}}}$$

Rearrange the equation to solve for the primary current.

$$
\begin{aligned}
I_{\text{primary}} &= \left(\frac{n_{\text{secondary}}}{n_{\text{primary}}} \right) I_{\text{secondary}} \\
&= \left(\frac{100}{5} \right)(2.5 \text{ A}) \\
&= \boxed{50 \text{ A}}
\end{aligned}
$$

The answer is (C).

2. Adjusting the no-load frequency set point of a generator controls real load. Adjusting the no-load voltage set point of a generator controls the reactive load.

The answer is (D).

3. Unbalanced faults are calculated using the method of symmetrical components. Generated voltage is assumed to be constant in stability studies, so small changes in the speed of the prime mover are considered not to impact the voltage. DC offsets and harmonics are ignored. That is, only synchronous frequency voltages and currents are considered.

The answer is (D).

4. Energy management is a broad term applying to any aspect of power system management.

A Modbus is a serial communications protocol common to many SCADA systems.

A power flow study is used to determine the stability of a system.

Scheduling is the process by which the SCADA system evaluates the load, load forecast, fuel cost, usage, and numerous other factors to determine the most economical usage.

The answer is (D).

5. Use the per unit system to convert the impedance values of the first transformer to the second transformer. The voltage base is the same on both transformers.

$$Z_{\text{pu,new}} = Z_{\text{pu,old}} \left(\frac{V_{\text{base,old}}}{V_{\text{base,new}}} \right)^2 \left(\frac{S_{\text{base,new}}}{S_{\text{base,old}}} \right)$$

$$= j0.03(1)^2 \left(\frac{4000 \text{ kVA}}{2000 \text{ kVA}} \right)$$

$$= j0.06$$

The transformers are now on the same base. The voltage base, regardless of the exact value of that voltage, is the same for both transformers. Taking the ratio of the full power gives

$$\frac{S_{T_1}}{S_{T_2}} = \frac{\dfrac{V^2}{Z_{T_1}}}{\dfrac{V^2}{Z_{T_2}}} = \frac{Z_{T_2}}{Z_{T_1}}$$

The transformers will add load in inverse proportion to their impedances. Substitute the same base values and solve for the full power of T_2 when T_1 is fully loaded.

$$\frac{S_{T_1}}{S_{T_2}} = \frac{Z_{T_2}}{Z_{T_1}}$$

$$S_{T_2} = S_{T_1} \left(\frac{Z_{T_1}}{Z_{T_2}} \right) = (2000 \text{ kVA}) \left(\frac{0.06}{0.045} \right)$$

$$= 2667 \text{ kVA}$$

The total load of the transformers is

$$S_{\text{total}} = S_{T_1} + S_{T_2}$$

$$= 2000 \text{ kVA} + 2667 \text{ kVA}$$

$$= \boxed{4667 \text{ kVA} \quad (4600 \text{ kVA})}$$

The answer is (C).

6. A properly coordinated curve is shown.

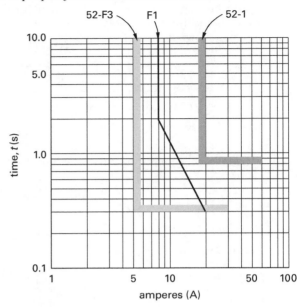

The result shown is achieved by adjusting or replacing F1 so that its curve falls between the two breaker curves 52-1 and 52-F3. (It can also be achieved by resetting breaker 52-1's characteristics so that its TCC curve moves to the right of the F1 curve.)

The answer is (B).

7. Curve D is the greatest hazard because it represents the greatest amount of *let-through energy* for any given time interval.

Though not strictly an energy unit, the I^2t term used in describing the let-through energy is called an *energy unit*.

The answer is (D).

8. A bus without any generators is called a load bus. A bus with a generator attached is called a generator bus, with one exception. A slack bus has a generator attached but is used in the studies as a reference bus. There are no variables to solve for on the reference bus. The slack bus will usually be the bus with the largest generator capability and will be designated bus 1.

Three popular software programs that can be used to perform power flow studies are ETAP, SKM, and EDSA.

The answer is (D).

9. The fuel efficiency is the inverse slope of the fuel usage curve at any given point. The generators are most efficient where the fuel usage curves are tangent to the origin. Therefore, generator 1 is most efficient at point 1.

Using data from point 2 (the most fuel-efficient point for generator 2), the fuel requirement for generator 2 is

$$R_{\text{fuel}} = \frac{E_{\text{input}}}{P_{\text{output}}}$$

$$= \frac{4 \times 10^9 \, \dfrac{\text{Btu}}{\text{h}}}{400 \, \text{MW}}$$

$$= 1 \times 10^7 \, \text{Btu/MW·h}$$

The answer is (A).

10. Arcing is occurring due to degraded insulation or out-of-alignment contacts. Current flow through the arc is being reestablished even with the main contacts open, which is known as a re-strike.

The answer is (D).

Topic XII: Measurement and Instrumentation

Chapter

Measurement/
Instrumentation

53 Measurement and Instrumentation

PRACTICE PROBLEMS

1. What instrumentation term can be defined as a measure of the reproducibility of observations?

(A) accuracy

(B) error

(C) precision

(D) resolution

2. A digital voltmeter is designed for a maximum ambient temperature of 65°C. On the 10 KHz scale, what is the expected noise voltage in a 100 kΩ series resistor?

(A) 4.3 nV

(B) 0.14 µV

(C) 1.9 µV

(D) 4.3 µV

3. What is the average value of the voltage signal $v(t) = 5 \sin 377t$?

(A) 0.0 V

(B) 3.2 V

(C) 3.5 V

(D) 5.6 V

4. The voltage signal in Prob. 3 is rectified and measured by a d'Arsonval meter movement. What is the voltage reading?

(A) 0.0 V

(B) 3.2 V

(C) 3.5 V

(D) 5.6 V

5. The manufacturer's data sheet lists the sensitivity of a DC voltmeter as 1000 Ω/V. What current is required for full-scale meter deflection?

(A) 1.0 nA

(B) 1.0 µA

(C) 1.0 mA

(D) 1.0 A

6. A DC voltmeter with a coil resistance of 150 Ω is to be designed with a sensitivity of 10,000 Ω/V and a voltage range of 0 V to 100 V. What type and what magnitude of resistance are required?

(A) parallel, 10 kΩ

(B) parallel, 1 MΩ

(C) series, 10 kΩ

(D) series, 1 MΩ

7. A d'Arsonval movement with a coil resistance of 150 Ω is to be designed for use as an ammeter. Most nearly, what is the full-scale deflection current?

(A) 0.3 mA

(B) 3.0 mA

(C) 0.3 A

(D) 3.0 A

8. A DC voltmeter with 20 kΩ of internal resistance measures the voltage across a resistance as 10.0 V. The current through the parallel combination of the resistor and voltmeter is measured as 1.00 mA by an ammeter with 100 Ω of internal resistance. Assume both meters are perfectly accurate. The resistor is fed from a source with a Thevenin resistance of 600 Ω. Determine the (a) resistance of the resistor and (b) current that will flow in the resistor if the meters are removed from the circuit.

9. A voltmeter is designed to measure voltages in the full-scale ranges of 3, 10, 30, and 100 V_{DC}. The meter movement has an internal resistance of 50 Ω and a full-scale current of 1 mA. Using a four-contact multiposition switch, design the voltmeter and specify the four resistor values used.

10. A digital meter uses the number system base b and has n digits. The noise is one-half the place value of the least significant digit. (a) Determine the signal-to-noise ratio for the case where the reading is half the full-scale value. (b) Obtain the signal-to-noise ratios for the following cases.

n	b	S/N (dB)
3	10	$20 \log (S/N)$
4	8	
8	2	
3	16	

SOLUTIONS

1. Precision is the range of values of a set of measurements. Therefore, it is a measure of the reproducibility of observations.

The answer is (C).

2. The thermal noise voltage is

$$V_{noise} = \sqrt{4\kappa TR(\mathrm{BW})}$$

$$= \sqrt{\begin{array}{c}(4)\left(1.3805 \times 10^{-23}\ \dfrac{\mathrm{J}}{\mathrm{K}}\right)(338\mathrm{K}) \\ \times (100 \times 10^3\ \Omega)(10 \times 10^3\ \mathrm{Hz})\end{array}}$$

$$= \boxed{4.3 \times 10^{-6}\ \mathrm{V} \quad (4.3\ \mu\mathrm{V})}$$

The answer is (D).

3. The average value of any unmodified sinusoid is zero.

The answer is (A).

4. A d'Arsonval meter responds to average values. The average value of a rectified sinusoid is

$$V_{ave} = \left(\frac{2}{\pi}\right)V_p$$

$$= \left(\frac{2}{\pi}\right)(5\ \mathrm{V})$$

$$= \boxed{3.18\ \mathrm{V} \quad (3.2\ \mathrm{V})}$$

The answer is (B).

5. The sensitivity is defined as

$$\mathrm{sensitivity} = \frac{1}{I_{fs}}$$

$$I_{fs} = \frac{1}{\mathrm{sensitivity}} = \frac{1}{1000\ \dfrac{\Omega}{\mathrm{V}}}$$

$$= \boxed{10^{-3}\ \mathrm{A} \quad (1.0\ \mathrm{mA})}$$

The answer is (C).

6. The voltmeter requires an external series resistance given by

$$\frac{1}{I_{fs}} = \frac{R_{ext} + R_{coil}}{V_{fs}}$$

$$R_{ext} = V_{fs}\left(\frac{1}{I_{fs}}\right) - R_{coil}$$

$$= (100 \text{ V})\left(10,000 \frac{\Omega}{\text{V}}\right) - 150 \text{ }\Omega$$

$$= \boxed{0.999 \times 10^6 \text{ }\Omega \quad (1 \text{ M}\Omega)}$$

The answer is (D).

7. For ammeters, the standard design voltage is 50 mV. The full-scale deflection current is

$$I_{fs} = \frac{V}{R_{coil}} = \frac{50 \times 10^{-3} \text{ V}}{150 \text{ }\Omega}$$

$$= \boxed{3.33 \times 10^{-4} \text{ A} \quad (0.3 \text{ mA})}$$

The answer is (A).

8.

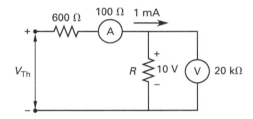

(a) The Thevenin equivalent voltage is

$$V_{Th} = 10 \text{ V} + \left(\frac{1 \text{ mA}}{1000 \frac{\text{mA}}{\text{A}}}\right)(700 \text{ }\Omega) = 10.7 \text{ V}$$

With 1 mA flowing into 10 V, the equivalent load resistance is

$$\left(\frac{10 \text{ V}}{1 \text{ mA}}\right)\left(1000 \frac{\text{mA}}{\text{A}}\right) = 10,000 \text{ }\Omega \quad (10 \text{ k}\Omega)$$

So,

$$10 \text{ k}\Omega = \frac{(20 \text{ k}\Omega)R}{20 \text{ k}\Omega + R}$$

$$(20 \text{ k}\Omega + R)(10 \text{ k}\Omega) = (20 \text{ k}\Omega)R$$

$$200 \times 10^3 \text{ k}\Omega + (10 \text{ k}\Omega)R = (20 \text{ k}\Omega)R$$

$$200 \times 10^3 \text{ k}\Omega = (10 \text{ k}\Omega)R$$

$$R = \boxed{20 \text{ k}\Omega}$$

(b) If the meters are removed, the current flowing will be

$$I = \frac{10.7 \text{ V}}{20,600 \text{ }\Omega} = \boxed{0.52 \times 10^{-3} \text{ A} \quad (0.52 \text{ mA})}$$

9. The meter configuration is shown.

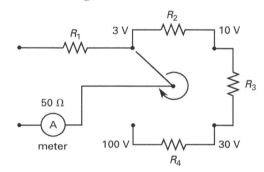

With the 1 mA full-scale current,

$$3 \text{ V: } 50 + R_1 = \frac{3 \text{ V}}{1 \times 10^{-3} \text{ A}} = 3000 \text{ }\Omega$$

$$R_1 = \boxed{2950 \text{ }\Omega}$$

$$10 \text{ V: } \quad R_2 = \frac{10 \text{ V} - 3 \text{ V}}{1 \times 10^{-3} \text{ A}}$$

$$= \boxed{7000 \text{ }\Omega}$$

$$30 \text{ V: } \quad R_3 = \frac{30 \text{ V} - 10 \text{ V}}{1 \times 10^{-3} \text{ A}}$$

$$= \boxed{20,000 \text{ }\Omega}$$

$$100 \text{ V: } \quad R_4 = \frac{100 \text{ V} - 30 \text{ V}}{1 \times 10^{-3} \text{ A}}$$

$$= \boxed{70,000 \text{ }\Omega}$$

10. (a) Full-scale is 999_{10} for the first case. The noise is 0.5. At half-scale the signal is 500_{10}, so the signal-to-noise ratio is $500/0.5 = 10^3$ or

$$\left.\text{S/N}\right|_{\text{dB}} = 20\log 10^3 = \boxed{60\text{ dB}}$$

(b) For the octal, full-scale in base 10 is

$$(7)(8)^3 + (7)(8)^2 + (7)(8)^1 + 7 = 4095$$

Half-scale is 2048.

$$\text{S/N} = 20\log \frac{2048}{0.5} = \boxed{72.2\text{ dB}}$$

For the binary, full-scale in base 10 is

$$(2)^7 + (2)^6 + (2)^5 + (2)^4 + (2)^3 + (2)^2 + (2)^1 + 1 = 255$$

Half-scale is 128.

$$\text{S/N} = 20\log \frac{128}{0.5} = \boxed{48.2\text{ dB}}$$

For the hexadecimal, full-scale is

$$(15)(16)^2 + (15)(16)^1 + 15 = 4095$$

For half-scale, the S/N ratio is the same as the octal half-scale.

$$\text{S/N} = 20\log \frac{2048}{0.5} = \boxed{72.2\text{ dB}}$$

Topic XIII: Electrical Materials

Electrical
Materials

54 Electrical Materials

PRACTICE PROBLEMS

1. Which of these types of materials generally has the highest thermal and electrical conductivity?

- (A) ceramic
- (B) metal
- (C) polymer
- (D) semiconductor

2. Which of these elements is a liquid at standard atmospheric temperature and pressure (STP)?

- (A) Ge
- (B) H
- (C) Hg
- (D) In

3. Which of these materials has the highest conductivity at 20°C?

- (A) aluminum
- (B) copper
- (C) gold
- (D) silver

4. Ceramic materials typically

- (A) contain no metallic elements
- (B) have low electrical resistivity
- (C) have low thermal conductivity
- (D) make poor insulators

5. Which type of bonding is responsible for the attraction of water molecules to one another?

- (A) covalent
- (B) ionic
- (C) metallic
- (D) van der Waals

SOLUTIONS

1. Several divisions of materials can occur during classification. One such division is into metals, polymers (that is, plastics), and ceramics. Metal has the highest thermal and electrical conductivity of the material types.

The answer is (B).

2. Of all the elements, only bromine (Br) and mercury (Hg) are liquid under standard atmospheric conditions.

The answer is (C).

3. The approximate values for electrical conductivity at 20°C for the listed materials are

metal	S/m
aluminium	38×10^6
copper	58×10^6
gold	41×10^6
silver	63×10^6

The answer is (D).

4. Ceramic materials typically have very low thermal conductivity.

Ceramics are brittle and contain both metallic and non-metallic elements. Since the 1980s, advanced ceramic materials have been developed that have low electrical resistivity within certain temperature ranges, but in general, ceramics typically have very high electrical resistivity, which makes them good insulators.

The answer is (C).

5. Ionic, covalent, and metallic bonds are the three primary bonding methods. Bonds between molecules of the same substance are called van der Waals forces. There are three types of van der Waals forces: dipole-to-dipole (between polar molecules), dispersion (between non-polar molecules), and hydrogen bond (also a dipole-to-dipole bond, but stronger).

> Water molecules form hydrogen bonds with each other. Therefore, the van der Waals bonding is responsible for the attraction of water molecules.

The answer is (D).

Topic XIV: Codes and Standards

Codes and
Standards

55 Biomedical Electrical Engineering

PRACTICE PROBLEMS

The following information is applicable to Prob. 1 and Prob. 2.

A medical incubator is designed to maintain temperature, humidity, and CO_2 at the proper levels for the sampled items. The humidifier uses the low-water-level alarm circuitry shown.

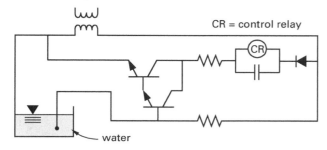

1. What term describes the configuration of the two transistors?

(A) common collector

(B) common base

(C) common source

(D) Darlington pair

2. Which of the following transistor parameters represents the gain used to drive the control relay?

(A) h_{fe}

(B) h_{ie}

(C) h_{oe}

(D) h_{re}

3. A Class II biological safety cabinet (hood) uses a 115 V single-phase AC motor to maintain the vacuum level within the hood that is necessary to prevent the spread of biological contaminants. With clean filters,

the motor draws 10 A at its rated power factor of 0.85 lagging. What is the power consumption of the motor?

(A) 980 W

(B) 1150 W

(C) 1700 W

(D) 1990 W

4. Which of the following items are used to reduce leakage current for 60 Hz biomedical devices?

(A) DC-DC inverters

(B) rechargeable batteries

(C) motor generators

(D) all of the above

5. Radio frequency units used for patient care may result in _____ human tissue.

(A) the reorientation of

(B) the heating of

(C) increased circulation within

(D) the reorganization of cellular structure of

SOLUTIONS

1. The two transistors are in a common emitter configuration, which is analogous to a common source configuration for field-effect transistors (FETs). This specific configuration is commonly referred to as a Darlington pair.

The answer is (D).

2. The Darlington pair has a high input resistance and a high current gain, h_{fe}. It is this forward current gain that enables the operation of the control relay.

The answer is (A).

3. The power in a single-phase motor is

$$
\begin{aligned}
P &= IV(\text{pf}) \\
&= (10\text{ A})(115\text{ V})(0.85) \\
&= \boxed{977.5\text{ W}\quad(980\text{ W})}
\end{aligned}
$$

The answer is (A).

4. Leakage current is minimized by using nonoscillatory power sources, such as DC-DC inverters and rechargeable batteries, or electrical items with nonelectromagnetic coupling, such as motor generators. For biomedical devices, leakage current is set by IEC 60601–1 and is generally limited to 0.01–10 μA depending on the testing method, classification of applied parts, and whether under normal or single-fault conditions.

The answer is (D).

5. Radio frequency fields of sufficient intensity can damage human cells by depositing energy through heat.

The answer is (B).

56 National Electrical Code

PRACTICE PROBLEMS

1. An unknown wire gage must be determined during an electrical inspection. The wire diameter is measured as 0.2591 cm. What is the AWG standard designation?

- (A) AWG 4/0
- (B) AWG 4
- (C) AWG 8
- (D) AWG 10

2. A given circuit is meant to carry a continuous lighting load of 16 A. In addition, four loads designed for permanent display stands are fastened in place and require 2 A each when operating. What is the rating of the overcurrent protective device (OCPD) on the branch circuit?

- (A) 20 A
- (B) 23 A
- (C) 28 A
- (D) 30 A

3. If standard three-conductor copper cable is used for the branch circuit in Prob. 2, what is the minimum conductor size?

- (A) AWG 8
- (B) AWG 10
- (C) AWG 12
- (D) AWG 14

4. A dwelling two-wire resistive heating circuit is designed for 240 V and 30 A. AWG 8 copper wire is used with a resistance of 0.809 Ω/1000 ft. What is the voltage drop for a one-way circuit length of 75 ft?

- (A) 1.0%
- (B) 1.5%
- (C) 3.6%
- (D) 5.0%

5. A three-wire feeder, AWG 1/0, carries 100 A of continuous load from a 480 V, 3ϕ (three-phase), three-wire source. The THHN copper conductors operate at 75°C and are enclosed within aluminum conduit. The expected power factor is 0.8. The total circuit length is 150 m. What is the voltage drop?

- (A) 4.0 V
- (B) 7.0 V
- (C) 11 V
- (D) 14 V

6. A 120 V dwelling branch circuit supplies four outlets, one of which has four receptacles. What is the total volt-ampere load for the circuit shown?

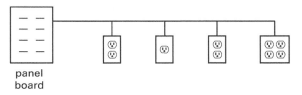

- (A) 180 VA
- (B) 360 VA
- (C) 720 VA
- (D) 900 VA

7. A three-phase, four-wire feeder with a full-sized neutral in a separate raceway carries 14 A continuous and 40 A noncontinuous loads. The feeder uses an overcurrent device with a terminal rating of 75°C. What is the minimum copper conductor size?

- (A) AWG 3
- (B) AWG 4
- (C) AWG 6
- (D) AWG 8

8. All of the following are properties of fuses rather than circuit breakers, with one exception. Which item is the property of circuit breakers?

   (A) adjustable time-current characteristics

   (B) low initial costs

   (C) longer downtime after operation

   (D) simplicity

9. Installed equipment within a dwelling is attached to an overcurrent device (breaker) rated at 15 A. What is the required size of the equipment's grounding copper conductor?

   (A) AWG 10

   (B) AWG 12

   (C) AWG 14

   (D) AWG 18

10. A three-phase, four-wire feeder with termination provisions designed to carry less than 100 A uses THWN AWG 8 copper wire in an ambient operating environment of 50°C. From NEC Table 310.15(b)(2)(a), the derating factor for four conductors in a single raceway is 0.8. What is the allowable ampacity?

   (A) 19 A

   (B) 24 A

   (C) 30 A

   (D) 40 A

11. A high-speed, single-phase, alternating-current motor has a nameplate rating of 15 A at 230 V. What is most nearly the horsepower rating?

   (A) 2.0 hp

   (B) 2.5 hp

   (C) 2.6 hp

   (D) 3.0 hp

12. A three-phase motor is supplied with power by two parallel sets of 300 kcmil copper cables in steel conduit. Each cable is 400 ft in length and is rated for 75°C. The NEC is applicable. What is most nearly the value of the phase-to-neutral reactance?

   (A) 5 mΩ

   (B) 6 mΩ

   (C) 8 mΩ

   (D) 10 mΩ

13. Which of the following most closely defines *continuous duty*?

   (A) large load for 3 hours or more

   (B) maximum load for 3 hours or more

   (C) steady large load for 3 hours or more

   (D) substantially constant load for an indefinite period

14. An electrical enclosure will be situated outdoors and exposed to rain, snow, and sleet, as well as corrosive agents. Which type of enclosure is acceptable?

   (A) Type 1

   (B) Type 2

   (C) Type 3RX

   (D) Type 4

15. Which dwelling-unit locations do NOT require the installation of ground-fault circuit interrupter (GFCI)-type receptacles?

   (A) bedrooms

   (B) boathouses

   (C) crawl spaces

   (D) garages

16. A 20A branch circuit supplies four 15A-rated receptacles in a work space where one receptacle will supply a plug-connected cutting machine and controller. What is the maximum current that can be drawn by the cutting machine so that it still meets NEC requirements?

   (A) 10 A

   (B) 12 A

   (C) 16 A

   (D) 20 A

17. A dwelling unit has an area of 1500 ft². What is most nearly the general lighting load?

   (A) 500 VA

   (B) 1500 VA

   (C) 4500 VA

   (D) 50,000 VA

Codes and Standards

18. A metal underground water pipe is considered adequate for grounding purposes if it is electrically continuous and in direct contact with the earth for how many meters?

(A) 2.4 m

(B) 3.0 m

(C) 6.0 m

(D) 10 m

19. In a 2×4 wood member, a hole is bored through which wire is run. What distance from the nearest edge of the wood should the hole be in order to be installed without special protection?

(A) 0.500 in

(B) 1.25 in

(C) 2.00 in

(D) 4.00 in

20. An electrical box is 10 in² and contains several of wires. What length of free conductor is required in the box?

(A) 3 in

(B) 6 in

(C) 8 in

(D) 12 in

21. An electrical cabinet enclosure has wires coming off a terminal block that remain within the enclosure. The wires are 12 AWG. What is the minimum wire bending space required?

(A) 1.5 in

(B) 2.0 in

(C) 3.0 in

(D) not specified

22. A box has a trade size of 75 mm × 50 mm × 38 mm. What is the maximum number of 10AWG wires allowed in the box?

(A) 1

(B) 2

(C) 3

(D) 4

23. Electrical metallic tubing (EMT) is used to protect wires in an exposed section of an office building. A single run of tubing is used between two widely spaced pull points. How many total degrees of bends are allowed?

(A) 90°

(B) 180°

(C) 270°

(D) 360°

24. A flexible cable of Thermoset Type E rated for 90°C is used in an electric vehicle. The cable is a two-conductor, 10 AWG cable used in a 50°C environment. What is most nearly the allowable ampacity?

(A) 8 A

(B) 24 A

(C) 25 A

(D) 30 A

25. The nameplate on a three-phase 115 V induction motor indicates a full-load current of 5.4 A. What is most nearly the horsepower rating?

(A) 0.2 hp

(B) 0.5 hp

(C) 0.6 hp

(D) 0.7 hp

26. The nameplate data of a 1 hp motor shows a locked-rotor letter code of A. Which of the following options is the expected power during a locked-rotor condition?

(A) 1.0 kVA

(B) 10 kVA

(C) 15 kVA

(D) 20 kVA

27. What is most nearly the maximum locked-rotor current for a three-phase, 15 hp, 230 V motor with a locked-rotor code of E?

(A) 0.20 A

(B) 75 A

(C) 170 A

(D) 190 A

Codes and
Standards

28. A garage is set up to charge two electric vehicles on a 20 A, single-phase branch circuit supplied with 208 V. What is the total minimum ventilation requirement in such a system?

(A) 37 cfm

(B) 49 cfm

(C) 85 cfm

(D) 170 cfm

29. A bipolar, solar photovoltaic system has an open-circuit voltage of 480 V (+240 V to ground and –240 V to ground) at 25°C. Per the *ASHRAE Handbook—Fundamentals*, the mean temperature of a given installation will be –20°C in the winter. What is the expected open-circuit voltage for this system in the winter?

(A) 240 V

(B) 480 V

(C) 490 V

(D) 570 V

30. According to the NEC, what is the recommended torque for a slotted No. 10 screw that will be used to hold a 10 AWG wire?

(A) 1.7 lbf-in

(B) 2.3 lbf-in

(C) 20 lbf-in

(D) 35 lbf-in

31. A square metal box will be used to house a number of 10 AWG wires. The box is 4 in × 1.5 in. No volume allowances are required beyond that for the wires themselves. How many 10 AWG wires can be routed through the box?

(A) 6

(B) 7

(C) 8

(D) 12

32. Which of the following options is NOT true of isolated grounding receptacles in a health care facility?

(A) limit electrical noise

(B) provide single path to ground

(C) use yellow stripes on insulation

(D) are allowed in patient care areas

33. The demand factor allowed for a marina with 10 shore power receptacles on a given pier is most nearly

(A) 50%

(B) 60%

(C) 70%

(D) 80%

34. What type of output cable to an electric vehicle is allowed?

(A) any cable 12 AWG or larger

(B) portable power cables (PPE)

(C) elevator cables for flexibility

(D) EV* family of flexible cables

35. A 200 hp, 460 V, three-phase, continuous-duty squirrel-cage induction motor is designed to operate at a 0.85 power factor. The supplying branch circuit is 30.5 m long and utilizes THHN copper conductors in a steel conduit. The voltage drop is limited to 3% in the branch circuit, leaving 2% for the feeder circuit. NEC Table 430.250 and NEC Chap. 9, Table 9 list the applicable data. What size cable should be used to meet the voltage drop and current requirements of the load?

(A) AWG 3

(B) AWG 3/0

(C) AWG 250 kcmil

(D) AWG 350 kcmil

Codes and Standards

SOLUTIONS

1. The wire diameter is

$$d = (0.2591 \text{ cm})\left(\frac{1 \text{ in}}{2.54 \text{ cm}}\right) = 0.1020 \text{ in}$$

Finding this value in a table of wire gages, such as the one in App. 56.A, indicates a 10 gage wire. That is, AWG 10.

The answer is (D).

2. From *National Electrical Code* (NEC), Art. 210.20(A), the overcurrent protective device (OCPD) must be rated at 100% of the noncontinuous load plus 125% of the continuous load.

$$\text{OCPD} = (1.00)(2 \text{ A} + 2 \text{ A} + 2 \text{ A} + 2 \text{ A}) + (1.25)(16 \text{ A})$$
$$= \boxed{28 \text{ A minimum}}$$

A standard fixed-trip circuit breaker or a fuse rated at 30 A can be used (see Sec. 240.6).

The answer is (D).

3. Conductor size is based on 100% of the noncontinuous load plus 125% of the continuous load [Sec. 21.19(A)(1)(a)]. From Prob. 2, the total load is 28 A. The ampacity cannot be less than the rating of the overcurrent device, 30 A in this case [Sec. 310.15(B) Informational Note(2)]. The ampacity of the conductor for a circuit of less than 100 A should be based on the 60°C column of NEC 310.15(B)(16) [Sec. 110.14(C)(1)(a)] (see App. 56.B).

Using App. 56.B, an AWG 10 conductor is chosen. (Checking Art. 240.4 of the *Code*, as is called for by the footnote in App. 56.B, indicates that 10 gage copper wire cannot have overcurrent protection greater than 30 A. All requirements are met.) Any derating of the conductor ampacity requires the use of smaller gage (larger diameter) conductor. See the conductor factors in NEC Tables 310.15(B)(2)(a), (B)(2)(b), and (B)(2)(c).

The answer is (B).

4. A two-wire circuit is either single phase or DC. The heating is resistive, so pf = 1 if the circuit is AC, which is the most likely scenario. Regardless, the voltage drop is given by Ohm's law as

$$V = IR$$
$$= IR_l l$$

The voltage drop occurs on both wires, so a factor of two must be used.

$$V = IR_l 2l$$

Substituting gives

$$V = 2IR_l l$$
$$= (2)(30 \text{ A})\left(\frac{0.809 \text{ }\Omega}{1000 \text{ ft}}\right)(75 \text{ ft})$$
$$= 3.64 \text{ V}$$

As a percentage of the 240 V nominal voltage, the voltage drop is

$$\frac{3.64 \text{ V}}{240 \text{ V}} \times 100 \% = \boxed{1.516\% \quad (1.5\%)}$$

The answer is (B).

The conductor ampacities can be found in App. 56.B (NEC Table 310.15(B)(16)). Conductor properties, such as resistance (Ω) per 1000 ft, are found in App. 56.E (NEC Chap. 9, Table 8).

5. Although the continuous load is 100 A, and the voltage drop is calculated using that value, the conductors and the overcurrent protector must be able to carry 125 A, that is, 125% of the continuous load [Sec. 215.2(A)(1)(a) and Sec. 215.3].

The AC voltage drop is

$$V_{\text{drop}} = IZ$$

The effective impedance, corrected for the power factor, is calculated using App. 56.F (NEC Chap. 9, Table 9). For a 1/0 gage feeder, the resistance to neutral per 1000 ft in aluminum conduit is 0.13 Ω.

The inductive reactance, X_L, to neutral per 1000 ft is 0.044 Ω. The columns titled "alternating-current resistance for uncoated copper wires" and "X_L (reactance) for all wires" are used. The uncoated copper wire column is used, although the THHN conductors are insulated. Also, the capacitive reactance is ignored. Therefore, the voltage drop calculated will be an approximate value.

Because the power factor is not 0.85, the effective impedance cannot be taken directly from the table. The corrected impedance, Z_c, is

$$Z_c = R_l(\text{pf}) + X_{L,l}\sin(\arccos \text{pf})$$
$$= \left(\frac{0.13 \text{ }\Omega}{1000 \text{ ft}}\right)(0.8)$$
$$+ \left(\frac{0.044 \text{ }\Omega}{1000 \text{ ft}}\right)\sin 36.86°$$
$$= 0.1304 \text{ }\Omega/1000 \text{ ft}$$

Codes and Standards

Because this is an "ohms-to-neutral" value, the line-to-neutral, or phase, voltage drop is

$$V_{\text{drop (line-to-neutral)}} = V_\phi = IZ_c$$
$$= (100 \text{ A})\left(\frac{0.1304 \ \Omega}{1000 \text{ ft}}\right)$$
$$= 13.04 \text{ V}/1000 \text{ ft}$$

Adjusting for the total circuit length gives

$$V_\phi = \left(\frac{13.04 \text{ V}}{1000 \text{ ft}}\right)(150 \text{ m})\left(\frac{1 \text{ ft}}{0.3048 \text{ m}}\right)$$
$$= 6.417 \text{ V}$$

Voltage drops are given in terms of line quantities for a three-phase, three-wire system.

$$V_{\text{line}} = V_\phi \sqrt{3}$$
$$= (6.417 \text{ V})\sqrt{3}$$
$$= \boxed{11.1 \text{ V} \quad (11 \text{ V})}$$

This is approximately 2.5% of the 480 V source and is within NEC recommendations.

The answer is (C).

6. The situation is as shown.

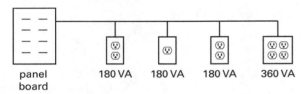

panel board — 180 VA — 180 VA — 180 VA — 360 VA

On any outlet with less than four receptacles, the VA load is 180 VA [Sec. 220.14(I)]. For four or more receptacles, the load is

$$\text{VA} = (\text{numbers of receptacles})\left(90 \ \frac{\text{VA}}{\text{receptacle}}\right)$$
$$= (4)(90 \text{ VA})$$
$$= 360 \text{ VA}$$
$$\text{VA}_{\text{total}} = 180 \text{ VA} + 180 \text{ VA} + 180 \text{ VA} + 360 \text{ VA}$$
$$= \boxed{900 \text{ VA}}$$

The answer is (D).

7. Feeder conductor size, before derating, is based on 100% of the noncontinuous load and 125% of the continuous load [Sec. 215.2(A)(1)(a)].

$$A_{\text{load}} = (1.00)(40 \text{ A}) + (1.25)(14 \text{ A})$$
$$= 57.5 \text{ A} \quad (58 \text{ A})$$

The total of 58 A is used because an ampacity of 0.5 A or greater is rounded up [Sec. 220.5(B)]. Using App. 56. B (NEC Table 310.15(B)(16)), select AWG 6 from the 75°C column. Always consider additional requirements or limits on the ratings. In this case, for circuits less than 100 A, the 60°C column is required [Sec. 110.14(C)(1)]. (Because of this article, the column primarily used in dwelling calculations is the one titled "60°C (140° F)".) Therefore, the AWG 6 conductor cannot be used as it is rated for 55 A at 60°C. Instead, $\boxed{\text{AWG 4}}$ with a 70 A rating is used.

The answer is (B).

8. The electrical characteristics of a fuse are set by the material used, and no external adjustments can occur. The time-current characteristics cannot change.

The answer is (A).

9. Equipment grounding conductor sizes are given in NEC Table 250.122. For a 15 A overcurrent device, the grounding conductor must be $\boxed{\text{AWG 14}}$ copper.

The answer is (C).

10. Because the circuit carries less than 100 A and the termination provisions are designed commensurate with this condition, the 60°C column of App. 56.B must be used [see Sec. 110.14(C)(1)(a)]. 40 A is the limiting allowable ampacity. The operating environment is above 30°C for which the values are calculated. THWN wire is rated for 75°C and 50 A. Using the correction factor table in App. 56.C, 0.75 is found in the 46–50°C row. The allowable ampacity in a given environment is determined by multiplying the rated ampacity in the table by all the correction factors—in this case, one for temperature and one for more than three conductors in a raceway.

$$A_{\text{total}} = A_{\text{rated}} F_{\text{temp}} F_{\text{cond}}$$
$$= (50 \text{ A})(0.75)(0.8)$$
$$= \boxed{30 \text{ A}}$$

Because 30 A is less than the 40 A maximum limit from the 60°C column, 30 A is the correct value.

The answer is (C).

11. From NEC Art. 430.6(A)(1), a high-speed motor is one with a speed of 1200 rpm or higher. When a high-speed motor's nameplate has an ampere rating and the horsepower rating is needed, interpolate the horsepower rating from the appropriate NEC table.

For a single-phase, alternating-current motor, use NEC Table 430.248. For a 230 V motor, 2 hp corresponds to 12 A and 3 hp corresponds to 17 A. Interpolate using the two-point form.

$$\frac{y - y_1}{x - x_1} = \frac{y_2 - y_1}{x_2 - x_1}$$

$$y = \frac{(y_2 - y_1)(x - x_1)}{x_2 - x_1} + y_1$$

$$= \frac{(3 \text{ hp} - 2 \text{ hp})(15 \text{ A} - 12 \text{ A})}{17 \text{ A} - 12 \text{ A}} + 2 \text{ hp}$$

$$= \boxed{2.6 \text{ hp}}$$

The answer is (C).

12. From App. 56.F (NEC Chap. 9, Table 9), the inductive reactance, X_L, for a 300 kcmil cable is 0.051 Ω/1000 ft.

The inductive reactance for a 400 ft, 300 kcmil cable is

$$X_{L,\text{cable}} = X_L D = \left(\frac{0.051 \text{ }\Omega}{1000 \text{ ft}}\right)(400 \text{ ft})$$

$$= 0.0204 \text{ }\Omega$$

Since the two sets of cables are used in parallel, the phase-to-neutral load reactance is half the cable reactance.

$$X_{L,\text{load}} = \frac{X_{L,\text{cable}} X_{L,\text{cable}}}{X_{L,\text{cable}} + X_{L,\text{cable}}} = \frac{(X_{L,\text{cable}})^2}{2X_{L,\text{cable}}} = \frac{X_{L,\text{cable}}}{2}$$

$$= \frac{0.0204 \text{ }\Omega}{2}$$

$$= \boxed{10.2 \text{ m}\Omega \quad (10 \text{ m}\Omega)}$$

The answer is (D).

13. The term *continuous duty* (in the NEC as duty, continuous) is defined in NEC Art. 100 as operation at a substantially constant load for an indefinitely long time.

(The term *continuous load* is defined in NEC Art. 100 as a load where the maximum current is expected to continue for 3 hours or more. The distinction between these definitions is important for a proper understanding of NEC requirements.)

The answer is (D).

14. The appropriate enclosure is found in NEC Art. 110.28, Table 110.28, in the For Outdoor Use section.

Types 1 and 2 do not qualify for exposure to rain, snow, and sleet, or to corrosive agents. Type 4 can be used in rain, snow, and sleet, but not with corrosive agents. Type 3RX meets both sets of requirements.

The answer is (C).

15. The dwelling unit locations requiring GFCIs are listed in NEC Sec. 210.8(A)(1)-(10) and include bathrooms, garages, outdoor spaces, crawl spaces, unfinished basements, kitchens, sinks (in areas other than kitchens), and boathouses. GFCIs are not required in bedrooms (though arc-fault circuit interrupter (AFCI) breakers are required—see NEC Art. 210.12).

The answer is (A).

16. Per NEC Sec. 210.21(B)(2) and Table 210.21(B)(2), the maximum cord-and-plug-connected load on a 15 A- or 20 A-rated circuit using 15 A-receptacles is 12 A.

The answer is (B).

17. The general lighting load for a dwelling unit, which is discussed in NEC Art. 220.12 and specified in Table 220.12, is 33 VA/m^2 (3 VA/ft^2). Calculate the total load.

$$S = A F_{\text{NEC}} = (1500 \text{ ft}^2)\left(3 \frac{\text{VA}}{\text{ft}^2}\right)$$

$$= \boxed{4500 \text{ VA}}$$

This calculation determines the number of lighting branch circuits required. The actual load on the branch circuits is calculated using the demand factors listed in Table 220.42.

The answer is (C).

18. Per NEC Sec. 250.52(A)(1), a metal underground water pipe must be in contact with the earth for 3.0 m (10 ft) or more to be adequate for grounding purposes.

The answer is (B).

19. Per NEC Sec. 300.4(A)(1), the minimum edge distnce for bored hole through wood members is 32 mm (1.25 in). If this requirement is not met, the wire or bored hole needs the protection of a steel plate or bushing that is at least 1.6 mm (0.0625 in) thick.

The answer is (B).

Codes and Standards

20. Per NEC Sec. 300.14, a box with an opening of at least 200 mm (8 in) must have at least $\boxed{150 \text{ mm (6 in)}}$ of *free conductor*. The term free conductor indicates the wire length available for use in the box, which is enough for proper connections to be made by an electrician.

The answer is (B).

21. The minimum bending space for wires not entering or leaving an enclosure through an opposite wall is covered in NEC Sec. 312.6(B)(1). See also Table 312.6(A). For AWG wire sizes 14–10, the bending space is $\boxed{\text{not specified}}$ for any number of wires per terminal, probably because the wire is of small enough diameter that damage due to bending is minimal.

The answer is (D).

22. The maximum number of wires allowed in a given box is found in NEC Art. 314.16 and Table 314.16(A). For a 75 mm × 50 mm × 38 mm box, and for 10 AWG wire, the table lists the maximum number of 10 AWG wires as $\boxed{3}$.

The answer is (C).

23. Per NEC Sec. 358.26 no more than four quarter bends (90° each, totalling $\boxed{360°}$) are the total degrees of bends allowed in a single run of EMT, that is, between pull points (e.g., conduits and boxes).

The answer is (D).

24. Thermoset Type E flexible cable is governed by NEC Art. 400.5(A) and Table 400.5(A)(1), Column B, for two current-carrying conductors per cable. For an ambient temperature of 30°C the table indicates an ampacity of 30 A. For temperatures other than 30°C, use the correction factors in NEC Table 310.15(B)(2)(a). For a 50°C ambient and a 90°C rating, the correction factor is 0.82. The allowable ampacity is

$$\begin{aligned} I_{\text{allowed}} &= I_{\text{rated}} F_{\text{correction}} \\ &= (30 \text{ A})(0.82) \\ &= \boxed{24.6 \text{ A} \quad (24 \text{ A})} \end{aligned}$$

The correct answer must be 24.6 A or less. Anything higher would exceed the allowable ampacity.

The answer is (B).

25. The horsepower rating is taken from NEC Table 430.250 by interpolating between the values for ½ hp (4.4 A) and ¾ hp (6.4 A).

$$\begin{aligned} P_{\text{hp}} &= P_{\text{hp},1/2} + P_{\text{hp},3/4} - P_{\text{hp},1/2}\left(\frac{I_{5.4} - I_{4.4}}{I_{6.4} - I_{4.4}}\right) \\ &= 0.50 \text{ hp} + (0.75 \text{ hp} - 0.50 \text{ hp}) \\ &\quad \times \left(\frac{5.4 \text{ A} - 4.4 \text{ A}}{6.4 \text{ A} - 4.4 \text{ A}}\right) \\ &= \boxed{0.63 \text{ hp} \quad (0.6 \text{ hp})} \end{aligned}$$

The answer is (C).

26. Locked-rotor letter codes are given in NEC Table 430.7(B). The letter A indicates a motor input range of 0 kVA/hp to 3.14 kVA/hp. Since the motor has a rating of 1hp, this is also the apparent power rating during a locked-rotor condition. $\boxed{1.0 \text{ kVA}}$ is the only option in that range.

The answer is (A).

27. From NEC Table 430.7(B), a motor with a locked-rotor code of E has a maximum motor input of 4.99 kVA/hp. For a 15 hp motor, the maximum apparent power is

$$\begin{aligned} S_{\text{locked-rotor}} &= P_{\text{hp}} F_{\text{code}} \\ &= (15 \text{ hp})\left(4.99 \frac{\text{kVA}}{\text{hp}}\right) \\ &= 74.85 \text{ kVA} \end{aligned}$$

The maximum locked-rotor current is

$$\begin{aligned} S_{\text{locked-rotor}} &= \sqrt{3} I_{\text{max}} V \\ I_{\text{max}} &= \frac{S_{\text{locked-rotor}}}{\sqrt{3} V} \\ &= \frac{74.85 \text{ kVA}}{\sqrt{3} (0.230 \text{ kV})} \\ &= \boxed{187.89 \text{ A} \quad (190 \text{ A})} \end{aligned}$$

The answer is (D).

28. Electric vehicle charging systems are covered in NEC Art. 625, and the ventilation requirements in cubic feet per minute (cfm) are found in Table 652.52(B)(1)(b). The table lists the values of ventilation required for a single vehicle. For a 20 A, single-phase branch circuit at 208 V, the requirement listed is 85 cfm. For two vehicles, the total minimum ventilation required is 2×85 cfm = $\boxed{170 \text{ cm.}}$

The answer is (D).

29. The voltage correction factors for temperatures under 25°C are given in NEC Table 690.7. For –20°C, the correction factor is 1.18. The design voltage for selection of conductors, disconnecting means, and over-current device settings should be

$$\begin{aligned} V_{\text{design}} &= V_{\text{rated}} F_{\text{ambient}} \\ &= (480 \text{ V})(1.18) \\ &= \boxed{566.40 \text{ V} \quad (570 \text{ V})} \end{aligned}$$

The answer is (D).

30. Recommended torques, used only in the absence of guidance from the manufacturer, are found in Annex I of the NEC, in Table I.1. Since the slot width and slot length were not provided, refer to the table note and select the largest torque associated with 10 AWG wire. In this case the correct value is $\boxed{35 \text{ lbf-in}}$, found in the first row of the table, in the second of the four columns associated with slotted head screws.

The answer is (D).

31. The number of wires allowed within a given box is specified in NEC Table 314.16(A) for standard trade sizes. The table is applicable as long as no additional fill requirements are specified in NEC Arts. 314.16(B)(2)–(B)(5).

The box trade size listed in the problem statement is located in the fifth row of NEC Table 314.16(A). For the fifth row, the maximum number of 10 AWG wires allowed is 8.

The answer is (C).

32. Isolated grounding receptacles are normally used to limit potential noise (interference) to a given piece of equipment.

Such receptacles provide no functional benefit from multiple equipment grounding paths available with other receptacles (See NEC 517.13). Such receptacles have a single path to ground, thus avoiding the electrical noise of other circuits.

The equipment grounded conductor is identified by green insulation or one or more yellow stripes along the entire conductor's length (See NEC 517.16(B)).

Such isolated grounding receptacles are identified with an orange triangle on the receptacle and should NOT be used within a patient care vicinity (See NEC 517.16(A)).

The answer is (D).

33. The demand factors for shore power receptacles are given in NEC Table 555.12. For an installation with 9–14 receptacles, the demand factor is 80%.

The answer is (D).

34. Electric vehicle cables are specified in NEC Art. 625.17(B) as EV, EVJ, EVE, EVJE, EVT or EVJT, which are flexible cables listed in NEC Table 400.4.

The answer is (D).

35. The key to determining the proper size conductors is calculating the maximum allowable effective impedance. The effective impedance is calculated from

$$\begin{aligned} Z_{0.85} &= \frac{VD_{\text{LN}}}{\left(\dfrac{\text{circuit length}}{1000 \text{ ft}}\right)(\text{circuit load in amps})} \\ &= \frac{VD_{\text{LN}}}{\left(\dfrac{L}{1000 \text{ ft}}\right) I_{\text{load}}} \quad [\text{I}] \end{aligned}$$

NEC Art. 210.19(A), Informational Note No. 4, recommends that conductors be sized to allow for a 3% voltage drop, with no more than 5% in the branch and feeder circuits combined. (NEC Art. 215.2(A)(1), Informational Note No. 2, recommends the same for a feeder circuit.) The branch-circuit voltage drop allowed was given as 3%. The voltage drop, line-to-line, is

$$\begin{aligned} VD_{\text{LL}} &= (0.03)(460 \text{ V}) \\ &= 13.8 \text{ V} \end{aligned}$$

The voltage drop, line-to-neutral, is

$$\begin{aligned} VD_{\text{LL}} &= \sqrt{3} \, VD_{\text{LN}} \\ VD_{\text{LN}} &= \frac{VD_{\text{LL}}}{\sqrt{3}} = \frac{13.8 \text{ V}}{\sqrt{3}} \\ &= 7.9674 \text{ V} \quad [\text{II}] \end{aligned}$$

The circuit length is 30.5 m. Converting to U.S. units, the circuit length is

$$L = \frac{30.5 \text{ m}}{0.3048 \; \frac{\text{m}}{\text{ft}}}$$
$$= 100.06 \text{ ft} \qquad \text{[III]}$$

The circuit load on the conductors must be a minimum of 125% of the continuous load per NEC Art. 210.19(A)(1). This requirement is echoed in NEC Art. 430.22(A) for a motor, which states that the supplying conductors must be rated for 125% of the motor's full-load rating. NEC Art. 430.6(A)(1) requires that the ampere ratings be determined from a series of tables unless both the horsepower rating and the ampere rating are given, in which case the nameplate ampere rating is used. The applicable table for three-phase alternating current motors is NEC Table 430.250, where the 200 hp, 460 V motor has a full-load ampere rating of 240 A. Therefore, the conductors must be sized for

$$I_{\text{load}} = 1.25 I_{\text{rated}} = (1.25)(240 \text{ A})$$
$$= 300 \text{ A} \qquad \text{[IV]}$$

Therefore, per Table 310.15(B)(16), using the 90°C THHN copper conductor column, a conductor of 300 kcmil (rated for 320 A) or larger is required to meet the full-load current requirement.

Substitute the values calculated in Eq. II, Eq. III, and Eq. IV into Eq. I.

$$Z_{0.85} = \frac{VD_{\text{LN}}}{\left(\dfrac{L}{1000 \text{ ft}}\right) I_{\text{load}}} = \frac{7.9674 \text{ V}}{\left(\dfrac{100.06 \text{ ft}}{1000 \text{ ft}}\right)(240 \text{ A})}$$
$$= 0.332 \; \Omega/1000 \text{ ft} \qquad \text{[V]}$$

The information calculated in Eq. V is now used to determine the conductor size using NEC Chap. 9, Table 9. The table is given in ohms-to-neutral per 1000 ft, hence the use of the line-to-neutral voltages and the per-1000 ft reference. Also, the reference power factor is 0.85. Should this not be the case, NEC Chap. 9, Table 9 provides the method for calculating the effective impedance. The effective impedance must be 0.332 Ω/ 1000 ft or less. For $Z_{0.85}$ for uncoated copper wires in steel conduit, a cable of AWG 4 or larger is required to meet the voltage drop requirement.

The only given size meeting both the voltage drop and current requirements is 350 kcmil.

(DC resistance is calculated in a similar manner to Eq. I without the need to convert to line-to-neutral voltage. Also, NEC Chap. 9, Table 8 is used to determine the conductor size instead of NEC Table 9.)

The answer is (D).

57 National Electrical Safety Code

PRACTICE PROBLEMS

1. Street and area lights under the control of utilities are covered under the rules of the

(A) American National Standards Institute (ANSI)

(B) Institute of Electrical Electronic Engineers (IEEE)

(C) *National Electrical Code* (NEC)

(D) *National Electrical Safety Code* (NESC)

2. The phase voltage of the live electrical parts of an alectric supply station is 13,800 V.

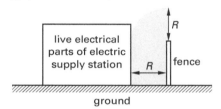

ground

Approximately what is the required minimum clearance distance, R, from the electric supply station to the fence?

(A) 1 m

(B) 4 m

(C) 10 m

(D) 30 m

3. Which of the following statements about the New England states are true?

(A) According to *National Electrical Safety Code* (NESC) Rule 250, the states are considered to be in the medium weather zone. Therefore, they must meet the clearance requirements of zone 1 of NESC Rule 230.

(B) According to NESC Rule 250, the states are considered to be in the medium weather zone. Therefore, they must meet the clearance requirements of zone 2 of NESC Rule 230.

(C) According to NESC Rule 250, the states are considered to be in the heavy weather zone. Therefore, they must meet the clearance requirements of zone 1 of NESC Rule 230.

(D) According to NESC Rule 250, the states are considered to be in the heavy weather zone. Therefore, they must meet the clearance requirements of zone 2 of NESC Rule 230.

4. A sail boat with an open supply conductor rated at 13.8 kV enters a body of water. The surface area of the water is more than 8 km². Most nearly, what vertical clearance is required for the open supply conductor?

(A) 4 m

(B) 8 m

(C) 12 m

(D) 30 m

5. The wind load, L_{SI}, on an overhead line structure is calculated in newtons as

$$L_{SI} = 0.613\, V_{m/s}^2 k_z\, G_{RF} I C_f A_{m^2}$$

In the formula given, k_z represents the

(A) force coefficient

(B) gust response factor

(C) velocity pressure exposure coefficient

(D) wind speed

SOLUTIONS

1. The NESC contains the governing requirements for lights under the exclusive control of utilities or their authorized contractors.

The answer is (D).

2. According to *National Electrical Safety Code* (NESC) Rule 110 and Table 110-1, for a voltage of 13,800 V, R must be at least 3.1 m (10.1 ft).

The answer is (B).

3. The New England states are in the heavy weather zone (called the heavy loading zone). The heavy, medium, light, and warm weather conditions of Rule 250 correspond to clearance zones 1, 2, 3, and 4 respectively, of Rule 230.

The answer is (C).

4. According to *National Electrical Safety Code* (NESC) Rule 232, Table 232-1 (metric) (row 7.d, column 5), the necessary clearance is 12.3 m based on the high water mark.

The answer is (C).

5. *National Electric Safety Code* (NESC) Rule 250.C.1, Table 250-2 defines each term used in the formula for calculating wind load. k_z is the velocity pressure coefficient.

The answer is (C).

Codes and Standards

Topic XV: Professional

Professional

58 Engineering Economic Analysis

PRACTICE PROBLEMS

1. At 6% effective annual interest, approximately how much will be accumulated if $1000 is invested for 10 years?

(A) $560

(B) $790

(C) $1600

(D) $1800

2. At 6% effective annual interest, the present worth of $2000 that becomes available in four years is most nearly

(A) $520

(B) $580

(C) $1600

(D) $2500

3. At 6% effective annual interest, approximately how much should be invested to accumulate $2000 in 20 years?

(A) $620

(B) $1400

(C) $4400

(D) $6400

4. At 6% effective annual interest, the year-end annual amount deposited over seven years that is equivalent to $500 invested now is most nearly

(A) $90

(B) $210

(C) $300

(D) $710

5. At 6% effective annual interest, the accumulated amount at the end of 10 years if $50 is invested at the end of each year for 10 years is most nearly

(A) $90

(B) $370

(C) $660

(D) $900

6. At 6% effective annual interest, approximately how much should be deposited at the start of each year for 10 years (a total of 10 deposits) in order to empty the fund by drawing out $200 at the end of each year for 10 years (a total of 10 withdrawals)?

(A) $190

(B) $210

(C) $220

(D) $250

7. At 6% effective annual interest, approximately how much should be deposited at the start of each year for five years to accumulate $2000 on the date of the last deposit?

(A) $350

(B) $470

(C) $510

(D) $680

8. At 6% effective annual interest, approximately how much will be accumulated in 10 years if three payments of $100 are deposited every other year for four years, with the first payment occurring at $t = 0$?

(A) $180

(B) $480

(C) $510

(D) $540

9. $500 is compounded monthly at a 6% nominal annual interest rate. Approximately how much will have accumulated in five years?

(A) $515

(B) $530

(C) $675

(D) $690

10. The effective annual rate of return on an $80 investment that pays back $120 in seven years is most nearly

(A) 4.5%

(B) 5.0%

(C) 5.5%

(D) 6.0%

11. A new machine will cost $17,000 and will have a resale value of $14,000 after five years. Special tooling will cost $5000. The tooling will have a resale value of $2500 after five years. Maintenance will be $2000 per year. The effective annual interest rate is 6%. The average annual cost of ownership during the next five years will be most nearly

(A) $2000

(B) $2300

(C) $4300

(D) $5500

12. An old covered wooden bridge can be strengthened at a cost of $9000, or it can be replaced for $40,000. The present salvage value of the old bridge is $13,000. It is estimated that the reinforced bridge will last for 20 years, will have an annual cost of $500, and will have a salvage value of $10,000 at the end of 20 years. The estimated salvage value of the new bridge after 25 years is $15,000. Maintenance for the new bridge would cost $100 annually. The effective annual interest rate is 8%. Which is the best alternative?

(A) Strengthen the old bridge.

(B) Build the new bridge.

(C) Both options are economically identical.

(D) Not enough information is given.

13. A firm expects to receive $32,000 each year for 15 years from sales of a product. An initial investment of $150,000 will be required to manufacture the product. Expenses will run $7530 per year. Salvage value is zero,

and straight-line depreciation is used. The income tax rate is 48%. The after-tax rate of return is most nearly

(A) 8.0%

(B) 9.0%

(C) 10%

(D) 11%

14. A public works project has initial costs of $1,000,000, benefits of $1,500,000, and disbenefits of $300,000.

(a) The benefit/cost ratio is most nearly

(A) 0.20

(B) 0.47

(C) 1.2

(D) 1.7

(b) The excess of benefits over costs is most nearly

(A) $200,000

(B) $500,000

(C) $700,000

(D) $800,000

15. A speculator in land pays $14,000 for property that he expects to hold for 10 years. $1000 is spent in renovation, and a monthly rent of $75 is collected from the tenants. (Use the year-end convention.) Taxes are $150 per year, and maintenance costs are $250 per year. In 10 years, the sale price needed to realize a 10% rate of return is most nearly

(A) $26,000

(B) $31,000

(C) $34,000

(D) $36,000

16. The effective annual interest rate for a payment plan of 30 equal payments of $89.30 per month when a lump sum payment of $2000 would have been an outright purchase is most nearly

(A) 27%

(B) 35%

(C) 43%

(D) 51%

17. A depreciable item is purchased for $500,000. The salvage value at the end of 25 years is estimated at $100,000.

(a) The depreciation in each of the first three years using the straight-line method is most nearly

(A) $4000

(B) $16,000

(C) $20,000

(D) $24,000

(b) The depreciation in each of the first three years using the sum-of-the-years' digits method is most nearly

(A) $16,000; $16,000; $16,000

(B) $30,000; $28,000; $27,000

(C) $31,000; $30,000; $28,000

(D) $32,000; $31,000; $30,000

(c) The depreciation in each of the first three years using the double-declining balance method is most nearly

(A) $16,000; $16,000; $16,000

(B) $31,000; $30,000; $28,000

(C) $37,000; $34,000; $31,000

(D) $40,000; $37,000; $34,000

18. Equipment that is purchased for $12,000 now is expected to be sold after 10 years for $2000. The estimated maintenance is $1000 for the first year, but it is expected to increase $200 each year thereafter. The effective annual interest rate is 10%.

(a) The present worth is most nearly

(A) $16,000

(B) $17,000

(C) $21,000

(D) $22,000

(b) The annual cost is most nearly

(A) $1100

(B) $2200

(C) $3600

(D) $3700

19. A new grain combine with a 20-year life can remove seven pounds of rocks from its harvest per hour. Any rocks left in its output hopper will cause $25,000 damage in subsequent processes. Several investments are available to increase the rock-removal capacity, as listed in the table. The effective annual interest rate is 10%. What should be done?

rock removal rate	annual probability of exceeding rock removal rate	required investment to achieve removal rate
7	0.15	0
8	0.10	$15,000
9	0.07	$20,000
10	0.03	$30,000

(A) Do nothing.

(B) Invest $15,000.

(C) Invest $20,000.

(D) Invest $30,000.

20. A mechanism that costs $10,000 has operating costs and salvage values as given. An effective annual interest rate of 20% is to be used.

year	operating cost	salvage value
1	$2000	$8000
2	$3000	$7000
3	$4000	$6000
4	$5000	$5000
5	$6000	$4000

(a) The cost of owning and operating the mechanism in year two is most nearly

(A) $3800

(B) $4700

(C) $5800

(D) $6000

(b) The cost of owning and operating the mechanism in year five is most nearly

(A) $5600

(B) $6500

(C) $7000

(D) $7700

(c) The economic life of the mechanism is most nearly

(A) one year

(B) two years

(C) three years

(D) five years

(d) Assuming that the mechanism has been owned and operated for four years already, the cost of owning and operating the mechanism for one more year is most nearly

(A) $6400

(B) $7200

(C) $8000

(D) $8200

21. (*Time limit: one hour*) A salesperson intends to purchase a car for $50,000 for personal use, driving 15,000 miles per year. Insurance for personal use costs $2000 per year, and maintenance costs $1500 per year. The car gets 15 miles per gallon, and gasoline costs $1.50 per gallon. The resale value after five years will be $10,000. The salesperson's employer has asked that the car be used for business driving of 50,000 miles per year and has offered a reimbursement of $0.30 per mile. Using the car for business would increase the insurance cost to $3000 per year and maintenance to $2000 per year. The salvage value after five years would be reduced to $5000. If the employer purchased a car for the salesperson to use, the initial cost would be the same, but insurance, maintenance, and salvage would be $2500, $2000, and $8000, respectively. The salesperson's effective annual interest rate is 10%.

(a) Is the reimbursement offer adequate?

(b) With a reimbursement of $0.30 per mile, approximately how many miles must the car be driven per year to justify the employer buying the car for the salesperson to use?

(A)   20,000 mi

(B)   55,000 mi

(C)   82,000 mi

(D)   150,000 mi

22. Alternatives A and B are being evaluated. The effective annual interest rate is 10%.

	alternative A	alternative B
first cost	$80,000	$35,000
life	20 years	10 years
salvage value	$7000	0
annual costs		
years 1–5	$1000	$3000
years 6–10	$1500	$4000
years 11–20	$2000	0
additional cost		
year 10	$5000	0

(a) The present worth for alternative A is most nearly

(A) $91,000

(B) $93,000

(C) $100,000

(D) $120,000

(b) The equivalent uniform annual cost (EUAC) for alternative A is most nearly

(A) $10,000

(B) $11,000

(C) $12,000

(D) $14,000

(c) The present worth for alternative B is most nearly

(A) $56,000

(B) $62,000

(C) $70,000

(D) $78,000

(d) The EUAC for alternative B is most nearly

(A) $9100

(B) $10,000

(C) $11,000

(D) $13,000

(e) Which alternative is economically superior?

(A) Alternative A is economically superior.

(B) Alternative B is economically superior.

(C) Alternatives A and B are economically equivalent.

(D) Not enough information is provided.

23. A car is needed for three years. Plans A and B for acquiring the car are being evaluated. An effective annual interest rate of 10% is to be used.

Plan A: lease the car for $0.25/mile (all inclusive)

Plan B: purchase the car for $30,000; keep the car for three years; sell the car after three years for $7200; pay $0.14 per mile for oil and gas; pay other costs of $500 per year

(a) What is the annual mileage of plan A?

(b) What is the annual mileage of plan B?

(c) For an equal annual cost, what is the annual mileage?

(d) Which plan is economically superior?

(A) Plan A is economically superior.

(B) Plan B is economically superior.

(C) Plans A and B are economically equivalent.

(D) Not enough information is provided.

24. Two methods are being considered to meet strict air pollution control requirements over the next 10 years. Method A uses equipment with a life of 10 years. Method B uses equipment with a life of five years that will be replaced with new equipment with an additional life of five years. Capacities of the two methods are different, but operating costs do not depend on the throughput. Operation is 24 hours per day, 365 days per year. The effective annual interest rate for this evaluation is 7%.

| | method A | method B | |
	years 1–10	years 1–5	years 6–10
installation cost	$13,000	$6000	$7000
equipment cost	$10,000	$2000	$2200
operating cost per hour	$10.50	$8.00	$8.00
salvage value	$5000	$2000	$2000
capacity (tons/yr)	50	20	20
life	10 years	5 years	5 years

(a) The uniform annual cost per ton for method A is most nearly

(A) $1800

(B) $1900

(C) $2100

(D) $2200

(b) The uniform annual cost per ton for method B is most nearly

(A) $3500

(B) $3600

(C) $4200

(D) $4300

(c) Over what range of throughput (in units of tons/yr) does each method have the minimum cost?

25. A transit district has asked for assistance in determining the proper fare for its bus system. An effective annual interest rate of 7% is to be used. The following additional information was compiled.

cost per bus	$60,000
bus life	20 years
salvage value	$10,000
miles driven per year	37,440
number of passengers per year	80,000
operating cost	$1.00 per mile in the first year, increasing $0.10 per mile each year thereafter

(a) If the fare is to remain constant for the next 20 years, the break-even fare per passenger is most nearly

(A) $0.51/passenger

(B) $0.61/passenger

(C) $0.84/passenger

(D) $0.88/passenger

(b) If the transit district decides to set the per-passenger fare at $0.35 for the first year, approximately how much should the passenger fare go up each year thereafter such that the district can break even in 20 years?

(A) $0.022 increase per year

(B) $0.036 increase per year

(C) $0.067 increase per year

(D) $0.072 increase per year

(c) If the transit district decides to set the per-passenger fare at $0.35 for the first year and the per-passenger fare goes up $0.05 each year thereafter, the additional government subsidy (per passenger) needed for the district to break even in 20 years is most nearly

(A) $0.11

(B) $0.12

(C) $0.16

(D) $0.21

26. Make a recommendation to your client to accept one of the following alternatives. Use the present worth comparison method. (Initial costs are the same.)

Alternative A: a 25-year annuity paying $4800 at the end of each year, where the interest rate is a nominal 12% per annum

Alternative B: a 25-year annuity paying $1200 every quarter at 12% nominal annual interest

(A) Alternative A is economically superior.

(B) Alternative B is economically superior.

(C) Alternatives A and B are economically equivalent.

(D) Not enough information is provided.

27. A firm has two alternatives for improvement of its existing production line. The data are as follows.

	alternative A	alternative B
initial installment cost	$1500	$2500
annual operating cost	$800	$650
service life	5 years	8 years
salvage value	0	0

Determine the best alternative using an interest rate of 15%.

(A) Alternative A is economically superior.

(B) Alternative B is economically superior.

(C) Alternatives A and B are economically equivalent.

(D) Not enough information is provided.

28. Two mutually exclusive alternatives requiring different investments are being considered. The life of both alternatives is estimated at 20 years with no salvage values. The minimum rate of return that is considered acceptable is 4%. Which alternative is best?

	alternative A	alternative B
investment required	$70,000	$40,000
net income per year	$5620	$4075
rate of return on total investment	5%	8%

(A) Alternative A is economically superior.

(B) Alternative B is economically superior.

(C) Alternatives A and B are economically equivalent.

(D) Not enough information is provided.

29. Compare the costs of two plant renovation schemes, A and B. Assume equal lives of 25 years, no salvage values, and interest at 25%.

	alternative A	alternative B
first cost	$20,000	$25,000
annual expenditure	$3000	$2500

(a) Determine the best alternative using the present worth method.

(A) Alternative A is economically superior.

(B) Alternative B is economically superior.

(C) Alternatives A and B are economically equivalent.

(D) Not enough information is provided.

(b) Determine the best alternative using the capitalized cost comparison.

(A) Alternative A is economically superior.

(B) Alternative B is economically superior.

(C) Alternatives A and B are economically equivalent.

(D) Not enough information is provided.

(c) Determine the best alternative using the annual cost comparison.

(A) Alternative A is economically superior.

(B) Alternative B is economically superior.

(C) Alternatives A and B are economically equivalent.

(D) Not enough information is provided.

30. A machine costs $18,000 and has a salvage value of $2000. It has a useful life of eight years. The interest rate is 8%.

(a) Using straight-line depreciation, its book value at the end of five years is most nearly

(A) $2000

(B) $3000

(C) $6000

(D) $8000

(b) Using the sinking fund method, the depreciation in the third year is most nearly

(A) $1500

(B) $1600

(C) $1800

(D) $2000

(c) Repeat part (a) using double declining balance depreciation. The balance value at the fifth year is most nearly

(A) $2000

(B) $4000

(C) $6000

(D) $8000

31. A chemical pump motor unit is purchased for $14,000. The estimated life is eight years, after which it will be sold for $1800. Find the depreciation in the first two years by the sum-of-the-years' digits method. The

after-tax depreciation recovery using 15% interest with 52% income tax is most nearly

(A) $3600

(B) $3900

(C) $4100

(D) $6300

32. A soda ash plant has the water effluent from processing equipment treated in a large settling basin. The settling basin eventually discharges into a river that runs alongside the basin. Recently enacted environmental regulations require all rainfall on the plant to be diverted and treated in the settling basin. A heavy rainfall will cause the entire basin to overflow. An uncontrolled overflow will cause environmental damage and heavy fines. The construction of additional height on the existing basic walls is under consideration.

Data on the costs of construction and expected costs for environmental cleanup and fines are shown. Data on 50 typical winters have been collected. The soda ash plant management considers 12% to be their minimum rate of return, and it is felt that after 15 years the plant will be closed. The company wants to select the alternative that minimizes its total expected costs.

additional basin height (ft)	number of winters with basin overflow	expense for environmental clean up per year	construction cost
0	24	$550,000	0
5	14	$600,000	$600,000
10	8	$650,000	$710,000
15	3	$700,000	$900,000
20	1	$800,000	$1,000,000
	50		

The additional height the basin should be built to is most nearly

(A) 5.0 ft

(B) 10 ft

(C) 15 ft

(D) 20 ft

33. A wood processing plant installed a waste gas scrubber at a cost of $30,000 to remove pollutants from the exhaust discharged into the atmosphere. The scrubber has no salvage value and will cost $18,700 to operate next year, with operating costs expected to increase at the rate of $1200 per year thereafter. Money can be borrowed at 12%. Approximately when should the company consider replacing the scrubber?

(A)   3 yr

(B)   6 yr

(C)   8 yr

(D)   10 yr

34. Two alternative piping schemes are being considered by a water treatment facility. Head and horsepower are reflected in the hourly cost of operation. On the basis of a 10-year life and an interest rate of 12%, what is most nearly the number of hours of operation for which the two installations will be equivalent?

	alternative A	alternative B
pipe diameter	4 in	6 in
head loss for required flow	48 ft	26 ft
size motor required	20 hp	7 hp
energy cost per hour of operation	$0.30	$0.10
cost of motor installed	$3600	$2800
cost of pipes and fittings	$3050	$5010
salvage value at end of 10 years	$200	$280

(A)   1000 hr

(B)   3000 hr

(C)   5000 hr

(D)   6000 hr

35. An 88% learning curve is used with an item whose first production time was six weeks.

(a) Approximately how long will it take to produce the fourth item?

(A)   4.5 wk

(B)   5.0 wk

(C)   5.5 wk

(D)   6.0 wk

(b) Approximately how long will it take to produce the sixth through fourteenth items?

(A)   35 wk

(B)   40 wk

(C)   45 wk

(D)   50 wk

36. A company is considering two alternatives, only one of which can be selected.

alternative	initial investment	salvage value	annual net profit	life
A	$120,000	$15,000	$57,000	5 yr
B	$170,000	$20,000	$67,000	5 yr

The net profit is after operating and maintenance costs, but before taxes. The company pays 45% of its year-end profit as income taxes. Use straight-line depreciation. Do not use investment tax credit.

(a) Determine whether each alternative has an ROR greater than the MARR.

(A)   Alternative A has ROR > MARR.

(B)   Alternative B has ROR > MARR.

(C)   Both alternatives have ROR > MARR.

(D)   Neither alternative has ROR > MARR.

(b) Find the best alternative if the company's minimum attractive rate of return is 15%.

(A)   Alternative A is economically superior.

(B)   Alternative B is economically superior.

(C)   Alternatives A and B are economically equivalent.

(D)   Not enough information is provided.

37. A company is considering the purchase of equipment to expand its capacity. The equipment cost is $300,000. The equipment is needed for five years, after which it will be sold for $50,000. The company's before-tax cash flow will be improved $90,000 annually by the purchase of the asset. The corporate tax rate is 48%, and straight-line depreciation will be used. The company will take an investment tax credit of 6.67%. What is the after-tax rate of return associated with this equipment purchase?

(A)   10.9%

(B)   11.8%

(C)   12.2%

(D)   13.2%

38. A 120-room hotel is purchased for $2,500,000. A 25-year loan is available for 12%. The year-end convention applies to loan payments. A study was conducted to determine the various occupancy rates.

occupancy	probability
65% full	0.40
70%	0.30
75%	0.20
80%	0.10

The operating costs of the hotel are as follows.

taxes and insurance	$20,000 annually
maintenance	$50,000 annually
operating	$200,000 annually

The life of the hotel is figured to be 25 years when operating 365 days per year. The salvage value after 25 years is $500,000. The new hotel owners want to receive an annual rate of return of 15% on their investment. Neglect tax credit and income taxes.

(a) The distributed profit is most nearly

(A) $300,000

(B) $320,000

(C) $340,000

(D) $380,000

(b) The annual daily receipts are most nearly

(A) $2300

(B) $2400

(C) $2500

(D) $2600

(c) The average occupancy is most nearly

(A) 0.65

(B) 0.70

(C) 0.75

(D) 0.80

(d) The average rate that should be charged per room per night is most nearly

(A) $27

(B) $29

(C) $30

(D) $31

39. A company is insured for $3,500,000 against fire and the insurance rate is $0.69/$1000. The insurance company will decrease the rate to $0.47/$1000 if fire sprinklers are installed. The initial cost of the sprinklers is $7500. Annual costs are $200; additional taxes are $100 annually. The system life is 25 years.

(a) The annual savings is most nearly

(A) $300

(B) $470

(C) $570

(D) $770

(b) The rate of return is most nearly

(A) 3.8%

(B) 5.0%

(C) 13%

(D) 16%

40. Heat losses through the walls in an existing building cost a company $1,300,000 per year. This amount is considered excessive, and two alternatives are being evaluated. Neither of the alternatives will increase the life of the existing building beyond the current expected life of six years, and neither of the alternatives will produce a salvage value. Improvements can be depreciated.

Alternative A: Do nothing, and continue with current losses.

Alternative B: Spend $2,000,000 immediately to upgrade the building and reduce the loss by 80%. Annual maintenance will cost $150,000.

Alternative C: Spend $1,200,000 immediately. Then, repeat the $1,200,000 expenditure three years from now. Heat loss the first year will be reduced 80%. Due to deterioration, the reduction will be 55% and 20% in the second and third years. (The pattern is repeated starting after the second expenditure.) There are no maintenance costs.

All energy and maintenance costs are considered expenses for tax purposes. The company's tax rate is 48%, and straight-line depreciation is used. 15% is regarded as the effective annual interest rate. Evaluate each alternative on an after-tax basis.

(a) The present worth of alternative A is most nearly

(A) −$5.9 million

(B) −$4.9 million

(C) −$2.6 million

(D) −$2.4 million

(b) The present worth of alternative B is most nearly

(A) −$3.4 million

(B) −$2.8 million

(C) −$2.6 million

(D) −$2.2 million

(c) The present worth of alternative C is most nearly

(A) −$3.2 million

(B) −$3.1 million

(C) −$2.4 million

(D) −$2.0 million

(d) Which alternative should be recommended?

(A) alternative A

(B) alternative B

(C) alternative C

(D) not enough information

41. You have been asked to determine if a 7-year-old machine should be replaced. Give a full explanation for your recommendation. Base your decision on a before-tax interest rate of 15%.

The existing machine is presumed to have a 10-year life. It has been depreciated on a straight-line basis from its original value of $1,250,000 to a current book value of $620,000. Its ultimate salvage value was assumed to be $350,000 for purposes of depreciation. Its present salvage value is estimated at $400,000, and this is not expected to change over the next three years. The current operating costs are not expected to change from $200,000 per year.

A new machine costs $800,000, with operating costs of $40,000 the first year, and increasing by $30,000 each year thereafter. The new machine has an expected life of 10 years. The salvage value depends on the year the new machine is retired.

year retired	salvage
1	$600,000
2	$500,000
3	$450,000
4	$400,000
5	$350,000
6	$300,000
7	$250,000
8	$200,000
9	$150,000
10	$100,000

42. A company estimates that the demand for its product will be 500,000 units per year. The product incorporates three identical valves. The acquisition cost per valve is $6.50, and the cost of keeping a valve in storage per year is 40% of the acquisition cost per valve per year. It costs the company an average of $49.50 to process an order. The company receives no quantity discounts and does not use inventory safety stocks. The most economical quantity of valves to order is most nearly

(A) 2700 valves

(B) 4400 valves

(C) 4800 valves

(D) 7600 valves

43. As production facilities move toward just-in-time manufacturing, it is important to minimize

(A) demand rate

(B) production rate

(C) inventory carrying cost

(D) setup cost

SOLUTIONS

1.

$i = 6\%$ a year

Using the formula from Table 58.1,

$$F = P(1+i)^n = (\$1000)(1+0.06)^{10}$$
$$= \boxed{\$1790.85 \quad (\$1800)}$$

Using the factor from App. 58.B, $(F/P, i, n) = 1.7908$ for $i = 6\%$ a year and $n = 10$ years.

$$F = P(F/P,6\%,10) = (\$1000)(1.7908)$$
$$= \boxed{\$1790.80 \quad (\$1800)}$$

The answer is (D).

2.

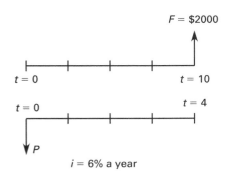

$i = 6\%$ a year

Using the formula from Table 58.1,

$$P = \frac{F}{(1+i)^n} = \frac{\$2000}{(1+0.06)^4}$$
$$= \boxed{\$1584.19 \quad (\$1600)}$$

Using the factor from App. 58.B, $(P/F, i, n) = 0.7921$ for $i = 6\%$ a year and $n = 4$ years.

$$P = F(P/F,6\%,4) = (\$2000)(0.7921)$$
$$= \boxed{\$1584.20 \quad (\$1600)}$$

The answer is (C).

3.

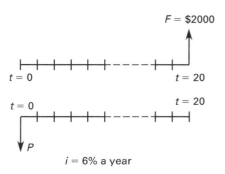

$i = 6\%$ a year

Using the formula from Table 58.1,

$$P = \frac{F}{(1+i)^n} = \frac{\$2000}{(1+0.06)^{20}}$$
$$= \boxed{\$623.61 \quad (\$620)}$$

Using the factor from App. 58.B, $(P/F, i, n) = 0.3118$ for $i = 6\%$ a year and $n = 20$ years.

$$P = F(P/F,6\%,20) = (\$2000)(0.3118)$$
$$= \boxed{\$623.60 \quad (\$620)}$$

The answer is (A).

4.

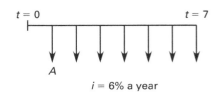

$i = 6\%$ a year

Professional

Using the formula from Table 58.1,

$$A = P\left(\frac{i(1+i)^n}{(1+i)^n - 1}\right) = (\$500)\left(\frac{(0.06)(1+0.06)^7}{(1+0.06)^7 - 1}\right)$$

$$= \boxed{\$89.57 \quad (\$90)}$$

Using the factor from App. 58.B, $(A/P, i, n) = 0.1791$ for $i = 6\%$ a year and $n = 7$ years.

$$A = P(A/P, 6\%, 7) = (\$500)(0.17914)$$

$$= \boxed{\$89.55 \quad (\$90)}$$

The answer is (A).

5.

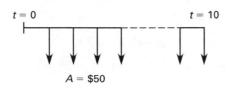

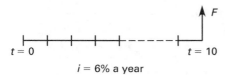

$i = 6\%$ a year

Using the formula from Table 58.1,

$$F = A\left(\frac{(1+i)^n - 1}{i}\right)$$

$$= (\$50)\left(\frac{(1+0.06)^{10} - 1}{0.06}\right)$$

$$= \boxed{\$659.04 \quad (\$660)}$$

Using the factor from App. 58.B, $(F/A, i, n) = 13.1808$ for $i = 6\%$ a year and $n = 10$ years.

$$F = A(F/A, 6\%, 10)$$

$$= (\$50)(13.1808)$$

$$= \boxed{\$659.04 \quad (\$660)}$$

The answer is (C).

6.

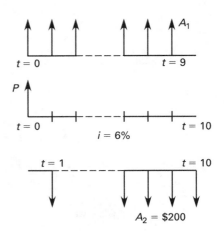

From Table 58.1, for each cash flow diagram,

$$P = A_1 + A_1\left(\frac{(1+0.06)^9 - 1}{(0.06)(1+0.06)^9}\right)$$

$$= A_2\left(\frac{(1+0.06)^{10} - 1}{(0.06)(1+0.06)^{10}}\right)$$

Therefore, for $A_2 = \$200$,

$$A_1 + A_1\left(\frac{(1+0.06)^9 - 1}{(0.06)(1+0.06)^9}\right)$$

$$= (\$200)\left(\frac{(1+0.06)^{10} - 1}{(0.06)(1+0.06)^{10}}\right)$$

$$7.80A_1 = \$1472.02$$

$$A_1 = \boxed{\$188.72 \quad (\$190)}$$

Using factors from App. 58.B,

$$(P/A, 6\%, 9) = 6.8017$$

$$(P/A, 6\%, 10) = 7.3601$$

$$A_1 + A_1(6.8017) = (\$200)(7.3601)$$

$$7.8017A_1 = \$1472.02$$

$$A_1 = \frac{\$1472.02}{7.8017}$$

$$= \boxed{\$188.68 \quad (\$190)}$$

The answer is (A).

7.

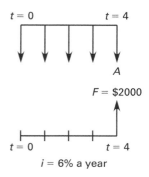

$i = 6\%$ a year

From Table 58.1,

$$F = A\left(\frac{(1+i)^n - 1}{i}\right)$$

Since the deposits start at the beginning of each year, five deposits are made that contribute to the final amount. This is equivalent to a cash flow that starts at $t = -1$ without a deposit and has a duration (starting at $t = -1$) of five years.

$$F = A\left(\frac{(1+i)^n - 1}{i}\right) = \$2000$$

$$= A\left(\frac{(1+0.06)^5 - 1}{0.06}\right)$$

$$\$2000 = 5.6371A = \frac{\$2000}{5.6371}$$

$$= \boxed{\$354.79 \quad (\$350)}$$

Using factors from App. 58.B,

$$F = A\big((F/P,6\%,4) + (F/A,6\%,4)\big)$$
$$\$2000 = A(1.2625 + 4.3746)$$
$$A = \boxed{\$354.79 \quad (\$350)}$$

The answer is (A).

8.

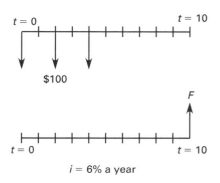

$i = 6\%$ a year

From Table 58.1, $F = P(1 + i)^n$. If each deposit is considered as P, each will accumulate interest for periods of 10, 8, and 6 years.

Therefore,

$$F = (\$100)(1 + 0.06)^{10} + (\$100)(1 + 0.06)^8$$
$$+ (\$100)(1 + 0.06)^6$$
$$= (\$100)(1.7908 + 1.5938 + 1.4185)$$
$$= \boxed{\$480.31 \quad (\$480)}$$

Using App. 58.B,

$$(F/P,i,n) = 1.7908 \text{ for } i = 6\% \text{ and } n = 10$$
$$= 1.5938 \text{ for } i = 6\% \text{ and } n = 8$$
$$= 1.4185 \text{ for } i = 6\% \text{ and } n = 6$$

By summation,

$$F = (\$100)(1.7908 + 1.5938 + 1.4185)$$
$$= \boxed{\$480.31 \quad (\$480)}$$

The answer is (B).

9.

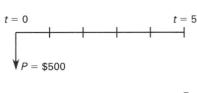

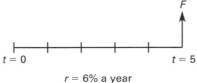

$r = 6\%$ a year

Since the deposit is compounded monthly, the effective interest rate should be calculated from Eq. 58.54.

$$i = \left(1 + \frac{r}{k}\right)^k - 1$$
$$= \left(1 + \frac{0.06}{12}\right)^{12} - 1$$
$$= 0.061678 \quad (6.1678\%)$$

From Table 58.1,

$$F = P(1+i)^n$$
$$= (\$500)(1 + 0.061678)^5$$
$$= \boxed{\$674.43 \quad (\$680)}$$

To use App. 58.B, interpolation is required.

$i\%$	factor F/P
6	1.3382
6.1678	desired
7	1.4026

$$\Delta(F/P) = \left(\frac{6.1678\% - 6\%}{7\% - 6\%}\right)(1.4026 - 1.3382)$$
$$= 0.0108$$

Therefore,

$$F/P = 1.3382 + 0.0108 = 1.3490$$
$$F = P(F/P,6.1677\%,5) = (\$500)(1.3490)$$
$$= \boxed{\$674.50 \quad (\$675)}$$

The answer is (C).

10.

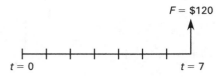

From Table 58.1,

$$F = P(1+i)^n$$

Therefore,

$$(1+i)^n = F/P$$
$$i = (F/P)^{1/n} - 1 = \left(\frac{\$120}{\$80}\right)^{1/7} - 1$$
$$= 0.0596 \approx \boxed{6.0\%}$$

From App. 58.B,

$$F = P(F/P,i\%,7)$$
$$(F/P,i\%,7) = F/P = \frac{\$120}{\$80} = 1.5$$

Searching App. 58.B,

$$(F/P,i\%,7) = 1.4071 \text{ for } i = 5\%$$
$$= 1.5036 \text{ for } i = 6\%$$
$$= 1.6058 \text{ for } i = 7\%$$

Therefore, $i \approx \boxed{6.0\%.}$

The answer is (D).

11.

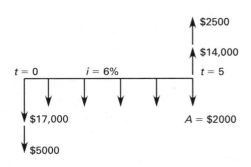

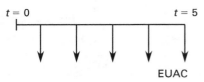

The annual cost of ownership, EUAC, can be obtained by the factors converting P to A and F to A.

$$P = \$17{,}000 + \$5000$$
$$= \$22{,}000$$
$$F = \$14{,}000 + \$2500$$
$$= \$16{,}500$$
$$\text{EUAC} = A + P(A/P,6\%,5) - F(A/F,6\%,5)$$
$$(A/P,6\%,5) = 0.2374$$
$$(A/F,6\%,5) = 0.1774$$
$$\text{EUAC} = \$2000 + (\$22{,}000)(0.2374)$$
$$- (\$16{,}500)(0.1774)$$
$$= \boxed{\$4295.70 \quad (\$4300)}$$

The answer is (C).

12.

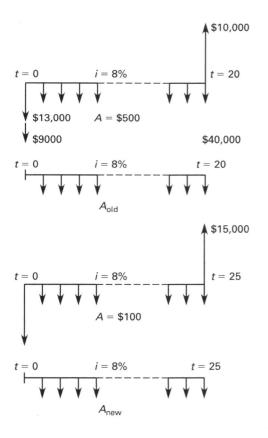

Consider the salvage value as a benefit lost (cost).

$$\text{EUAC}_{\text{old}} = \$500 + (\$9000 + \$13{,}000)(A/P,8\%,20)$$
$$- (\$10{,}000)(A/F,8\%,20)$$
$$(A/P,8\%,20) = 0.1019$$
$$(A/F,8\%,20) = 0.0219$$
$$\text{EUAC}_{\text{old}} = \$500 + (\$22{,}000)(0.1019)$$
$$- (\$10{,}000)(0.0219)$$
$$= \$2522.80$$

Similarly,

$$\text{EUAC}_{\text{new}} = \$100 + (\$40{,}000)(A/P,8\%,25)$$
$$- (\$15{,}000)(A/F,8\%,25)$$
$$(A/P,8\%,25) = 0.0937$$
$$(A/F,8\%,25) = 0.0137$$
$$\text{EUAC}_{\text{new}} = \$100 + (\$40{,}000)(0.0937)$$
$$- (\$15{,}000)(0.0137)$$
$$= \$3642.50$$

Therefore, the new bridge is more costly.

$$\boxed{\text{The best alternative is to strengthen the old bridge.}}$$

The answer is (A).

13.

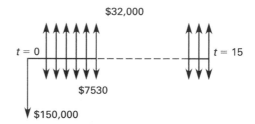

The annual depreciation is

$$D = \frac{C - S_n}{n} = \frac{\$150{,}000}{15} = \$10{,}000/\text{year}$$

The taxable income is

$$\$32{,}000 - \$7530 - \$10{,}000 = \$14{,}470/\text{year}$$

Taxes paid are

$$(\$14{,}470)(0.48) = \$6945.60/\text{year}$$

The after-tax cash flow is

$$\$32{,}000 - \$7530 - \$6945.60 = \$17{,}524.40$$

Professional

The present worth of the alternate is zero when evaluated at its ROR.

$$0 = -\$150{,}000 + (\$17{,}524.40)(P/A,i\%,15)$$

Therefore,

$$(P/A,i\%,15) = \frac{\$150{,}000}{\$17{,}524.40} = 8.55949$$

Searching App. 58.B, this factor matches $i = 8\%$.

$$\boxed{\text{ROR} = 8.0\%}$$

The answer is (A).

14. (a) The conventional benefit/cost ratio is

$$B/C = \frac{B-D}{D}$$

The benefit/cost ratio will be

$$B/C = \frac{\$1{,}500{,}000 - \$300{,}000}{\$1{,}000{,}000} = \boxed{1.2}$$

The answer is (C).

(b) The excess of benefits over cost is $\boxed{\$200{,}000.}$

The answer is (A).

15. The annual rent is

$$(\$75)\left(12\ \frac{\text{months}}{\text{year}}\right) = \$900$$

$$P = P_1 + P_2 = \$15{,}000$$

$$A_1 = -\$900$$

$$A_2 = \$250 + \$150 = \$400$$

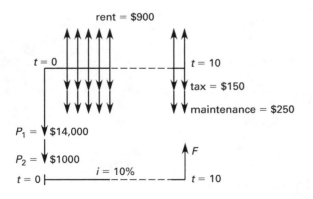

Use App. 58.B.

$$F = (\$15{,}000)(F/P,10\%,10)$$
$$+(\$400)(F/A,10\%,10)$$
$$-(\$900)(F/A,10\%,10)$$
$$(F/P,10\%,10) = 2.5937$$
$$(F/A,10\%,10) = 15.9374$$
$$F = (\$15{,}000)(2.5937) + (\$400)(15.9374)$$
$$-(\$900)(15.9374)$$
$$= \boxed{\$30{,}937 \quad (\$31{,}000)}$$

The answer is (B).

16.

```
t = 0                          t = 30
|___|___|___ _ _ _ ___|___|___|
    |   |   |          |   |   |
    v   v   v          v   v   v A = $89.30


t = 0                          t = 30
|___|___|___ _ _ _ _ _ ___|___|
|
v P = $2000
```

From Table 58.1,

$$P = A\left(\frac{(1+i)^n - 1}{i(1+i)^n}\right)$$

$$\frac{(1+i)^{30} - 1}{i(1+i)^{30}} = \frac{\$2000}{\$89.30} = 22.40$$

By trial and error,

$i\%$	$(1+i)^{30}$	$\dfrac{(1+i)^{30} - 1}{i(1+i)^{30}}$
10	17.45	9.43
6	5.74	13.76
4	3.24	17.29
2	1.81	22.40

2% per month is close.

$$i = (1+0.02)^{12} - 1 = \boxed{0.2682 \quad (27\%)}$$

The answer is (A).

17. (a) Use the straight-line method, Eq. 58.25.

$$D = \frac{C - S_n}{n}$$

Each year depreciation will be the same.

$$D = \frac{\$500{,}000 - \$100{,}000}{25} = \boxed{\$16{,}000}$$

The answer is (B).

(b) Use Eq. 58.27.

$$T = \tfrac{1}{2}n(n+1) = \left(\frac{1}{2}\right)(25)(25+1) = 325$$

Sum-of-the-years' digits (SOYD) depreciation can be calculated from Eq. 58.28.

$$D_j = \frac{(C - S_n)(n - j + 1)}{T}$$

$$D_1 = \frac{(\$500{,}000 - \$100{,}000)(25 - 1 + 1)}{325}$$

$$= \boxed{\$30{,}769 \quad (\$31{,}000)}$$

$$D_2 = \frac{(\$500{,}000 - \$100{,}000)(25 - 2 + 1)}{325}$$

$$= \boxed{\$29{,}538 \quad (\$30{,}000)}$$

$$D_3 = \frac{(\$500{,}000 - \$100{,}000)(25 - 3 + 1)}{325}$$

$$= \boxed{\$28{,}308 \quad (\$28{,}000)}$$

The answer is (C).

(c) The double declining balance (DDB) method can be used. By Eq. 58.32,

$$D_j = dC(1 - d)^{j-1}$$

Use Eq. 58.31.

$$d = \frac{2}{n} = \frac{2}{25}$$

$$D_1 = \left(\frac{2}{25}\right)(\$500{,}000)\left(1 - \frac{2}{25}\right)^0 = \boxed{\$40{,}000}$$

$$D_2 = \left(\frac{2}{25}\right)(\$500{,}000)\left(1 - \frac{2}{25}\right)^1 = \boxed{\$36{,}800 \quad (\$37{,}000)}$$

$$D_3 = \left(\frac{2}{25}\right)(\$500{,}000)\left(1 - \frac{2}{25}\right)^2 = \boxed{\$33{,}856 \quad (\$34{,}000)}$$

The answer is (D).

18.

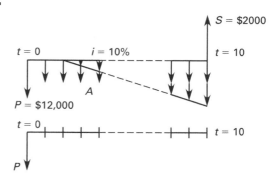

(a) A is $1000, and G is $200 for $t = n - 1 = 9$ yr. With $F = S = \$2000$, the present worth is

$$P = \$12{,}000 + A(P/A,10\%,10) + G(P/G,10\%,10)$$
$$\quad - F(P/F,10\%,10)$$
$$= \$12{,}000 + (\$1000)(6.1446) + (\$200)(22.8913)$$
$$\quad - (\$2000)(0.3855)$$
$$= \boxed{\$21{,}952 \quad (\$22{,}000)}$$

The answer is (D).

(b) The annual cost is

$$A = (\$12{,}000)(A/P,10\%,10) + \$1000$$
$$\quad + (\$200)(A/G,10\%,10)$$
$$\quad - (\$2000)(A/F,10\%,10)$$
$$= (\$12{,}000)(0.1627) + \$1000 + (\$200)(3.7255)$$
$$\quad - (\$2000)(0.0627)$$
$$= \boxed{\$3572.10 \quad (\$3600)}$$

The answer is (C).

19. An increase in rock removal capacity can be achieved by a 20-year loan (investment). Different cases available can be compared by equivalent uniform annual cost (EUAC).

$$\begin{aligned}
\text{EUAC} &= \text{annual loan cost} \\
&\quad + \text{expected annual damage} \\
&= \text{cost}\,(A/P,10\%,20) \\
&\quad + (\$25{,}000)(\text{probability})
\end{aligned}$$

$$(A/P,10\%,20) = 0.1175$$

Professional

A table can be prepared for different cases.

An effective annual interest rate	cost ($)	annual loan cost ($)	expected annual damage ($)	EUAC ($)
7	0	0	3750	3750.00
8	15,000	1761.90	2500	4261.90
9	20,000	2349.20	1750	4099.20
10	30,000	3523.80	750	4273.80

> It is cheapest to do nothing.

The answer is (A).

20. (a) Calculate the cost of owning and operating for years one and two.

$$A_1 = (\$10,000)(A/P,20\%,1) + \$2000$$
$$- (\$8000)(A/F,20\%,1)$$
$$(A/P,20\%,1) = 1.2$$
$$(A/F,20\%,1) = 1.0$$
$$A_1 = (\$10,000)(1.2) + \$2000 - (\$8000)(1.0)$$
$$= \$6000$$
$$A_2 = (\$10,000)(A/P,20\%,2) + \$2000$$
$$+ (\$1000)(A/G,20\%,2)$$
$$- (\$7000)(A/F,20\%,2)$$
$$(A/P,20\%,2) = 0.6545$$
$$(A/G,20\%,2) = 0.4545$$
$$(A/F,20\%,2) = 0.4545$$
$$A_2 = (\$10,000)(0.6545) + \$2000$$
$$+ (\$1000)(0.4545) - (\$7000)(0.4545)$$
$$= \boxed{\$5818 \quad (\$5800)}$$

The answer is (C).

(b) Calculate the cost of owning and operating for years three through five.

$$A_3 = (\$10,000)(A/P,20\%,3) + \$2000$$
$$+ (\$1000)(A/G,20\%,3)$$
$$- (\$6000)(A/F,20\%,3)$$
$$(A/P,20\%,3) = 0.4747$$
$$(A/G,20\%,3) = 0.8791$$
$$(A/F,20\%,3) = 0.2747$$

$$A_3 = (\$10,000)(0.4747) + \$2000$$
$$+ (\$1000)(0.8791)$$
$$- (\$6000)(0.2747)$$
$$= \$5977.90$$

$$A_4 = (\$10,000)(A/P,20\%,4)$$
$$+ \$2000 + (\$1000)(A/G,20\%,4)$$
$$- (\$5000)(A/F,20\%,4)$$

$$(A/P,20\%,4) = 0.3863$$
$$(A/G,20\%,4) = 1.2742$$
$$(A/F,20\%,4) = 0.1863$$
$$A_4 = (\$10,000)(0.3863) + \$2000$$
$$+ (\$1000)(1.2742) - (\$5000)(0.1863)$$
$$= \$6205.70$$

$$A_5 = (\$10,000)(A/P,20\%,5) + \$2000$$
$$+ (\$1000)(A/G,20\%,5)$$
$$- (\$4000)(A/F,20\%,5)$$
$$(A/P,20\%,5) = 0.3344$$
$$(A/G,20\%,5) = 1.6405$$
$$(A/F,20\%,5) = 0.1344$$
$$A_5 = (\$10,000)(0.3344) + \$2000$$
$$+ (\$1000)(1.6405) - (\$4000)(0.1344)$$
$$= \boxed{\$6446.90 \quad (\$6500)}$$

The answer is (B).

(c) Since the annual owning and operating cost is smallest after two years of operation, it is advantageous to sell the mechanism after the second year.

> The economic life is two years.

The answer is (B).

(d) After four years of operation, the owning and operating cost of the mechanism for one more year will be

$$A = \$6000 + (\$5000)(1 + i) - \$4000$$
$$i = 0.2 \quad (20\%)$$
$$A = \$6000 + (\$5000)(1.2) - \$4000$$
$$= \boxed{\$8000}$$

The answer is (C).

Professional

21. (a) To find out if the reimbursement is adequate, calculate the business-related expense.

Charge the company for business travel.

$$\text{insurance: } \$3000 - \$2000 = \$1000$$
$$\text{maintenance: } \$2000 - \$1500 = \$500$$
$$\text{drop in salvage value: } \$10,000 - \$5000 = \$5000$$

The annual portion of the drop in salvage value is

$$A = (\$5000)(A/F,10\%,5)$$
$$(A/F,10\%,5) = 0.1638$$
$$A = (\$5000)(0.1638)$$
$$= \$819/\text{yr}$$

The annual cost of gas is

$$\left(\frac{50,000 \text{ mi}}{15 \dfrac{\text{mi}}{\text{gal}}}\right)\left(\frac{\$1.50}{\text{gal}}\right) = \$5000$$

$$\text{EUAC per mile} = \frac{\$1000 + \$500 + \$819 + \$5000}{50,000 \text{ mi}}$$
$$= \$0.14638/\text{mi}$$

Since the reimbursement per mile was \$0.30 and since $\$0.30 > \$0.14638$, the reimbursement is $\boxed{\text{adequate.}}$

(b) Determine (with reimbursement) how many miles the car must be driven to break even.

If the car is driven M miles per year,

$$\left(\frac{\$0.30}{1 \text{ mi}}\right)M = (\$50,000)(A/P,10\%,5) + \$2500$$
$$+ \$2000 - (\$8000)(A/F,10\%,5)$$
$$+ \left(\frac{M}{15 \dfrac{\text{mi}}{\text{gal}}}\right)(\$1.50)$$

$$(A/P,10\%,5) = 0.2638$$
$$(A/F,10\%,5) = 0.1638$$

$$0.3M = (\$50,000)(0.2638) + \$2500 + \$2000$$
$$- (\$8000)(0.1638) + 0.1M$$
$$0.2M = \$16,379.60$$
$$M = \frac{\$16,379.60}{\dfrac{\$0.20}{1 \text{ mi}}}$$
$$= \boxed{81,898 \text{ mi} \quad (82,000 \text{ mi})}$$

The answer is (C).

22. The present worth of alternative A is

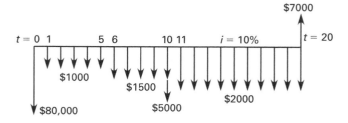

$$P_A = \$80,000 + (\$1000)(P/A,10\%,5)$$
$$+ (\$1500)(P/A,10\%,5)(P/F,10\%,5)$$
$$+ (\$2000)(P/A,10\%,10)(P/F,10\%,10)$$
$$+ (\$5000)(P/F,10\%,10)$$
$$- (\$7000)(P/F,10\%,20)$$

$$(P/A,10\%,5) = 3.7908$$
$$(P/F,10\%,5) = 0.6209$$
$$(P/A,10\%,10) = 6.1446$$
$$(P/F,10\%,10) = 0.3855$$
$$(P/F,10\%,20) = 0.1486$$
$$P_A = \$80,000 + (\$1000)(3.7908)$$
$$+ (\$1500)(3.7908)(0.6209)$$
$$+ (\$2000)(6.1446)(0.3855)$$
$$+ (\$5000)(0.3855)$$
$$- (\$7000)(0.1486)$$
$$= \boxed{\$92,946.15 \quad (\$93,000)}$$

The answer is (B).

(b) Since the lives are different, compare by EUAC.

$$\text{EUAC(A)} = (\$92,946.14)(A/P,10\%,20)$$
$$= (\$92,946.14)(0.1175)$$
$$= \boxed{\$10,921 \quad (\$11,000)}$$

The answer is (B).

(c) Evaluate alternative B.

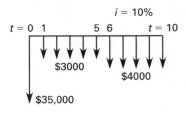

$$P_B = \$35,000 + (\$3000)(P/A,10\%,5)$$
$$+(\$4000)(P/A,10\%,5)(P/F,10\%,5)$$
$$(P/A,10\%,5) = 3.7908$$
$$(P/F,10\%,5) = 0.6209$$
$$P_B = \$35,000 + (\$3000)(3.7908)$$
$$+(\$4000)(3.7908)(0.6209)$$
$$= \boxed{\$55,787.23 \quad (\$56,000)}$$

The answer is (A).

(d) Since the lives are different, compare by EUAC.

$$\text{EUAC(B)} = (\$55,787.23)(A/P,10\%,10)$$
$$= (\$55,787.23)(0.1627)$$
$$= \boxed{\$9077 \quad (\$9100)}$$

The answer is (A).

(e) Since EUAC(B) < EUAC(A),

$$\boxed{\text{Alternative B is economically superior.}}$$

The answer is (B).

23. (a) If the annual cost is compared with a total annual mileage of M, for plan A,

$$A_A = \boxed{\$0.25M}$$

(b) For plan B,

$$A_B = (\$30,000)(A/P,10\%,3) + \$0.14M$$
$$+\$500 - (\$7200)(A/F,10\%,3)$$
$$(A/P,10\%,3) = 0.4021$$
$$(A/F,10\%,3) = 0.3021$$
$$A_B = (\$30,000)(0.4021) + \$0.14M + \$500$$
$$-(\$7200)(0.3021)$$
$$= \boxed{\$10,387.88 + \$0.14M}$$

(c) For an equal annual cost $A_A = A_B$,

$$\$0.25M = \$10,387.88 + \$0.14M$$
$$\$0.11M = \$10,387.88$$
$$M = 94,435 \quad (94,000)$$

An annual mileage would be $\boxed{M = 94,000 \text{ mi.}}$

(d) For an annual mileage less than that, $A_A < A_B$.

$$\boxed{\begin{array}{l}\text{Plan A is economically superior until 94,000 mi is}\\\text{exceeded.}\end{array}}$$

The answer is (A).

24. (a) Method A:

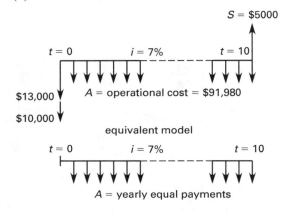

$$24 \text{ hr/day}$$
$$365 \text{ days/yr}$$
$$\text{total of } (24)(365) = 8760 \text{ hr/yr}$$
$$\$10.50 \text{ operational cost/hr}$$
$$\text{total of } (8760)(\$10.50) = \$91,980 \text{ operational cost/yr}$$

$$A = \$91,980 + (\$23,000)(A/P,7\%,10)$$
$$-(\$5000)(A/F,7\%,10)$$
$$(A/P,7\%,10) = 0.1424$$
$$(A/F,7\%,10) = 0.0724$$
$$A = \$91,980 + (\$23,000)(0.1424)$$
$$-(\$5000)(0.0724)$$
$$= \$94,893.20/\text{yr}$$

Therefore, the uniform annual cost per ton each year will be

$$\frac{\$94,893.20}{50 \text{ ton}} = \boxed{\$1897.86 \quad (\$1900)}$$

The answer is (B).

(b) Method B:

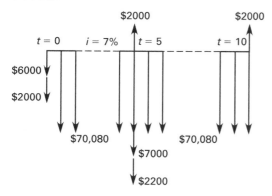

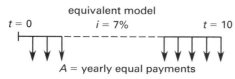

8760 hr/yr

$8 operational cost/hr

total of $70,080 operational cost/yr

$$A = \$70,080 + (\$6000 + \$2000)$$
$$\times (A/P,7\%,10)$$
$$+ (\$7000 + \$2200 - \$2000)$$
$$\times (P/F,7\%,5)(A/P,7\%,10)$$
$$- (\$2000)(A/F,7\%,10)$$

$$(A/P,7\%,10) = 0.1424$$
$$(A/F,7\%,10) = 0.0724$$
$$(P/F,7\%,5) = 0.7130$$
$$A = \$70,080 + (\$8000)(0.1424)$$
$$+ (\$7200)(0.7130)(0.1424)$$
$$- (\$2000)(0.0724)$$
$$= \$71,805.42/\text{yr}$$

Therefore, the uniform annual cost per ton each year will be

$$\frac{\$71,805.42}{20 \text{ ton}} = \boxed{\$3590.27 \quad (\$3600)}$$

The answer is (B).

(c) The following table can be used to determine which alternative is least expensive for each throughput range.

tons/yr	cost of using A		cost of using B		cheapest
0–20	$94,893	(1×)	$71,805	(1×)	B
20–40	$94,893	(1×)	$143,610	(2×)	A
40–50	$94,893	(1×)	$215,415	(3×)	A
50–60	$189,786	(2×)	$215,415	(3×)	A
60–80	$189,786	(2×)	$287,220	(4×)	A

25.

$$A_e = (\$60,000)(A/P,7\%,20) + A$$
$$+ G(P/G,7\%,20)(A/P,7\%,20)$$
$$- (\$10,000)(A/F,7\%,20)$$
$$(A/P,7\%,20) = 0.0944$$

$$A = (37,440 \text{ mi})\left(\frac{\$1.0}{1 \text{ mi}}\right) = \$37,440$$

$$G = 0.1A = (0.1)(\$37,440) = \$3744$$
$$(P/G,7\%,20) = 77.5091$$
$$(A/F,7\%,20) = 0.0244$$
$$A_e = (\$60,000)(0.0944) + \$37,440$$
$$+ (\$3744)(77.5091)(0.0944)$$
$$- (\$10,000)(0.0244)$$
$$= \$70,254.32$$

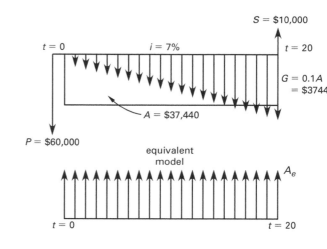

(a) With 80,000 passengers a year, the break-even fare per passenger would be

$$\text{fare} = \frac{A_e}{80{,}000}$$

$$= \frac{\$70{,}254.32}{80{,}000}$$

$$= \boxed{\$0.878/\text{passenger} \quad (\$0.88/\text{passenger})}$$

The answer is (D).

(b) The passenger fare should go up each year by

$$\$0.878 = \$0.35 + G(A/G,7\%,20)$$

$$G = \frac{\$0.878 - \$0.35}{7.3163}$$

$$= \boxed{\$0.072 \text{ increase per year}}$$

The answer is (D).

(c) As in part (b), the subsidy should be

$$\text{subsidy} = \text{cost} - \text{revenue}$$
$$P = \$0.878 - (\$0.35 + (\$0.05)(A/G,7\%,20))$$
$$= \$0.878 - (\$0.35 + (\$0.05)(7.3163))$$
$$= \boxed{\$0.162 \quad (\$0.16)}$$

The answer is (C).

26. Use the present worth comparison method.

$$P(\text{A}) = (\$4800)(P/A,12\%,25) = (\$4800)(7.8431)$$
$$= \$37{,}646.88$$

$$(4 \text{ quarters})(25 \text{ years}) = 100 \text{ compounding periods}$$

$$P(\text{B}) = (\$1200)(P/A,3\%,100) = (\$1200)(31.5989)$$
$$= \$37{,}918.68$$

$$\boxed{\text{Alternative B is economically superior.}}$$

The answer is (B).

27. Use the equivalent uniform annual cost method.

$$\text{EUAC}(\text{A}) = (\$1500)(A/P,15\%,5) + \$800$$
$$= (\$1500)(0.2983) + \$800$$
$$= \$1247.45$$

$$\text{EUAC}(\text{B}) = (\$2500)(A/P,15\%,8) + \$650$$
$$= (\$2500)(0.2229) + \$650$$
$$= \$1207.25$$

$$\boxed{\text{Alternative B is economically superior.}}$$

The answer is (B).

28. The data given imply that both investments return 4% or more. However, the increased investment of $30,000 may not be cost effective. Do an incremental analysis.

$$\text{incremental cost} = \$70{,}000 - \$40{,}000 = \$30{,}000$$

$$\text{incremental income} = \$5620 - \$4075 = \$1545$$

$$0 = -\$30{,}000 + (\$1545)(P/A,i\%,20)$$
$$(P/A,i\%,20) = 19.417$$
$$i \approx 0.25\% < 4\%$$

$$\boxed{\text{Alternative B is economically superior.}}$$

(The same conclusion could be reached by taking the present worths of both alternatives at 4%.)

The answer is (B).

29. (a) The present worth comparison is

$$P(\text{A}) = (-\$3000)(P/A,25\%,25) - \$20{,}000$$
$$= (-\$3000)(3.9849) - \$20{,}000$$
$$= -\$31{,}954.70$$
$$P(\text{B}) = (-\$2500)(3.9849) - \$25{,}000$$
$$= -\$34{,}962.25$$

$$\boxed{\text{Alternative A is economically superior.}}$$

The answer is (A).

(b) The capitalized cost comparison is

$$\text{CC}(\text{A}) = \$20{,}000 + \frac{\$3000}{0.25} = \$32{,}000$$

$$\text{CC}(\text{B}) = \$25{,}000 + \frac{\$2500}{0.25} = \$35{,}000$$

$$\boxed{\text{Alternative A is economically superior.}}$$

The answer is (A).

(c) The annual cost comparison is

$$\text{EUAC(A)} = (\$20{,}000)(A/P,25\%,25) + \$3000$$
$$= (\$20{,}000)(0.2509) + \$3000$$
$$= \$8018.00$$

$$\text{EUAC(B)} = (\$25{,}000)(0.2509) + \$2500$$
$$= \$8772.50$$

$$\boxed{\text{Alternative A is economically superior.}}$$

The answer is (A).

30. (a) The depreciation in the first five years is

$$\text{BV} = \$18{,}000 - (5)\left(\frac{\$18{,}000 - \$2000}{8}\right)$$
$$= \boxed{\$8000}$$

The answer is (D).

(b) With the sinking fund method, the basis is

$$(\$18{,}000 - \$2000)(A/F,8\%,8)$$
$$= (\$18{,}000 - \$2000)(0.0940)$$
$$= \$1504$$
$$D_1 = (\$1504)(1.000) = \$1504$$
$$D_2 = (\$1504)(1.0800) = \$1624$$
$$D_3 = (\$1504)(1.0800)^2 = \boxed{\$1754 \quad (\$1800)}$$

The answer is (C).

(c) Using double declining balance depreciation, the first five years' depreciation is

$$D_1 = \left(\frac{2}{8}\right)(\$18{,}000) = \$4500$$

$$D_2 = \left(\frac{2}{8}\right)(\$18{,}000 - \$4500) = \$3375$$

$$D_3 = \left(\frac{2}{8}\right)(\$18{,}000 - \$4500 - \$3375) = \$2531$$

$$D_4 = \left(\frac{2}{8}\right)(\$18{,}000 - \$4500 - \$3375 - \$2531)$$
$$= \$1898$$

$$D_5 = \left(\frac{2}{8}\right)(\$18{,}000 - \$4500 - \$3375$$
$$\qquad - \$2531 - \$1898)$$
$$= \$1424$$

The balance value at the fifth year is

$$\text{BV} = \$18{,}000 - \$4500 - \$3375 - \$2531$$
$$\qquad - \$1898 - \$1424$$
$$= \boxed{\$4272 \quad (\$4000)}$$

The answer is (B).

31. The after-tax depreciation recovery is

$$T = \left(\frac{1}{2}\right)(8)(9) = 36$$

$$D_1 = \left(\frac{8}{36}\right)(\$14{,}000 - \$1800) = \$2711$$

$$\Delta D = \left(\frac{1}{36}\right)(\$14{,}000 - \$1800) = \$339$$

$$D_2 = \$2711 - \$339 = \$2372$$

$$\text{DR} = (0.52)(\$2711)(P/A,15\%,8)$$
$$\qquad - (0.52)(\$339)(P/G,15\%,8)$$
$$= (0.52)(\$2711)(4.4873)$$
$$\qquad - (0.52)(\$339)(12.4807)$$
$$= \boxed{\$4125.74 \quad (\$4100)}$$

The answer is (C).

32. Use the equivalent uniform annual cost method to find the best alternative.

$$(A/P,12\%,15) = 0.1468$$

$$\text{EUAC}_{5\,\text{ft}} = (\$600{,}000)(0.1468)$$
$$+ \left(\frac{14}{50}\right)(\$600{,}000) + \left(\frac{8}{50}\right)(\$650{,}000)$$
$$+ \left(\frac{3}{50}\right)(\$700{,}000) + \left(\frac{1}{50}\right)(\$800{,}000)$$
$$= \$418{,}080$$

$$\text{EUAC}_{10\,\text{ft}} = (\$710{,}000)(0.1468)$$
$$+ \left(\frac{8}{50}\right)(\$650{,}000) + \left(\frac{3}{50}\right)(\$700{,}000)$$
$$+ \left(\frac{1}{50}\right)(\$800{,}000)$$
$$= \$266{,}228$$

$$\text{EUAC}_{15\,\text{ft}} = (\$900{,}000)(0.1468)$$
$$+ \left(\frac{3}{50}\right)(\$700{,}000) + \left(\frac{1}{50}\right)(\$800{,}000)$$
$$= \$190{,}120$$
$$\text{EUAC}_{20\,\text{ft}} = (\$1{,}000{,}000)(0.1468)$$
$$+ \left(\frac{1}{50}\right)(\$800{,}000)$$
$$= \$162{,}800$$

$$\boxed{\text{Build to 20 ft.}}$$

The answer is (D).

33. Assume replacement after one year.

$$\text{EUAC}(1) = (\$30{,}000)(A/P,12\%,1) + \$18{,}700$$
$$= (\$30{,}000)(1.12) + \$18{,}700$$
$$= \$52{,}300$$

Assume replacement after two years.

$$\text{EUAC}(2) = (\$30{,}000)(A/P,12\%,2)$$
$$+ \$18{,}700 + (\$1200)(A/G,12\%,2)$$
$$= (\$30{,}000)(0.5917) + \$18{,}700$$
$$+ (\$1200)(0.4717)$$
$$= \$37{,}017$$

Assume replacement after three years.

$$\text{EUAC}(3) = (\$30{,}000)(A/P,12\%,3)$$
$$+ \$18{,}700 + (\$1200)(A/G,12\%,3)$$
$$= (\$30{,}000)(0.4163) + \$18{,}700$$
$$+ (\$1200)(0.9246)$$
$$= \$32{,}299$$

Similarly, calculate to obtain the numbers in the following table.

years in service	EUAC
1	$52,300
2	$37,017
3	$32,299
4	$30,207
5	$29,152
6	$28,602
7	$28,335
8	$28,234
9	$28,240
10	$28,312

$$\boxed{\text{Replace after 8 yr.}}$$

The answer is (C).

34. Since the head and horsepower data are already reflected in the hourly operating costs, there is no need to work with head and horsepower.

Let N = no. of hours operated each year.

$$\text{EUAC(A)} = (\$3600 + \$3050)(A/P,12\%,10)$$
$$- (\$200)(A/F,12\%,10) + 0.30N$$
$$= (\$6650)(0.1770) - (\$200)(0.0570) + 0.30N$$
$$= 1165.65 + 0.30N$$
$$\text{EUAC(B)} = (\$2800 + \$5010)(A/P,12\%,10)$$
$$+ (\$280)(A/F,12\%,10) + 0.10N$$
$$= (\$7810)(0.1770) - (\$280)(0.0570) + 0.10N$$
$$= 1366.41 + 0.10N$$

$$\text{EUAC(A)} = \text{EUAC(B)}$$
$$1165.65 + 0.30N = 1366.41 + 0.10N$$
$$N = \boxed{1003.8 \text{ hr} \quad (1000 \text{ hr})}$$

The answer is (A).

35. (a) From Eq. 58.89,

$$\frac{T_2}{T_1} = 0.88 = 2^{-b}$$
$$\log 0.88 = -b \log 2$$
$$-0.0555 = -(0.3010)b$$
$$b = 0.1843$$

$$T_4 = (6)(4)^{-0.1843} = \boxed{4.65 \text{ wk} \quad (4.5 \text{ wk})}$$

The answer is (A).

(b) From Eq. 58.90,

$$T_{6-14} = \left(\frac{T_1}{1-b}\right)\left((n_2 + \tfrac{1}{2})^{1-b} - (n_1 - \tfrac{1}{2})^{1-b}\right)$$
$$= \left(\frac{6}{1 - 0.1843}\right)$$
$$\times \left(\left(14 + \frac{1}{2}\right)^{1-0.1843} - \left(6 - \frac{1}{2}\right)^{1-0.1843}\right)$$
$$= \left(\frac{6}{0.8157}\right)(8.857 - 4.017)$$
$$= \boxed{35.6 \text{ wk} \quad (35 \text{ wk})}$$

The answer is (A).

36. (a) First check that both alternatives have an ROR greater than the MARR. Work in thousands of dollars. Evaluate alternative A.

$$P(A) = -\$120 + (\$15)(P/F,i\%,5)$$
$$+(\$57)(P/A,i\%,5)(1 - 0.45)$$
$$+\left(\frac{\$120 - \$15}{5}\right)(P/A,i\%,5)(0.45)$$
$$= -\$120 + (\$15)(P/F,i\%,5)$$
$$+(\$40.8)(P/A,i\%,5)$$

Try 15%.

$$P(A) = -\$120 + (\$15)(0.4972) + (\$40.8)(3.3522)$$
$$= \$24.23$$

Try 25%.

$$P(A) = -\$120 + (\$15)(0.3277) + (\$40.8)(2.6893)$$
$$= -\$5.36$$

Since $P(A)$ goes through 0,

$$(\text{ROR})_A > \text{MARR} = 15\%$$

Next, evaluate alternative B.

$$P(B) = -\$170 + (\$20)(P/F,i\%,5)$$
$$+(\$67)(P/A,i\%,5)(1 - 0.45)$$
$$+\left(\frac{\$170 - \$20}{5}\right)(P/A,i\%,5)(0.45)$$
$$= -\$170 + (\$20)(P/F,i\%,5)$$
$$+(\$50.35)(P/A,i\%,5)$$

Try 15%.

$$P(B) = -\$170 + (\$20)(0.4972) + (\$50.35)(3.3522)$$
$$= \$8.73$$

Since $P(B) > 0$ and will decrease as i increases,

$$(\text{ROR})_B > 15\%$$

$$\boxed{\text{ROR} > \text{MARR for both alternatives.}}$$

The answer is (C).

(b) Do an incremental analysis to see if it is worthwhile to invest the extra $170 - $120 = $50.

$$P(B-A) = -\$50 + (\$20 - \$15)(P/F,i\%,5)$$
$$+(\$50.35 - \$40.8)(P/A,i\%,5)$$

Try 15%.

$$P(B-A) = -\$50 + (\$5)(0.4972)$$
$$+(\$9.55)(3.3522)$$
$$= -\$15.50$$

Since $P(B - A) < 0$ and would become more negative as i increases, the ROR of the added investment is greater than 15%.

$$\boxed{\text{Alternative A is superior.}}$$

The answer is (A).

37. Use the year-end convention with the tax credit. The purchase is made at $t = 0$. However, the credit is received at $t = 1$ and must be multiplied by $(P/F, i\%, 1)$.

$$P = -\$300,000 + (0.0667)(\$300,000)(P/F,i\%,1)$$
$$+(\$90,000)(P/A,i\%,5)(1 - 0.48)$$
$$+\left(\frac{\$300,000 - \$50,000}{5}\right)(P/A,i\%,5)(0.48)$$
$$+(\$50,000)(P/F,i\%,5)$$
$$= -\$300,000 + (\$20,010)(P/F,i\%,1)$$
$$+(\$46,800)(P/A,i\%,5)$$
$$+(\$24,000)(P/A,i\%,5)$$
$$+(\$50,000)(P/F,i\%,5)$$

By trial and error,

i	P
10%	$17,625
15%	-$20,409
12%	$1456
13%	-$6134
12¼%	-$472

$$\boxed{i \text{ is between } 12\% \text{ and } 12\tfrac{1}{4}\%.}$$

The answer is (C).

38. (a) The distributed profit is

$$\text{distributed profit} = (0.15)(\$2,500,000)$$
$$= \boxed{\$375,000 \quad (\$380,000)}$$

The answer is (D).

(b) Find the annual loan payment.

$$\text{payment} = (\$2,500,000)(A/P,12\%,25)$$
$$= (\$2,500,000)(0.1275)$$
$$= \$318,750$$

After paying all expenses and distributing the 15% profit, the remainder should be 0.

$$0 = \text{EUAC}$$
$$= \$20,000 + \$50,000 + \$200,000$$
$$+\$375,000 + \$318,750 - \text{annual receipts}$$
$$-(\$500,000)(A/F,15\%,25)$$
$$= \$963,750 - \text{annual receipts}$$
$$-(\$500,000)(0.0047)$$

This calculation assumes $i = 15\%$, which equals the desired return. However, this assumption only affects the salvage calculation, and since the number is so small, the analysis is not sensitive to the assumption.

$$\text{annual receipts} = \$961,400$$

The average daily receipts are

$$\frac{\$961,400}{365} = \boxed{\$2634 \quad (\$2600)}$$

The answer is (D).

(c) Use the expected value approach. The average occupancy is

$$(0.40)(0.65) + (0.30)(0.70) + (0.20)(0.75)$$
$$+(0.10)(0.80) = \boxed{0.70}$$

The answer is (B).

(d) The average number of rooms occupied each night is

$$(0.70)(120 \text{ rooms}) = 84 \text{ rooms}$$

The minimum required average daily rate per room is

$$\frac{\$2634}{84} = \boxed{\$31.36 \quad (\$31)}$$

The answer is (D).

39. (a) The annual savings are

$$\begin{matrix}\text{annual} \\ \text{savings}\end{matrix} = \left(\frac{0.69 - 0.47}{1000}\right)(\$3,500,000) = \boxed{\$770}$$

The answer is (D).

(b)

$$P = -\$7500 + (\$770 - \$200 - \$100)$$
$$\times (P/A,i\%,25) = 0$$
$$(P/A,i\%,25) = 15.957$$

Searching the tables and interpolating, the rate of return is

$$i \approx \boxed{3.75\% \quad (3.8\%)}$$

The answer is (A).

40. (a) Evaluate alternative A, working in millions of dollars.

$$P(A) = -(\$1.3)(1 - 0.48)(P/A,15\%,6)$$
$$= -(\$1.3)(0.52)(3.7845)$$
$$= \boxed{-\$2.56 \quad (-\$2.6) \quad [\text{millions}]}$$

The answer is (C).

(b) Use straight-line depreciation to evaluate alternative B.

$$D_j = \frac{\$2}{6} = \$0.333$$
$$P(B) = -\$2 - (0.20)(\$1.3)(1 - 0.48)(P/A,15\%,6)$$
$$-(\$0.15)(1 - 0.48)(P/A,15\%,6)$$
$$+(\$0.333)(0.48)(P/A,15\%,6)$$
$$= -\$2 - (0.20)(\$1.3)(0.52)(3.7845)$$
$$-(\$0.15)(0.52)(3.7845)$$
$$+(\$0.333)(0.48)(3.7845)$$
$$= \boxed{-\$2.202 \quad (-\$2.2) \quad [\text{millions}]}$$

The answer is (D).

Professional

(c) Evaluate alternative C.

$$D_j = \frac{1.2}{3} = 0.4$$

$$
\begin{aligned}
P(\text{C}) = &-(\$1.2)\big(1 + (P/F,15\%,3)\big) \\
&-(\$0.20)(\$1.3)(1-0.48) \\
&\times\big((P/F,15\%,1) + (P/F,15\%,4)\big) \\
&-(\$0.45)(\$1.3)(1-0.48) \\
&\times\big((P/F,15\%,2) + (P/F,15\%,5)\big) \\
&-(\$0.80)(\$1.3)(1-0.48) \\
&\times\big((P/F,15\%,3) + (P/F,15\%,6)\big) \\
&+(\$0.4)(\$0.48)(P/A,15\%,6) \\
= &-(\$1.2)(1.6575) \\
&-(\$0.20)(\$1.3)(0.52)(0.8696 + 0.5718) \\
&-(\$0.45)(\$1.3)(0.52)(0.7561 + 0.4972) \\
&-(\$0.80)(\$1.3)(0.52)(0.6575 + 0.4323) \\
&+(\$0.4)(0.48)(3.7845) \\
= &\boxed{-\$2.428 \quad (-\$2.4) \quad [\text{millions}]}
\end{aligned}
$$

The answer is (C).

(d) From parts (a) through (c), $\boxed{\text{alternative B}}$ $\boxed{\text{is superior.}}$

The answer is (B).

41. This is a replacement study. Since production capacity and efficiency are not a problem with the defender, the only question is when to bring in the challenger.

Since this is a before-tax problem, depreciation is not a factor, nor is book value.

The cost of keeping the defender one more year is

$$
\begin{aligned}
\text{EUAC}(\text{defender}) &= \$200,000 + (0.15)(\$400,000) \\
&= \$260,000
\end{aligned}
$$

For the challenger,

$$
\begin{aligned}
&\text{EUAC}(\text{challenger}) \\
&= (\$800,000)(A/P,15\%,10) + \$40,000 \\
&\quad +(\$30,000)(A/G,15\%,10) \\
&\quad -(\$100,000)(A/F,15\%,10) \\
&= (\$800,000)(0.1993) + \$40,000 \\
&\quad +(\$30,000)(3.3832) \\
&\quad -(\$100,000)(0.0493) \\
&= \$296,006
\end{aligned}
$$

Since the defender is cheaper, keep it. The same analysis next year will give identical answers. Therefore, keep the defender for the next three years, at which time the decision to buy the challenger will be automatic.

Having determined that it is less expensive to keep the defender than to maintain the challenger for 10 years, determine whether the challenger is less expensive if retired before 10 years.

If retired in nine years,

$$
\begin{aligned}
\text{EUAC}(\text{challenger}) &= (\$800,000)(A/P,15\%,9) + \$40,000 \\
&\quad +(\$30,000)(A/G,15\%,9) \\
&\quad -(\$150,000)(A/F,15\%,9) \\
&= (\$800,000)(0.2096) \\
&\quad +\$40,000 + (\$30,000)(3.0922) \\
&\quad -(\$150,000)(0.0596) \\
&= \$291,506
\end{aligned}
$$

Similar calculations yield the following results for all the retirement dates.

n	EUAC
10	\$296,000
9	\$291,506
8	\$287,179
7	\$283,214
6	\$280,016
5	\$278,419
4	\$279,909
3	\$288,013
2	\$313,483
1	\$360,000

Since none of these equivalent uniform annual costs are less than that of the defender, it is not economical to buy and keep the challenger for any length of time.

$$\boxed{\text{Keep the defender.}}$$

42. The annual demand for valves is

$$
\begin{aligned}
D &= (500,000 \text{ products})\left(3 \frac{\text{valves}}{\text{product}}\right) \\
&= 1,500,000 \text{ valves}
\end{aligned}
$$

The cost of holding (storing) a valve for one year is

$$
h = \left(0.4 \frac{1}{\text{yr}}\right)(\$6.50) = \$2.60 \; 1/\text{yr}
$$

The most economical quantity of valves to order is

$$Q^* = \sqrt{\frac{2aK}{h}} = \sqrt{\frac{(2)\left(1{,}500{,}000 \ \dfrac{1}{\text{yr}}\right)(\$49.50)}{\$2.60 \ \dfrac{1}{\text{yr}}}}$$

$$= \boxed{7557 \text{ valves} \quad (7600 \text{ valves})}$$

The answer is (D).

43. With just-in-time manufacturing, production is one-at-a-time, according to demand. When one is needed, one is made. In order to make the EOQ approach zero, the $\boxed{\text{setup cost}}$ must approach zero.

The answer is (D).

59 Engineering Law

PRACTICE PROBLEMS

1. List the different forms of company ownership. What are the advantages and disadvantages of each?

2. Define the requirements for a contract to be enforceable.

3. What standard features should a written contract include?

4. Describe the ways a consulting fee can be structured.

5. What is a retainer fee?

6. Which of the following organizations is NOT a contributor to the standard design and construction contract documents developed by the Engineers Joint Contract Documents Committee (EJCDC)?

(A) National Society of Professional Engineers

(B) Construction Specifications Institute

(C) Associated General Contractors of America

(D) American Institute of Architects

7. To be affected by the Fair Labor Standards Act (FLSA) and be required to pay minimum wage, construction firms working on bridges and highways must generally have

(A) 1 or more employees

(B) 2 or more employees and annual gross billings of $500,000

(C) 10 or more employees and be working on federally funded projects

(D) 50 or more employees and have been in business for longer than 6 months

8. A "double-breasted" design firm

(A) has errors and omissions as well as general liability insurance coverage

(B) is licensed to practice in both engineering and architecture

(C) serves both union and nonunion clients

(D) performs post-construction certification for projects it did not design

9. The phrase "without expressed authority" means which of the following when used in regards to partnerships of design professionals?

(A) Each full member of a partnership is a general agent of the partnership and has complete authority to make binding commitments, enter into contracts, and otherwise act for the partners within the scope of the business.

(B) The partnership may act in a manner that it considers best for the client, even though the client has not been consulted.

(C) Only plans, specifications, and documents that have been signed and stamped (sealed) by the authority of the licensed engineer may be relied upon.

(D) Only officers to the partnership may obligate the partnership.

10. A limited partnership has one managing general partner, two general partners, one silent partner, and three limited partners. If all partners cast a single vote when deciding on an issue, how many votes will be cast?

(A) 1

(B) 3

(C) 4

(D) 7

11. Which of the following are characteristics of a limited liability corporation (LLC)?

 I. limited liability for all members

 II. no taxation as an entity (no double taxation)

 III. more than one class of stock

 IV. limited to fewer than 25 members

 V. fairly easy to establish

 VI. no "continuity of life" like regular corporation

 (A) I, II, and IV

 (B) I, II, III, and VI

 (C) I, II, III, IV, and V

 (D) I, III, IV, V, and VI

12. Which of the following statements is FALSE in regard to joint ventures?

 (A) Members of a joint venture may be any combination of sole proprietorships, partnerships, and corporations.

 (B) A joint venture is a business entity separate from its members.

 (C) A joint venture spreads risk and rewards, and it pools expertise, experience, and resources. However, bonding capacity is not aggregated.

 (D) A joint venture usually dissolves after the completion of a specific project.

13. Which of the following construction business types can have unlimited shareholders?

 I. S corporation

 II. LLC

 III. corporation

 IV. sole proprietorship

  (A) II and III only

  (B) I, II, and III

  (C) I, III, and IV

  (D) I, II, III, and IV

14. The phrase "or approved equal" allows a contractor to

 (A) substitute one connection design for another

 (B) substitute a more expensive feature for another

 (C) replace an open-shop subcontractor with a union subcontractor

 (D) install a product whose brand name and model number are not listed in the specifications

15. Cities, other municipalities, and departments of transportation often have standard specifications, in addition to the specifications issued as part of the construction document set, that cover such items as

 (A) safety requirements

 (B) environmental requirements

 (C) concrete, fire hydrant, manhole structures, and curb requirements

 (D) procurement and accounting requirements

16. What is intended to prevent a contractor from bidding on a project and subsequently backing out after being selected for the project?

 (A) publically recorded bid

 (B) property lien

 (C) surety bond

 (D) proposal bond

17. Which of the following is illegal, in addition to being unethical?

 (A) bid shopping

 (B) bid peddling

 (C) bid rigging

 (D) bid unbalancing

18. Which of the following is NOT normally part of a construction contract?

- I. invitation to bid
- II. instructions to bidders
- III. general conditions
- IV. supplementary conditions
- V. liability insurance policy
- VI. technical specifications
- VII. drawings
- VIII. addenda
- IX. proposals
- X. bid bond
- XI. agreement
- XII. performance bond
- XIII. labor and material payment bond
- XIV. nondisclosure agreement

- (A) I
- (B) II
- (C) V
- (D) XIV

19. Once a contract has been signed by the owner and contractor, changes to the contract

- (A) cannot be made
- (B) can be made by the owner, but not by the contractor
- (C) can be made by the contractor, but not by the owner
- (D) can be made by both the owner and the contractor

20. A constructive change is a change to the contract that can legally be construed to have been made, even though the owner did not issue a specific, written change order. Which of the following situations is normally a constructive change?

- (A) request by the engineer-architect to install OSHA-compliant safety features
- (B) delay caused by the owner's failure to provide access
- (C) rework mandated by the building official
- (D) expense and delay due to adverse weather

21. A bid for foundation construction is based on owner supplied soil borings showing sandy clay to a depth of 12 ft. However, after the contract has been assigned and during construction, the backhoe encounters large pieces of concrete buried throughout the construction site. This situation would normally be referred to as

- (A) concealed conditions
- (B) unexplained features
- (C) unexpected characteristics
- (D) hidden detriment

22. If a contract has a value engineering clause and a contractor suggests to the owner that a feature or method be used to reduce the annual maintenance cost of the finished project, what will be the most likely outcome?

- (A) The contractor will be able to share one time in the owner's expected cost savings.
- (B) The contractor will be paid a fixed amount (specified by the contract) for making the suggestion, but only if the suggestion is accepted.
- (C) The contract amount will be increased by some amount specified in the contract.
- (D) The contractor will receive an annuity payment over some time period specified in the contract.

23. A contract has a value engineering clause that allows the parties to share in improvements that reduce cost. The contractor had originally planned to transport concrete on site for a small pour with motorized wheelbarrows. On the day of the pour, however, a concrete pump is available and is used, substantially reducing the contractor's labor cost for the day. This is an example of

- (A) value engineering whose benefit will be shared by both contractor and owner
- (B) efficient methodology whose benefit is to the contractor only
- (C) value engineering whose benefit is to the owner only
- (D) cost reduction whose benefit will be shared by both contractor and laborers

24. A material breach of contract occurs when the

- (A) contractor uses material not approved by the contract to use
- (B) contractor's material order arrives late
- (C) owner becomes insolvent
- (D) contractor installs a feature incorrectly

Professional

25. When an engineer stops work on a job site after noticing unsafe conditions, the engineer is acting as a(n)

    I.   agent

    II.   local official

    III.   OSHA safety inspector

    IV.   competent person

    (A)   I only

    (B)   I and IV

    (C)   II and III

    (D)   III and IV

26. While performing duties pursuant to a contract for professional services with an owner/developer, a professional engineer gives erroneous instruction to a subcontractor hired by the prime contractor. To whom would the subcontractor look for financial relief?

    I.   professional engineer

    II.   owner/developer

    III.   subcontractor's bonding company

    IV.   prime contractor

    (A)   III only

    (B)   IV only

    (C)   I and II only

    (D)   I, II, III, and IV

27. A professional engineer employed by a large, multinational corporation is the lead engineer in designing and producing a consumer product that proves injurious to some purchasers. What can the engineer expect in the future?

    (A)   legal action against him/her from consumers

    (B)   termination and legal action against him/her from his/her employer

    (C)   legal action against him/her from the U.S. Consumer Protection Agency

    (D)   thorough review of his/her work, legal support from the employer, and possible termination

28. A professional engineer in a private consulting practice makes a calculation error that causes the collapse of a structure during the construction process. What can the engineer expect in the future?

    (A)   cancellation of his/her Errors & Omissions insurance

    (B)   criminal prosecution

    (C)   loss of his/her professional engineering license

    (D)   incarceration

29. A professional engineer is hired by a homeowner to design a septic tank and leach field. The septic tank fails after 18 years of operation. What sentence best describes what comes next?

    (A)   The homeowner could pursue a tort action claiming a septic tank should retain functionality longer than 18 years.

    (B)   The engineer will be protected by a statute of limitations law.

    (C)   The homeowner may file a claim with the engineer's original bonding company.

    (D)   The engineer is ethically bound to provide remediation services to the homeowner.

Professional

SOLUTIONS

1. The three different forms of company ownership are the (1) sole proprietorship, (2) partnership, and (3) corporation.

A *sole proprietor* is his or her own boss. This satisfies the proprietor's ego and facilitates quick decisions, but unless the proprietor is trained in business, the company will usually operate without the benefit of expert or mitigating advice. The sole proprietor also personally assumes all the debts and liabilities of the company. A sole proprietorship is terminated upon the death of the proprietor.

A *partnership* increases the capitalization and the knowledge base beyond that of a proprietorship, but offers little else in the way of improvement. In fact, the partnership creates an additional disadvantage of one partner's possible irresponsible actions creating debts and liabilities for the remaining partners.

A *corporation* has sizable capitalization (provided by the stockholders) and a vast knowledge base (provided by the board of directors). It keeps the company and owner liability separate. It also survives the death of any employee, officer, or director. Its major disadvantage is the administrative work required to establish and maintain the corporate structure.

2. To be legal, a contract must contain an *offer*, some form of *consideration* (which does not have to be equitable), and an *acceptance* by both parties. To be enforceable, the contract must be voluntarily entered into, both parties must be competent and of legal age, and the contract cannot be for illegal activities.

3. A written contract will identify both parties, state the purpose of the contract and the obligations of the parties, give specific details of the obligations (including relevant dates and deadlines), specify the consideration, state the boilerplate clauses to clarify the contract terms, and leave places for signatures.

4. A consultant will either charge a fixed fee, a variable fee, or some combination of the two. A one-time fixed fee is known as a *lump-sum fee*. In a *cost plus fixed fee* contract, the consultant will also pass on certain costs to the client. Some charges to the client may depend on other factors, such as the salary of the consultant's staff, the number of days the consultant works, or the eventual cost or value of an item being designed by the consultant.

5. A *retainer* is a (usually) nonreturnable advance paid by the client to the consultant. While the retainer may be intended to cover the consultant's initial expenses until the first big billing is sent out, there does not need to be any rational basis for the retainer. Often, a small retainer is used by the consultant to qualify the client

(i.e., to make sure the client is not just shopping around and getting free initial consultations) and as a security deposit (to make sure the client does not change consultants after work begins).

6. The Engineers Joint Contract Documents Committee (EJCDC) consists of the National Society of Professional Engineers, the American Council of Engineering Companies (formerly the American Consulting Engineers Council), the American Society of Civil Engineers, Construction Specifications Institute, and the Associated General Contractors of America. The American Institute of Architects is not a member, and it has its own standardized contract documents.

The answer is (D).

7. A business in the construction industry must have two or more employees and a minimum annual gross sales volume of $500,000 to be subject to the Fair Labor Standards Act (FLSA). Individual coverage also applies to employees whose work regularly involves them in commerce between states (i.e., interstate commerce). Any person who works on, or otherwise handles, goods moving in interstate commerce, or who works on the expansion of existing facilities of commerce, is individually subject to the protection of the FLSA and the current minimum wage and overtime pay requirements, regardless of the sales volume of the employer.

The answer is (B).

8. A *double-breasted* design firm serves both union and nonunion clients. When union-affiliated companies find themselves uncompetitive in bidding on nonunion projects, the company owners may decide to form and operate a second company that is *open shop* (i.e., employees are not required to join a union as a condition of employment). Although there are some restrictions requiring independence of operation, the common ownership of two related firms is legal.

The answer is (C).

9. *Without expressed authority* means each member of a partnership has full authority to obligate the partnership (and the other partners).

The answer is (A).

10. Only general partners can vote in a partnership. Both silent and limited partners share in the profit and benefits of the partnership, but they only contribute financing and do not participate in the management. The identities of silent partners are often known only to a few, whereas limited partners are known to all.

The answer is (B).

11. Limited liability corporations have limited liability for all members, no double taxation, more than one type of stock, and no continuity of life. They are not limited in members, and they are comparatively fairly difficult to establish.

The answer is (B).

12. One of the reasons for forming joint ventures is that the bonding capacity is aggregated. Even if each contractor cannot individually meet the minimum bond requirements, the total of the bonding capacities may be sufficient.

The answer is (C).

13. Both normal corporations and limited liability corporations (LLCs) can have unlimited shareholders. S corporations and sole proprietorships are for individuals.

The answer is (A).

14. When the specifications include a nonstructural, brand-named article and the accompanying phrase "or approved equal," the contractor can substitute something with the same functionality, even though it is not the brand-named article. "Or approved equal" would not be used with a structural detail such as a connection design.

The answer is (D).

15. Municipalities that experience frequent construction projects within their boundaries have standard specifications that are included by reference in every project's construction document set. This document set would cover items such as concrete, fire hydrants, manhole structures, and curb requirements.

The answer is (C).

16. *Proposal bonds*, also known as *bid bonds*, are insurance policies payable to the owner in the event that the contractor backs out after submitting a qualified bid.

The answer is (D).

17. *Bid rigging*, also known as *price fixing*, is an illegal arrangement between contractors to control the bid prices of a construction project or to divide up customers or market areas. *Bid shopping* before or after the bid letting is where the general contractor tries to secure better subcontract proposals by negotiating with the subcontractors. *Bid peddling* is done by the subcontractor to try to lower its proposal below the lowest proposal. *Bid unbalancing* is where a contractor pushes the payment for some expense items to prior construction phases in order to improve cash flow.

The answer is (C).

18. The construction contract includes many items, some explicit only by reference. The contractor may be required to carry liability insurance, but the policy itself is between the contractor and its insurance company, and is not normally part of the construction contract.

The answer is (C).

19. Changes to the contract can be made by either the owner or the contractor. The method for making such changes is indicated in the contract. Almost always, the change must be agreed to by both parties in writing.

The answer is (D).

20. A *constructive change* to the contract is the result of an action or lack of action of the owner or its agent. If the project is delayed by the owner's failure to provide access, the owner has effectively changed the contract.

The answer is (B).

21. *Concealed conditions* are also known as *changed conditions* and *differing site conditions*. Most, but not all, contracts have provisions dealing with changed conditions. Some place the responsibility to confirm the site conditions before bidding on the contractor. Others detail the extent of changed conditions that will trigger a review of reimbursable expenses.

The answer is (A).

22. Changes to a structure's performance, safety, appearance, or maintenance that benefit the owner in the long run will be covered by the value engineering clause of a contract. Normally, the contractor is able to share in cost savings in some manner by receiving a payment or credit to the contract.

The answer is (A).

23. The problem gives an example of efficient methodology, where the benefit is to the contractor only. It is not an example of value engineering, as the change affects the contractor, not the owner. Performance, safety, appearance, and maintenance are unaffected.

The answer is (B).

24. *A material breach of the contract* is a significant event that is grounds for cancelling the contract entirely. Typical triggering events include failure of the owner to make payments, the owner causing delays, the owner declaring bankruptcy, the contractor abandoning the job, or the contractor getting substantially off schedule.

The answer is (C).

Professional

25. An engineer with the authority to stop work gets his/her authority from the agency clause of the contract for professional services with the owner/developer. It is unlikely that the local building department or OSHA would have authorized the engineer to act on their behalves. It is also possible that the engineer may have been designated as the jobsite's "competent person" (as required by OSHA) for one or more aspects of the job, with authority to implement corrective action.

The answer is (B).

26. In the absence of other information, the subcontractor only has a contractual relationship with the prime contractor, so the request for financial relief would initially go to the prime contractor. The prime contractor would ask for relief from the owner/developer, who in turn may want relief from the professional engineer. The subcontractor's bonding company's role is to guarantee the subcontractor's work product against subcontractor errors, not engineering errors.

The answer is (B).

27. Manufacturers of consumer products are generally held responsible for the safety of their products. Individual team members are seldom held personally accountable, and when they are, there is evidence of intentional wrongdoing and/or gross incompetence. The manufacturer will organize and absorb the costs of the legal defense. If the engineer is fired, it is likely this will occur after the case is closed.

The answer is (D).

28. A professional engineer is not held to the standard of perfection, so making a calculation error is neither a criminal act nor sufficient evidence of incompetence to result in loss of license. Without criminal prosecution for fraud or other wrongdoing, it is unlikely that the engineer would ever see the inside of a jail. However, it is likely that the engineer's bonding company will cancel his/her policy (after defending the case).

The answer is (A).

29. Satisfactory operation for 18 years is proof that the design was partially, if not completely, adequate. There is no ethical obligation to remediate normal wear-and-tear, particularly when the engineer was not involved in how the septic system was used or maintained. While the homeowner can threaten legal action and even file a complaint, in the absence of a warranty to the contrary, the engineer is probably well-protected by a statute of limitations.

The answer is (B).

Professional

60 Engineering Ethics

INTRODUCTION TO ETHICS PROBLEMS

Case studies in law and ethics can be interpreted in many ways. The problems presented are simple thumbnail outlines. In most real cases, there will be more facts to influence a determination than are presented in the case scenarios. In some cases, a state may have specific laws affecting the determination; in other cases, prior case law will have been established.

The determination of whether an action is legal can be made in two ways. The obvious interpretation of an illegal action is one that violates a specific law or statute. An action can also be *found to be illegal* if it is judged in court to be a breach of a written, verbal, or implied contract. Both of these approaches are used in the following solutions.

These answers have been developed to teach legal and ethical principles. While being realistic, they are not necessarily based on actual incidents or prior case law.

PRACTICE PROBLEMS

(Each problem has two parts. Determine whether the situation is (or can be) permitted legally. Then, determine whether the situation is permitted ethically.)

1. (a) Was it legal and/or ethical for an engineer to sign and seal plans that were not prepared by him or prepared under his responsible direction, supervision, or control?

(b) Was it legal and/or ethical for an engineer to sign and seal plans that were not prepared by him but were prepared under his responsible direction, supervision, and control?

2. Under what conditions would it be legal and/or ethical for an engineer to rely on the information (e.g., elevations and amounts of cuts and fills) furnished by a grading contractor?

3. Was it legal and/or ethical for an engineer to alter the soils report prepared by another engineer for his client?

4. Under what conditions would it be legal and/or ethical for an engineer to assign work called for in his contract to another engineer?

5. A licensed professional engineer was convicted of a felony totally unrelated to his consulting engineering practice.

(a) What actions would you recommend be taken by the state registration board?

(b) What actions would you recommend be taken by the professional or technical society (e.g., ASCE, ASME, IEEE, NSPE, and so on)?

6. An engineer came across some work of a predecessor. After verifying the validity and correctness of all assumptions and calculations, the engineer used the work. Under what conditions would such use be legal and/or ethical?

7. A building contractor made it a policy to provide cell phones to the engineers of projects he was working on. Under what conditions could the engineers accept the phones?

8. An engineer designed a tilt-up slab building for a client. The design engineer sent the design out to another engineer for checking. The checking engineer sent the plans to a concrete contractor for review. The concrete contractor made suggestions that were incorporated into the recommendations of the checking contractor. These recommendations were subsequently incorporated into the plans by the original design engineer. What steps must be taken to keep the design process legal and/or ethical?

9. A consulting engineer registered his corporation as "John Williams, P.E. and Associates, Inc." even though he had no associates. Under what conditions would this name be legal and/or ethical?

10. When it became known that a chemical plant was planning on producing a toxic product, an engineer employed by the plant wrote to the local newspaper condemning the chemical plant's action. Under what conditions would the engineer's action be legal and/or ethical?

11. An engineer signed a contract with a client. The fee the client agreed to pay was based on the engineer's estimate of time required. The engineer was able to complete the contract satisfactorily in half the time he expected. Under what conditions would it be legal and/or ethical for the engineer to keep the full fee?

12. After working on a project for a client, the engineer was asked by a competitor of the client to perform design services. Under what conditions would it be legal and/or ethical for the engineer to work for the competitor?

13. Two engineers submitted bids to a prospective client for a design project. The client told engineer A how much engineer B had bid and invited engineer A to beat the amount. Under what conditions could engineer A legally/ethically submit a lower bid?

14. A registered civil engineer specializing in well-drilling, irrigation pipelines, and farmhouse sanitary systems took a booth at a county fair located in a farming town. By a random drawing, the engineer's booth was located next to a hog-breeder's booth, complete with live (prize) hogs. The engineer gave away helium balloons with his name and phone number to all visitors to the booth. Did the engineer violate any laws/ethical guidelines?

15. While in a developing country supervising construction of a project an engineer designed, the engineer discovered the client's project manager was treating local workers in an unsafe and inhumane (but, for that country, legal) manner. When the engineer objected, the client told the engineer to mind his own business. Later, the local workers asked the engineer to participate in a walkout and strike with them.

(a) What legal/ethical positions should the engineer take?

(b) Should it have made any difference if the engineer had or had not yet accepted any money from the client?

16. While working for a client, an engineer learns confidential knowledge of a proprietary production process being used by the client's chemical plant. The process is clearly destructive to the environment, but the client will not listen to the objections of the engineer. To inform the proper authorities will require the engineer to release information that was gained in confidence. Is it legal and/or ethical for the engineer to expose the client?

17. While working for an engineering design firm, an engineer was moonlighting as a soils engineer. At night, the engineer used the employer's facilities to analyze and plot the results of soils tests. He then used his employer's computers to write his reports. The equipment and computers would otherwise be unused. Under what conditions could the engineer's actions be considered legal and/or ethical?

18. Ethical codes and state legislation forbidding competitive bidding by design engineers are

(A) enforceable in some states

(B) not enforceable on public (nonfederal) projects

(C) enforceable for projects costing less than $5 million

(D) not enforceable

SOLUTIONS

1. (a) Stamping plans for someone else is illegal. The registration laws of all states permit a registered engineer to stamp/sign/seal only plans that were prepared by him personally or were prepared under his direction, supervision, or control. This is sometimes called being in *responsible charge*. The stamping/signing/sealing, for a fee or gratis, of plans produced by another person, whether that person is registered or not and whether that person is an engineer or not, is illegal.

An illegal act, being a concealed act, is intrinsically unethical. In addition, stamping/signing/sealing plans that have not been checked violates the rule contained in all ethical codes that requires an engineer to protect the public.

(b) This is both ethical and legal. Consulting engineering firms typically operate in this manner, with a senior engineer being in responsible charge for the work of subordinate engineers.

2. Unless the engineer and contractor worked together such that the engineer had personal knowledge that the information was correct, accepting the contractor's information is illegal. Not only would using unverified data violate the state's registration law (for the same reason that stamping/signing/sealing unverified plans in Sec. 126.1 was illegal), but the engineer's contract clause dealing with assignment of work to others would probably be violated.

The act is unethical. An illegal act, being a concealed act, is intrinsically unethical. In addition, using unverified data violates the rule contained in all ethical codes that requires an engineer to protect the client.

3. It is illegal to alter a report to bring it "more into line" with what the client wants unless the alterations represent actual, verified changed conditions. Even when the alterations are warranted, however, use of the unverified remainder of the report is a violation of the state registration law requiring an engineer only to stamp/sign/seal plans developed by or under him. Furthermore, this would be a case of fraudulent misrepresentation unless the originating engineer's name was removed from the report.

Unless the engineer who wrote the original report has given permission for the modification, altering the report would be unethical.

4. Assignment of engineering work is legal (1) if the engineer's contract permitted assignment, (2) all prerequisites (e.g., notifying the client) were met, and (3) the work was performed under the direction of another licensed engineer.

Assignment of work is ethical (1) if it is not illegal, (2) if it is done with the awareness of the client, and (3) if the assignor has determined that the assignee is competent in the area of the assignment.

5. (a) The registration laws of many states require a hearing to be held when a licensee is found guilty of unrelated, but nevertheless unforgivable, felonies (e.g., moral turpitude). The specific action (e.g., suspension, revocation of license, public censure, and so on) taken depends on the customs of the state's registration board.

(b) By convention, it is not the responsibility of technical and professional organizations to monitor or judge the personal actions of their members. Such organizations do not have the authority to discipline members (other than to revoke membership), nor are they immune from possible retaliatory libel/slander lawsuits if they publicly censure a member.

6. The action is legal because, by verifying all the assumptions and checking all the calculations, the engineer effectively does the work. Very few engineering procedures are truly original; the fact that someone else's effort guided the analysis does not make the action illegal.

The action is probably ethical, particularly if the client and the predecessor are aware of what has happened (although it is not necessary for the predecessor to be told). It is unclear to what extent (if at all) the predecessor should be credited. There could be other extenuating circumstances that would make referring to the original work unethical.

7. Gifts, per se, are not illegal. Unless accepting the phones violates some public policy or other law, or is in some way an illegal bribe to induce the engineer to favor the contractor, it is probably legal to accept the phones.

Ethical acceptance of the phones requires (among other considerations) that (1) the phones be required for the job, (2) the phones be used for business only, (3) the phones are returned to the contractor at the end of the job, and (4) the contractor's and engineer's clients know and approve of the transaction.

8. There are two issues: (1) the assignment and (2) the incorporation of work done by another. To avoid a breach, the contracts of both the design and checking engineers must permit the assignments. To avoid a violation of the state registration law requiring engineers to be in responsible charge of the work they stamp/sign/seal, both the design and checking engineers must verify the validity of the changes.

To be ethical, the actions must be legal and all parties (including the design engineer's client) must be aware that the assignments have occurred and that the changes have been made.

Professional

9. The name is probably legal. If the name was accepted by the state's corporation registrar, it is a legally formatted name. However, some states have engineering registration laws that restrict what an engineering corporation may be named. For example, all individuals listed in the name (e.g., "Cooper, Williams, and Somerset—Consulting Engineers") may need to be registered. Whether having "Associates" in the name is legal depends on the state.

Using the name is unethical. It misleads the public and represents unfair competition with other engineers running one-person offices.

10. Unless the engineer's accusation is known to be false or exaggerated, or the engineer has signed an agreement (confidentiality, nondisclosure, and so on) with his employer forbidding the disclosure, the letter to the newspaper is probably not illegal.

The action is probably unethical. (If the letter to the newspaper is unsigned it is a concealed action and is definitely unethical.) While whistle-blowing to protect the public is implicitly an ethical procedure, unless the engineer is reasonably certain that manufacture of the toxic product represents a hazard to the public, he has a responsibility to the employer. Even then, the engineer should exhaust all possible remedies to render the manufacture nonhazardous before blowing the whistle. Of course, the engineer may quit working for the chemical plant and be as critical as the law allows without violating engineer-employer ethical considerations.

11. Unless the engineer's payment was explicitly linked in the contract to the amount of time spent on the job, taking the full fee would not be illegal or a breach of the contract.

An engineer has an obligation to be fair in estimates of cost, particularly when the engineer knows no one else is providing a competitive bid. Taking the full fee would be ethical if the original estimate was arrived at logically and was not meant to deceive or take advantage of the client. An engineer is permitted to take advantage of economies of scale, state-of-the-art techniques, and break-through methods. (Similarly, when a job costs more than the estimate, the engineer may be ethically bound to stick with the original estimate.)

12. In the absence of a nondisclosure or noncompetition agreement or similar contract clause, working for the competitor is probably legal.

Working for both clients is unethical. Even if both clients know and approve, it is difficult for the engineer not to "cross-pollinate" his work and improve one client's position with knowledge and insights gained at the expense of the other client. Furthermore, the mere appearance of a conflict of interest of this type is a violation of most ethical codes.

13. In the absence of a sealed-bid provision mandated by a public agency and requiring all bids to be opened simultaneously (and the award going to the lowest bidder), the action is probably legal.

It is unethical for an engineer to undercut the price of another engineer. Not only does this violate a standard of behavior expected of professionals, it unfairly benefits one engineer because a similar chance is not given to the other engineer. Even if both engineers are bidding openly against each other (in an auction format), the client must understand that a lower price means reduced service. Each reduction in price is an incentive to the engineer to reduce the quality or quantity of service.

14. It is generally legal for an engineer to advertise his services. Unless the state has relevant laws, the engineer probably did not engage in illegal actions.

Most ethical codes prohibit unprofessional advertising. The unfortunate location due to a random drawing might be excusable, but the engineer should probably refuse to participate. In any case, the balloons are a form of unprofessional advertising, and as such, are unethical.

15. (a) As stated in the scenario statement, the client's actions are legal for that country. The fact that the actions might be illegal in another country is irrelevant. Whether or not the strike is legal depends on the industry and the laws of the land. Some or all occupations (e.g., police and medical personnel) may be forbidden to strike. Assuming the engineer's contract does not prohibit participation, the engineer should determine the legality of the strike before making a decision to participate.

If the client's actions are inhumane, the engineer has an ethical obligation to withdraw from the project. Not doing so associates the profession of engineering with human misery.

(b) The engineer has a contract to complete the project for the client. (It is assumed that the contract between the engineer and client was negotiated in good faith, that the engineer had no knowledge of the work conditions prior to signing, and that the client did not falsely induce the engineer to sign.) Regardless of the reason for withdrawing, the engineer is breaching his contract. In the absence of proof of illegal actions by the client, withdrawal by the engineer requires a return of all fees received. Even if no fees have been received, withdrawal exposes the engineer to other delay-related claims by the client.

16. A contract for an illegal action cannot be enforced. Therefore, any confidentiality or nondisclosure agreement that the engineer has signed is unenforceable if the production process is illegal, uses illegal chemicals, or violates laws protecting the environment. If the production process is not illegal, it is not legal for the engineer to expose the client.

Society and the public are at the top of the hierarchy of an engineer's responsibilities. Obligations to the public take precedence over the client. If the production process is illegal, it would be ethical to expose the client.

17. It is probably legal for the engineer to use the facilities, particularly if the employer is aware of the use. (The question of whether the engineer is trespassing or violating a company policy cannot be answered without additional information.)

Moonlighting, in general, is not ethical. Most ethical codes prohibit running an engineering consulting business while receiving a salary from another employer. The rationale is that the moonlighting engineer is able to offer services at a much lower price, placing other consulting engineers at a competitive disadvantage. The use of someone else's equipment compounds the problem since the engineer does not have to pay for using the equipment, and so does not have to charge any clients for it. This places the engineer at an unfair competitive advantage compared to other consultants who have invested heavily in equipment.

18. Ethical bans on competitive bidding are not enforceable. The National Society of Professional Engineers' (NSPE) ethical ban on competitive bidding was struck down by the U.S. Supreme Court in 1978 as a violation of the Sherman Antitrust Act of 1890.

The answer is (D).

61 Electrical Engineering Frontiers

PRACTICE PROBLEMS

1. What force particles are responsible for electromagnetic interaction?

- (A) bosons
- (B) gluons
- (C) gravitons
- (D) photons

SOLUTIONS

1. Photons are the force particles responsible for electromagnetic interaction. Photons are sometimes called virtual photons because they cannot be detected without interfering with the electromagnetic interaction.

The answer is (D).

62 Engineering Licensing in the United States

PRACTICE PROBLEMS

1. Obtaining a professional engineer (PE) license will help you gain

  I. respect

  II. a higher salary

  III. proof of your engineering knowledge

  (A) I only

  (B) I and II only

  (C) III only

  (D) I, II, and III

SOLUTIONS

1. A PE license provides your peers and employers with proof of your engineering knowledge and self-discipline. Earning a PE license will engender greater respect and have the potential for a positive impact on your salary.

The answer is (D).

Professional